INFORMATION TECHNOLOGY (IT) IN THE INDIAN ECONOMY

Policies, Prospects and Challenges

Information Technology (IT) in the Indian Economy

Policies, Prospects and Challenges

Edited by

M.S. Bhatt and Asheref Illiyan
Department of Economics, Jamia Millia Islamia
(A Central University)
New Delhi, India

New Century Publications
New Delhi, India

NEW CENTURY PUBLICATIONS
4800/24, Bharat Ram Road,
Ansari Road, Daryaganj,
New Delhi – 110 002 (India)

Tel.: 011-2324 7798, 4358 7398, 6539 6605
Fax: 011-4101 7798
E-mail: indiatax@vsnl.com • info@newcenturypublications.com
www.newcenturypublications.com

Editorial office:
LG–7, Aakarshan Bhawan
4754-57/23, Ansari Road, Daryaganj,
New Delhi – 110 002

Tel.: 011-4356 0919

First Published: June 2009

ISBN: 978-81-7708-205-0

Published by New Century Publications and printed at Salasar Imaging Systems, New Delhi

Designs: Patch Creative Unit, New Delhi

PRINTED IN INDIA

About the Book

India has emerged over the last 20 years as the most preferred destination for outsourcing of Information Technology (IT) services. The vibrant IT industry is contributing immensely by providing information about latest technology and international business practices. Various sectors of the Indian economy, such as industry, finance, insurance, communications and transport, have adopted IT in a big way. Trade in IT and IT-enabled services has been the main driver of growth in India's trade in services in recent years.

Starting out with basic data entry tasks, the industry has rapidly acquired a reputation as the primary low-cost destination for voice-based customer contact/support services, finance and accounting, and a range of back-office processing activities. The past few years have seen the scope of these services expanding to include increasingly more complex processes involving rule-based decision making and research/analytics services requiring informed individual judgment and domain/vertical knowledge.

Indian companies adopt all the modes for supply of IT services abroad. The larger companies have set up wholly-owned subsidiaries in USA, UK and the European continent or have opened branches to facilitate the supply of services. Some of them have adopted the strategy of acquiring local companies or entering into strategic partnerships with them. They use the quality processes, innovative technologies and delivery models of acquired companies to penetrate the local market. They have permanent employees or hire consultants based in the host country.

However, there are certain areas of concern which need to be given the desired policy attention and weightage. Some of these are: (a) sustainability of high growth rate of software exports in future, (b) shortage of skilled labour, (c) low diffusion of information technology in the domestic market, (d) lagging hardware sector, (e) regional concentration and (f) excessive dependence on USA. Hence, there is a need for recasting of approach to the growth of IT sector. Dispassionate debate/discussion on the various aspects of Indian IT sector is a necessary requirement for such a rethink. It is against this backdrop that the present volume seeks to provide answers to some of the key issues and questions. It consists of 16 research papers, authored by eminent scholars in the field, which have been categorised in the following 3 theme parts: Part I: Performance, Challenges and Prospects of Indian IT Sector, Part II: Diffusion (Use) of IT into Other Sectors of the Economy and Part III: Social Aspects of IT Sector.

Editors' Profile

M.S. Bhatt is Professor and Head, Department of Economics, Jamia Millia Islamia (A Central University), New Delhi, India. He has been associated, among others, with the University of Kashmir, Srinagar; Jawaharlal Nehru University (JNU), New Delhi; Institute of Economic Growth (IEG), Delhi; Institute for Social and Economic Change (ISEC), Bangalore and Centre for Rural and Industrial Research, Chandigarh. His major sphere of interest comprises development economics, agrarian economics, environmental economics and Kashmir studies. Professor Bhatt was a member of the expert committee for economic reforms (Godbole Committee) appointed by the Government of Jammu and Kashmir. He has also acted as a principal investigator in a number of research projects. He is currently directing a number of research projects including University Grants Commission's Special Assistance Programme in Environmental Economics. He has to his credit 4 books and over 70 published articles in referred journals.

Asheref Illiyan is Senior Assistant Professor in the Department of Economics, Jamia Millia Islamia (A Central University), New Delhi. His main areas of interest include economics of information technology, agriculture and rural development, planning and decentralized governance and industrial economics. He has published a number of articles in reputed national and international journals on planning and rural development, economics of information technology and engineering goods industries. He has to his credit 1 edited book.

Acknowledgements

We would like to place on record our gratitude to all those persons and institutions for their assistance and cooperation in this venture. In particular, the following deserve a special mention: Professor Mushirul Hasan, Vice-Chancellor, Jamia Millia Islamia, for his encouragement and excellent leadership at various stages regarding the publication of this book; all contributors to the present volume who took lot of pains to write their papers as per our requirements and our colleagues in the Department of Economics, Jamia Millia Islamia, for their support.

Thanks are also due to the students of the Department of Economics for their valuable support and assistance; technical team of the Department, viz. M.M.A Baig, Ashfaque and Muqeem for their technical assistance; Ministerial staff of the Department, namely Abida Begum, Nazim, Shamsad Khan and Arif Ali for their back up services and Mr. Sandeep Sury of New Century Publications, New Delhi for his commendable efforts in getting the book printed in an excellent way.

Finally, we are grateful to the Ministry of Communications and Information Technology, Government of India; Ministry of Science and Technology, Government of India and Indian Council of Social Science Research (ICSSR), New Delhi, for their financial help and collaboration.

M.S. Bhatt
Asheref Illiyan

Contributors

Name	Affiliation	E-mail
Babu P. Remesh	Associate Fellow, V.V. Giri National Labour Institute, NOIDA, UP, India.	neetbabu@gmail.com
D.N. Rao	Professor, CESP, SIS, Jawaharlal Nehru University, New Delhi, India	rao_dn@hotmail.com
Harinam Singh	M.Phil. Fellow, Bharatiar University, Coimbatore, Tamil Nadu, India.	harinam_4a41@rediffmail.com
K.J. Joseph	Professor, Centre for Development Studies, Trivandrum, Kerala, India.	kjjoseph@cds.ac.in
Krishan Kumar	Lecturer in Economics, Directorate of Education, Government of NCT of Delhi, India.	krishan_kr_y@hotmail.com
M.F.M Nusry	Technical Lead, Integrated Systems International (Pvt) Ltd., First Floor, No 106, Dutugemunu Street, Dehiwala, Sri Lanka.	nusry@mubasher.net
M.Y. Khan	Chief Engineer, R&D Division, ITI Limited, Naini, Allahabad, UP, India.	khanmy_nni@itiltd.co.in
Neha Aggrawal	Department of Economics, Jamia Millia Islamia, New Delhi, India.	…
Pham Van Bo	Director, Centre for Training Consultancy and International Relations College of Training and Retraining Construction Staff, Hanoi, Vietnam.	phamvanbo@yahoo.com
Prathiba Wasan	Senior Lecturer, IEC College of Engineering and Technology, Greater NOIDA, UP, India.	pratibhawasan@rediffmail.com
P. Vigneswara Ilavarasan	Assistant Professor, Department of Humanities and Social Sciences, IIT, Hauz Khas, New Delhi, India.	evignesh@gmail.com

Ravinder Goel	Senior Lecturer, Department of Commerce, Satyawati College (Evening), Ashok Vihar, Delhi, India.	ravi_goel2001@yahoo.com
Santhosh Rebello	Professor and Head, Department of MCA, MSc (Bio-Informatics, Software Technology) St. Aloysius College, Mangalore, Karnataka, India.	santhosh.rebello@hotmail.com
Shelly Tara	Ph.D. Scholar, Department of Humanities and Social Sciences, Indian Institute of Technology, Hauz Khas, New Delhi, India.	shelly.tara@rediffmail.com
Shishir Saxena	Department of Economics, Monash University, Caulfield East, Victoria 3145, Australia.	shishir80@gmail.com
Somesh Kumar Mathur	Assistant Professor of Economics, HSS Department, IIT Kanpur, UP, India.	skmathur@iitk.ac.in
Vrushal T. Ghoble	Ph.D. Fellow, Centre for West Asian and African Studies, SIS, JNU, New Delhi, India	vrushal_jnu@yahoo.co.in
Zakir Hussain	Researcher, Institute for Defence Studies and Analysis, New Delhi, India.	shahabzakir@gmail.com

Contents

Part II: Diffusion (Use) of IT into Other Sectors of the Economy

Part III: Social Aspects of IT Sector

Editors' Note on the Book

The presence of India's Information Technology (IT) capabilities in the world market for IT software and services is well-documented. IT and IT-Enabled Services (ITES) have emerged as a promising sector for Indian economy. It has generated revenue to the tune of US$ 64 billion during 2007-08 and is 'expected to touch US$ 71.7 billion in 2008-09. [1] Its contribution to the Gross Domestic Product (GDP) has nearly quadruped from 1.2 percent in the year 1997-98 to 5.8 percent in the year 2008-09. The sector has also become the biggest employment generator with the number of jobs added almost doubling each year. The total IT software and services employment increased from 0.28 million in 1999-2000 to 1.63 million in 2006-07 and was expected to reach the 2.0 million mark in 2007-08 (excluding employment in the hardware sector). The indirect employment attributed to the sector was estimated at about 8.0 million for the year 2007-08. This translates into the creation of about 10 million job opportunities. Above all, IT has spawned a number of ancillary businesses such as transportation, real estate and catering, and has contributed to a rising class of young consumers with high disposable incomes as well as a rise in direct tax collection. [2]

Exports remain the mainstay of the Indian IT-BPO growth story. Software and services exports are likely to reach US$ 47 billion in 2008-09. [3 and 4] Thus, export-oriented software and service sector is the driving force of Indian IT industry. Its share in total software industry increased from 32 percent in 1985-86 to 79.94 percent in 2004-05. At an annual growth rate of 50 percent over the last decade (1990-2000), the Indian software and service sector expanded faster than in any other country of the world of comparable size. [5] Such a growth rate has been unparalleled in any other sector of the Indian economy since Independence in 1947. According to a World Bank-funded study, its share in the global cross-country customized software development market grew from 11.9 percent in 1991 to 19.5 percent in 2000. [6] Further, with foresighted policies it could become a major force, capturing 5 to 8 percent of the world market. [7] Moreover, India's exports of software and services are reaching to more than 130 countries and 'India continues to be the nerve-centre for global sourcing with over 2/3rd of the Fortune 500 and a majority of the Global 2,000 firms leveraging global service delivery—now sourcing from India.' [8]

Moreover, majority of the IT firms in the world with CMM-level 5 certification (highest quality certification in the world) are from India. Significantly, software exports are poised to emerge as the country's largest exporting sector with its share expected to rise from 26 percent in 2007-08 to 36 percent in 2008-09. Due to all these reasons it is firmly believed that 'this is one of the few areas where India has potential to become a Global Powerhouse. [9]

The phenomenal growth of Indian software export can be attributed to the following factors:

1. Growing respect for Indian software industry in the international market.
2. Continued rise in the offshore services. [10]
3. Quality service, timely delivery.
4. Entry into new markets.
5. Y2K data conversion business.
6. International linkages.
7. Various steps taken by the Government to promote software export (such as simplifying procedures, tax concessions, establishments of software technology parks, more liberal foreign investment policies).
8. A number of comparative advantages India possess (such as 2nd largest pool of scientific and skilled manpower which is also English speaking, low cost of labour, investment-friendly economic climate and policy, locational time difference with the western world enabling round the clock development).
9. Building up of national level institutions and continued expansion of their capacity.
10. Promotion of national innovation system.
11. Pro-active role by Nasscom (the software industry association).
12. Market diversification. [11]

Be that as it may there are certain areas of concern, which need to be given the desired policy attention and weightage. Some of these are as under:

1. A very important aspect that merits discussion is *sustainability of high growth rate of software exports in future*. The available evidence suggests that while software export registered an annual compound growth of above 45 percent during the period 1990-00, the compound annual growth has slipped down to about 25 percent since 2001. This is attributed mainly to slowdown in US economy since 2000. Further, a number of countries such as China, Russia, Philippines, Canada and Ireland are emerging as competitors to India in the international market for software and services. Therefore, we have to find ways for sustaining high growth rate of software export.

2. *Shortage of skilled labour* for software industry is another serious area of concern. There is growing demand for skilled labour in software industry. Despite the fact that country is known for its human resource for IT industry it has been struggling hard to meet the growing demand for IT professionals. Added to this is the problem of migration of skilled manpower to developed countries. Are our institutions capable of meeting the growing demand for IT workforce?.

The answer is stunt no. Therefore, steps have to be taken to increase the flow of skilled manpower.

3. *Low diffusion of information technology in the domestic market* is a matter of serious concern. While software export is growing very fast domestic software could not keep pace with software exports. A strong and vibrant domestic market is essential for sustained software exports and to insulate it from any fluctuation in international market. Because of many reasons domestic market is very weak. Therefore, Government has to formulate appropriate policies to promote the diffusion of IT into various sectors of the economy.

4. While India has marched ahead and demonstrated its competitive strength in the field of software in international market but unfortunately she has been *lagging behind in the filed of hardware production and export*. The booming software sector with lagging hardware is a major issue of concern particularly in view of the fact that China is emerging as a major competitor to India in the filed of software. Given, the solid hardware base it is perceived to have high potential for growth, which in turn calls for strengthening India's hardware base to sustain its high compound annual growth rate.

5. Undoubtedly one of the contributory factors for the success of the software industry has been the *pro-active role played* by the Government. But government policies have not been free from limitations and gray areas. There is need for recasting these policies.

6. *Concentration on software services rather than software products* is a critical challenge to be addressed quickly. So far we have been able to concentrate only on low-end software services rather than high-end software products. For instance, 'in 2004-05 software packaged products accounted for only 5 percent of Indian software Industry's revenue while software services made up 68 percent of the industry and BPO brought in another 27 percent' [12] But in China more than 50 percent of the software output is in the form of products. India has to increase its share in software products by innovation, research and development, copy right protection, developing niche products etc.

7. There is *high concentration of software industry* in seven metropolitan cities (Tier 1) across India. Easy availability of skilled manpower, better infrastructure and investment friendly climate are the main factors responsible for this skewed development. This is not without a cost. It has created not only the uneven development of the regions but also there are heavy strains on infrastructure along traffic jams, high operating costs, migration of resources etc. These issues need to be addressed on a priority. Literally these cities are running out of resources. Companies will have to shift their operations to Tier 2 cities to gain competitive advantage. It is a matter of consolation that the software

companies have started moving to Tier 2 cities such as Ahmedabad, Kochi, Coimbatore, Indore, Lucknow, Jaipur etc. But the biggest issue in Tier 2 cities is the infrastructure. It requires huge investment from government and private sector. [13]

8. *US financial crisis* is one of the most important challenges facing Indian IT Industry today. The crisis has created a ripple effect across industries and nations. Major developed nations are already in recession or in the verge of recession. India being a globalized economy cannot be totally insulated from this crisis. The immediate effect is seen in the form of crumbling of major stock markets in the country. Further, GDP growth has dropped from 9 to 7 percent in 2008-09 and estimate for 2009-10 are roughly between 5 to 6.5 percent. This crisis is expected further adversely impact IT/ITES sector, which has heavy dependence on US market as 60 percent of our exports are sourced to US. In addition to IT sector, Automobile, Air Travel and Real Estate are also badly affected.

In the times to come IT sector will have a challenging time ahead. Spending on IT services has come down. New customers will be more focused on cost reduction and existing ones will be keen to renegotiate the contracts at the lower prices. There has been cancellation of discretionary projects and reduction in existing contracts. Nasscom has scaled down the export growth target for 2008-09. Indeed the crisis has affected profitability, recruitment and growth of IT industry. However, there is a silver lining at the end of the tunnel. Some IT experts see present crisis as an opportunity for Indian IT sector. Companies are adopting cost cutting measures, increase the productivity of its employees. The attrition level has also come down drastically. It has in turn helped IT companies to explore the great potential existing in the domestic market. Even during this crisis period the IT export has recorded a double digit growth during 2008-09. This is a no mean achievement.

9. In addition to the above mentioned issues there are other areas that merit elaborate discussion. These areas are: socio-economic, psychological and health implications of IT industry, digital divide, IPR (Intellectual Property Rights) issues, Research and Development in IT industry, regional variations in IT development, changes in the family dynamics and so on.

In view of the above, there is need for a recasting of our approach to the growth and development of IT sector. Dispassionate debates and discussions on the various aspects of Indian IT sector and its role and diffusion into other sectors of the economy is a necessary requirement for such a rethink. We believe that these debates would help to understand the salient features and challenges facing the Indian IT sector. These will also generate an informed opinion about the significance of IT in enhancing productivity and efficiency of domestic sectors of the economy.

We are conscious that many crucial theoretical and policy issues could not find a place in this volume. With all humility, we do accept the primacy of these issues. But due to obvious reasons, these could not be accommodated. Whatever we have presented is expected to prove a useful addition to the existing literature on IT Industry and its policies. It will serve, in some way, as a fruitful guide to effective formulation and implementation of appropriate policies.

The volume has been organized in three parts. Part I deals with performance, challenges and prospects of Indian IT sector in macro context. Part II is devoted to the role and diffusion of IT into other sector of the economy. Part III relates to social aspects of IT sector. These Parts are interrelated and are in a continuum as Joseph [14] notes: 'Analytically, the contribution of IT towards development could be viewed at two, different but interrelated, levels; (a) on account of the growth of IT industries and (b) on account of the diffusion/use of IT. The former refers to the contribution in output, employment, export earning etc on account of the production of IT goods and services and is often more visible than those from use. The latter refers to IT-induced development through enhanced productivity, competitiveness, growth and human welfare on account of the use of this technology by the different sectors of the economy and society'. As a corollary to the above mentioned aspects, there is one more dimension which flows from the above two aspects. This relates to the social dimensions of IT sector which, *inter alia*, include digital divide, regional divide, changes in the culture and outlook of people and society, changes in the family dynamics, interaction among different technology group members, psychological impact etc. In the backdrop of this context, the present volume has been arranged in three parts. A chapter-wise resume of these parts is presented below.

Part I: Performance, Challenges and Prospects of Indian IT sector

This part presents a macro perspective of Indian IT sector and its performance, challenges and opportunities. In a bunch of 9 chapters it is difficult to deal with all the relevant issues. Given this we have tried to put together some macro themes which in our considered opinion form the bedrock for policy formulation and implementation.

As mentioned earlier, India has emerged as a leading force in the area of IT especially software and related services exports in international market. Software exports have recorded about 50 percent annual compound growth rate in the last decade. This unparalleled success story of India is unfolded by Professor K.J. Joseph in Chapter 1. His paper sheds light on India's performance with respect to production and exports of IT, the underlying factors in India's success and the limits and challenges to India's achievement. He argues that despite non-availability of intellectual property rights, infant industry protection and non-supportive trade regime, Indian IT industry has been moving up the value chain and has embarked on diversification. Professor Joseph emphasizes the pro-active role of government in capacity building in the innovation system to sustain the present growth and harness the benefits in future. The author concludes:

> The experience of India in the IT sector is a clear indication of the possibility of developing countries to participate successfully in the international division of labour in knowledge intensive sectors. Though India is one among the pioneering developing countries to initiate policy measures and institutional interventions towards evolving a broad-based electronics industry, the performance in the IT hardware sector has been less remarkable because of its failure to reap the linkages between trade and investment in an assembly oriented industry like electronics. While hardware lagged behind, the software sector emerged as a major source of export earnings and employment generation and there are indications to show that the industry has been moving up the value chain. The existence of a vibrant sectoral innovation system built up over the years, as evident from the highly competitive production and export network led by local firms with active participation of foreign firms, network of R&D institutions and firms in a conducive policy environment, seems to have facilitated an upward movement along the value chain. This is evident from the country's increasing participation in the global division of labour in R&D, embedded software development and other high value adding activities.

Heeks (1996) [15] argued that single most important factor that led to the emergence of internationally competitive software and service sector in India has been the availability of skilled manpower at highly competitive rates. However, this labour cost arbitrage cannot be sustained as some of the low cost developing countries such as China, Sri Lanka, Ireland, Philippines etc. are emerging as strong competitors for India in the international software and related services market. A plausible strategy adopted by Indian software firms to counter and overcome the competition unleashed by these countries is 'Economic Upgrading'. Economic Upgrading in a broader sense means over all improvement in the Economic capability of a firm or a service provider. Dr. Babu P. Remesh in Chapter 2 tries to explore the kind of economic upgrading strategies being pursued by the enterprises in the IT sector, within globally integrated supply chains. Determinants and constraints of upgrading and possible impact of upgrading on the social aspects and welfare of labour engaged in the sector (such as labour standards and promotion of decent work etc.) are also attempted in the paper. He concludes:

> The firms in IT Services and ITES/BPO Sector in India are currently in an intensive phase of economic upgrading. From the perspective of the entrepreneurs and employer bodies in the industry, this process of economic upgrading is expected to bring certain favourable results in the economy as well as labour market, which include: enhanced contributions to GDP and export earnings, advancements in technology and economic infrastructure, gains in employment levels and quality of employment, increase in labour standards (in terms of better work environments, improved working conditions, higher salaries, perks etc.), increased allocation of funds towards training and skill upgradation of workforce, greater earmarking of resources towards social spending and so on. In this context, it is important to have more detailed and socially/worker oriented enquiries that confirm whether the perceived advantages of economic upgrading under global production chains are ultimately realized by the workers in the sector as well as the society at large.

The positive impact of investment in plant and machinery and R&D on productivity, long run growth, generation of learning effects (which spill to other firms) have been extensively discussed in theoretical and empirical works (see for instance, DeLong and Summers 1991; [16] Greenwood et al. 1997; [17] Shaw 1992; [18] Romer 1986; [19] Griliches 1979 etc.). [20] Dr. Shishir Saxena in Chapter 3 re-tests some of this related proposition by taking Indian electronics industry as a case study. This study is a useful addition to the existing literature. Dr. Saxena attempted to verify whether a firm's own R&D and investment in plant and machinery has led to substantial output and productivity growth of these firms and whether knowledge spill over could explain exemplary growth of Indian electronics industry. He contends:

> Not only does knowledge generated by a firm's internal efforts, but also, intra-industry pullovers generated through all other firms' investment efforts, favourably affected the output of Indian Electronics firm. The analysis suggests that the recently purchased equipment stock of a firm favourably affects output. This stock may comprise domestically purchased and/or imported machinery. Since the latest machinery embodies contemporary technology, it is suggested that firms, including India's large number of labour-intensive firms, should ensure that their equipment stock incorporates the latest design and technology. There is strong evidence linking a firm's productivity to the "learning by watching" effects from investments in equipment, made by all other firms. This further underscores the need for using equipment of a new vintage, which would positively affect the output of all firms in that industry. Furthermore, this spill over suggests that a firm enjoys the "level of development" effects of the industry to which it belongs. In a labour-abundant country such as India, the need for training the workforce, therefore, assumes even more significance. It is

also found, albeit only partially, that a firm's own R&D, makes a significant contribution to output.

With companies across the globe becoming more cost and quality conscious, the need for maintenance of quality at every level is being felt. Therefore, many companies have gone for different quality standards such as ISO 9000: 2000, ISO 900 and 14000, SEI-CMM, PCMM and Six Sigma etc. India can boast of the largest number of companies in the world with SEI-CMM level 5, the highest quality certification in the world. Indian companies have realized that there is no short cut to success except acquiring quality standards. Hence acquiring and maintaining quality standard is one of the solid strengths developed by Indian software companies to have edge over its competitors. Undoubtedly, cheap availability of skilled labour, world class quality along with business skills and re-programmability has transformed India into a 'global powerhouse' in the IT software and service sector. Chapter 4 by Professor Santhosh Rebello assumes significance in this backdrop. It examines how quality (along with other factors) has helped in the emergence of India Inc Software Pvt Ltd. Professor Rebello argues:

> India has moved rapidly on the quality front. Indian software organizations first adopted the ISO model and then shifted to the Capability Maturity Model (CMM. Though no doubt the success of Indian companies with these frameworks is due to the exam taking abilities that our people have, still these models have been by-and-large absorbed quite well. This has given the Indian companies solid project and process management strength these models are supposed to bring. In fact, outside of the developed countries, it is only India that has companies that have been capable of successfully adopt global quality models. The second strength is the business expertise that now exists in the industry. With time, experience, projects and dealings with clients across the world, many managers have developed good understanding of business practices in other countries and have acquired domain knowledge and an understanding of the cultural context of the customer. Finally, Indian software companies have developed the ability to change rapidly with technology. This has been amply demonstrated by the quick movement to Java, E-commerce related technologies etc.

India presents a skewed picture of booming software and lagging hardware industry. The hardware revolution is also essential for the continued high growth of the software industry. As Vinnie Mehta, director of MAIT, puts it: "India can lose out on the software advantage it has already built up, and the future potential, if it does not concentrate on the hardware front. [21] Again, the hardware industry has the potential to create large scale employment opportunities even to the uneducated youth of the country. But unfortunately hardware industry is lagging behind and software industry is booming. Professor D.N. Rao in Chapter 5 highlights the consequence of this divide. He expresses concern over low growth in domestic hardware production and exports. He opines that imports of hardware are likely to increase as compared to its exports. In the electronic sector we are nowhere as compared to Asian Tigers like China and Taiwan. He is of the view that our commitment to zero duty import on hardware by WTO agreement spelled disaster for the Indian hardware industry. He visualizes a strong oligopolistic market structure and hence rules out any meaningful competition in the global hardware market. He suggests that for sustainable development of this sector we will have to set up production through continuous technological innovations. Some clear deep thinking is needed in this area at different levels.

Trade Related Intellectual Property Rights (TRIPS) is an important Agreement under the aegis of World Trade Organization (WTO). The objectives of TRIPS are: to promote technical innovation, R&D, transfer and dissemination of technology and so on. This in turn entails monopoly to product owners and is expected to promote R&D and offer improved technology and product to users. Strengthening of Intellectual Property (IP) is recommended on the ground that in the long run it is beneficial to the society. Even though in the short run it reduces competition and increases price in the market in which IP protected product is sold, it is claimed that in the long run by providing rent or monopoly profits, it increases incentive to undertake R&D, by allowing the fixed costs of R&D to be recouped. This in turn confirms long-run gains in terms of improved technology and better products.

Contrary to the expectations, Neha Aggrawal and Ravinder Goel in Chapter 6 find that Intellectual Property protection does not lead to these perceived benefits as claimed by the protagonists of Intellectual Property Rights. Elaborating their arguments with reference to Microsoft Corporation of USA and Infosys of India, the authors attempt to see whether or not IP protection spurs R&D and offers improved technology and product to users. According to them Copyright laws confers monopoly on its owners and monopoly leads to waste, profit maximization etc. The authors argue:

> For developing countries, the case of strong IPR protection has always been debatable. Firstly, they are net users of research and development-intensive products and they do not benefit from the monopoly profits that are created by IP protection. Their consumers suffer from the higher prices. Second, because their markets are small in relation to global demand actions by them to strengthen IP protection have little impact on the incentive to undertake additional R&D. Thus, a combination of higher costs in the short-run and the likely absence of dynamic

> gains over time mean that raising levels of protection would not be beneficial to them. Further, the claims of strong IP laws promoting R&D can also be questioned. In fact, in the case of software, at least it seems that guarding or maintaining monopoly, which is technically not feasible, is leading to heavy wasteful expenditure on sales and marketing only.

Dr. Zakir Hussain in Chapter 7 attempts to discern and delineate the deep rooted nexus between globalization, technology and international migration. He argues that the role of technology in international migration has become more emphatic and decisive since the mid-20th century as the world witnessed the ICT (Information and Communication Technology) revolution. The cheap and widely accessed network of the IT has not only helped the far flung people to connect themselves but enabled them to pass on the relevant information as well as link up within no time. The real phase of globalization, in fact, took place when 'internet' entering common man's life. This enabled the global creative forces to access information and move to those regions and countries where such skills and talents were either scare or welcome. This is more likely to increase in the future. The author does not consider movement of skilled professionals especially IT professionals as 'Brain Drain' rather treats it as 'Brain Gain' and 'Brain Circulation'. In this connection Dr. Hussain contends:

> The implication of the movement of skilled professionals from the abundant countries like India has resulted into both kinds of debates, positive as well as negative. The positive side of the debate underscores that the movement of skilled personnel from India to the developed countries has transferred the much needed capital, new ideas, technology know-how and more over working as a 'bridge community' in fostering trade and commerce between India and the developed world. So far the negative side of skilled emigration is concerned though many consider that the drain of the skilled personnel would surely produce cascading effects on our own systems, but this notion is not so popular now as it was during the 1960s and 1970s. Today, the receiving countries are themselves conscious to return back the origin countries the fruits of their talent pools (brain gain). They have resorted to a number of policies in this direction. These measures will surely encourage the current as well as potential migrants to utilize their hard-earned income for productive purposes. In order to maximize the net effects of the emigration of the skilled workers, at policy front a country like India needs to emphasize on: (i) promoting mostly the returnable migrants through mapping out the emerging skills in developed countries and matching it with our potential supply; (ii) harnessing the skills of the return migrants by maintaining proper records and their skill specificities; (iii) utilizing the Diaspora platform for developmental purposes; and (iv) reorganizing our emigration policies in view of the transforming immigration policies of the developed countries.

Demand for and supply of IT resources has been one of the main areas of concern for IT sector. Demand means volume of services and goods required, which users are willing to acquire at a given price in a given period of time. Supply of IT unlike the common concept of supply used in economic terms is the availability of entire skill to the field at a given period of time. The general perception is that there has been a shortage of IT resources especially skilled and trained labour. This is because of many reasons: Faster growth of IT Industry and inability of educational institutions to keep pace with the requirement of the industry. Moreover, all the available labours are not fit for employment in IT sector. A Nasscom study [22] shows among the 3 million graduates and post graduates added to the Indian workforce, only 25 percent of technical graduates and 10 to 15 percent of general graduate are considered employable by growing IT/ITES segment. Added to this is the movement of best of the talents to other countries. Dr. Krishan Kumar in Chapter 8 focuses on these crucial issues of demand and supply of IT resources and suggests a viable way for the creation of stable demand-supply framework for IT resources through splitting of business organization into demand and supply units. He argues:

> In most of our discussions on the evolution and development of information technology, the United States is always talked about. According to much of what one hears in media reports and policy debates, the US never had enough high-tech workers. Warnings of shortage of engineers, scientists, and now information technology specialists have appeared regularly since the 1950s. On most the occasions, such professionals have been sourced from other countries and India has always been a big contributor. With the advent of globalization the trend has spread over the developing world. Emergence of countries like India and China to the fore of the Info-Tech has brought in new issues. While China has emerged as a strong global production base for the hardware, the emergence of India as a software master and producer of human IT talent has drawn everyone's attention. Technological strength is seen as a crucial ingredient of economic prosperity. Now the country itself feels being short of trained and skilled IT talent in many areas. The major problem has been the movement of the best of the talent to other countries. Students from the premier institutes are lured by the overseas hirers. Other factors add to the continual pressure to maximize the country's supply of technical talent. Large labour pools help employers contain their payroll costs and improve the odds for a steady supply of occasional brilliant performers, the technical wizards whose contributions have been shown to be crucial. Training all these specialists keeps academics busy; large academic establishments also contribute to research and development capabilities, fuelling the emergent technologies.

The emergence of Sri Lanka in the world Business Process Outsourcing (BPO) map provides both challenges and opportunities for Indian BPO sector. Today, Sri Lanka's IT and BPO sectors are drawing the attention of foreign investors in a moderate way with latter wanting to put up an alternate business hub to their Indian operations in case of any exigency. At present Sri Lanka may not be a threat to Indian BPO sector as it is focusing on niche markets (such as financial accounting work, analytical works etc.) rather than trying to compete directly with the general services that are being offered in India and elsewhere. Above all, Sri Lanka's IT and BPO is a less than 1 billion business while India's IT and ITES sector is a 70 billion plus industry and it has established a brad name in the international market. Moreover, the prospects of Sri Lankan IT sector are marred by civil and political instability, lack of mass transport, high telecom cost, constraints on capacity creation in education etc. But in near future Sri Lanka may be able to overcome these problems and attract investment in IT sector on a big way as it has one of the lowest labour cost in the region. These are some of the interesting issues discussed by M.F.M. Nusry in Chapter 9. The author argues:

> Sri Lanka's vicinity to India, frequent transport facilities, lower taxations, tax holidays, cheap floor prices, high literacy rate, lower labour costs, low attrition rates and attractive investment offerings by the government for foreign investments are making Sri Lanka an attractive destination. With the appreciation of Indian currency, increasing wages in the BPO sector and tax revisions, the sector is getting hampered and players are looking at Tier-2 cities and alternatives. In contrast the appreciation of the US dollar and the British Pound make the industry even more attractive in the Island country. Also due to the vicinity and close ties with India some players are considering Sri Lanka as a second operational and backup site. With a limited labour force Sri Lanka is not comparable to the Indian market but its niche segmental offerings for: financial accounting research, analytical work as well as medical and clinical sectors (due to the availability of British certified accountants, doctors and researchers) may help it expand, attract and create more BPO players.

Part II: Diffusion (Use) of IT into Other Sectors of the Economy

IT being a general purpose technology can be applied in each and every aspects of human activity and it does not know barriers of race, religion, sector, states, fields, occupation etc. IT is revolutionizing life on earth in such a way that no other technology has done in the human history. It has potential to transform the world into a 'Global Village'. Against this backdrop this section presents a sample of 5 chapters to highlight the significance of IT in sector specific development strategies. It is true that this section does not include all of the key sectors of the economy. But given the central focus of the section we understand that the illustrations under reference are enough to elaborate the forward and backward linkages of IT.

Information and Communication Technologies can play a pivotal role in enhancing cooperation among developing nations particularly among South Asian Countries. The full development potential of the regions could be materialized by mutual cooperation. The policy of give and take can indeed work to the benefit of all. The South Asian cooperation in ICT can cover: (1) Education and Training through building institutions of research and higher learning; (2) Joint development of hardware and software in local languages and open source platform like LINUX; (3) Exchange of experiences in Electronic Governance programs for poor and the under-privileged. However, ICT sector in these countries are facing numerous hurdles. Dr. Somesh Kumar Mathur in Chapter 10 discusses some of these issues and suggests ways for policy formulation. Dr. Mathur argues:

> There are common set of problems faced by ICT sectors in the South Asian countries. These problems include limited access to ICT technologies and low tele-density especially in rural areas, inadequate quality and skills of graduates, rising salaries not commensurate with increase in productivity, weak infrastructure which results in frequent power shortages, low level of PC (personal computer) use and internet penetration, low level of domestic technology development, limited bandwidth, inadequate availability of venture capital for the small scale units, high attrition rates in BPO companies, e-security, improper content design of e-governance programs and limited domestic market for knowledge based technology and products. The South Asian governments have to address them in times to come and they need to pool their resources to overcome these problems.

Introduction of IT and internet technologies have revolutionized the way stock markets function and retail stock investment is made. Various empirical studies have confirmed that introduction of information technology has helped stock markets to improve performance and efficiency, reduce the cost and enable the quick transaction and innumerable other benefits to various stakeholders. One important area of stock market wherein internet technology has played a vital role is the retail trading (E-trading/Online trading). E-trading now accounts for more than half of retail stock trades in most of the developed countries. Considering the enormous potential of E-trading SEBI approved E-trading in India since 2001. However, it has failed to enthuse and gain popularity with majority of the retail Indian investors. There exists a sharp digital divide among the retails stock traders due to non-E-comfortability and retail trading is still dominated by the presence of traditional traders who trade via the broker's interventions. Prathiba

Wasan in Chapter 11 examines the prospects and problems of this emerging area. She argues:

> E-trading in India is not evolving along the global lines. The electronic markets of India have taken a deviation from the rest of the world. There is a huge digital divide and the restricted or no use of Internet for the purpose of transactions is self-imposed by the traders. This self-exclusion is resulting from poor scores on E-comfortability. If the E-comfortability deficits are overcome, one can hope for a better prospect for E-trading in India and this would require the active involvement of all the market participants including the investors. All hope is not lost as players like Reliance and Bharti have started laying optic fibre across the length and breadth of the country and in a matter of years, the last mile connectivity issue will be solved. The solution lies in balancing a technology-centric approach to transactions with the human touch factor which will be key determinant in making the transition from the traditional method of broking to e-broking. Brand building, assurances of security, developing multiple delivery channels with anytime telephonic grievance redressal options is some directions which may be of use for the immediate future. Online trading firms can also market themselves aggressively to students who are entering the professional arena, ensuring that their entry into equity happens online.

Energy Security and Energy self- sufficiency is a global concern today. Due to the ever increasing demand for oil and gas the energy industry is unable to meet global demand for oil and gas. A major challenge in energy market today is locating where the oil is, processing it and distributing it to different consumers at minimum cost. Adoption of IT into energy market can greatly benefit both consumers and producers. IT solutions of Seismic mapping, deeper onshore and offshore drilling production technique, reservoir stimulators and other advances have increased the accessibility of oil and gas. Above all, technology has diversified supply sources by allowing more energy to be accessed in new locations around the world. India being a net importer of oil and gas, enhancement of domestic production is need of the hour. Adoption of IT technology in exploration, production and distribution can play a significant role in this direction. Chapter 12 by Vrushal T. Ghoble is an attempt to study some of the IT technologies like Broadband Satellite Technology, Supervisory Control and Data Acquisition Solutions, Intelligent oil field etc. and their benefits to Indian energy market. Ghoble contends:

> The Communications and IT (Information Technologies) industries are rapidly converging with the energy industry. IT solutions to the oil and gas industry could tremendously enhance the hidden potentials of the emerging market. The four major business issues in the oil and gas industry (vis-à-vis Information Technology) are speed to market, convergence, consolidation and cost reduction.

> Each of these areas calls for specific integrated IT industry solutions. India has become highly dependent on oil and gas. In this context, IT can prove pivotal in sufficing to its energy needs. In a situation, where changes in the global demand are bringing pressure to increase production and revenue, it would be topical to probe, how new Information Technologies can address to the global oil and gas demand.

E-Governance has become a buzz word now. It is considered as an alternative system of governance to conventional system of governance and has the potential to ensure good governance. Studies have revealed that governments which use ICT effectively are able to lower their unit cost of operations and produce same (or higher) output with less money and less time. [23] It ensures Simple, Moral, Accountable, Responsive and Transparent governance (SMART). Harinam Singh and Pham Van Bo in Chapters 13 and 14 respectively throw light on the various aspects of e-governance in India and Vietnam respectively. Although India ranks slightly better (54th rank in the e-readiness ranking of Economist Intelligence Unit, 2007 as compared to 65th of Vietnam) than Vietnam there are many parallels, similarities and differences in the adoption of e-governance in these countries. Both can learn lessons from each other's experience. Singh's study focuses on various e-government projects across India with a view to explore nature of implementation of these projects, benefits imparted from them to citizens and their up and down sides. The application and implementation of IT in the area of self-government i.e. Panjayath Raj Institutions are also attempted in the paper. He is of the opinion that e-governance is a tool for governance not an object in itself. So while declaring agenda for e-governance, it must be defined what kind of good governance will be achieved through e-governance. In India implementation of e-governance is marred by a number of problems. Singh opines:

> E-Governance can use IT as a vehicle to create a knowledge society leading to the improvement in life of a common man. The only handicap in the e-governance system is that in India, there are a large number of illiterates and even a large number of computer illiterates. In the modern age, the latest technology and scientific knowledge have become imperative for maximizing social welfare of the people. Though a lot of progress has been made within the last decade or so in the realm of innovations and their applications, much is left to be done in our rural areas. We should not lose sight of the fact that unless life in the rural areas is improved with the help of modern technology, it would not be correct to say that India has been progressing in maximizing its social welfare.

Dr. Pham Van Bo recasts Vietnam's experience and experiments with e-governance. The main focus of Bo's paper is to explore Vietnamese policy framework for e-governance and Public Administration Reform (PAR) and assess Vietnam's e-readiness and compare it with other countries' e-readiness for governance. He argues that despite many policies and efforts to implement e-governance Vietnam still lags behind in the e-readiness index. The author proposes a number of policy prescriptions. Dr. Bo notes:

> The Vietnamese government has long recognized the important role of information communication and technology (ICT). Since 2000, a number of strategies, laws, decrees and decisions have been approved and a number of government and party committees and boards have been established to implement these plans. In 2001, the Prime Minister approved the Public Administration Reform (PAR) master plan 2001-2010. Public Administration Reform is an essential element for e-government success and will also impact project 112. Both state administrative management computerization (SAMcom) or project 112 and PAR are mainly supported to buy hardware. Vietnam lagging behind not only its Asian neighbours but also much of the world. However, she has gained progress of ICT development in the year 2007-08.

Part III: Social Aspects of IT Sector

New Technology not only affects and alters the prevailing modes of production and distribution but also impacts Socio-Psychological life of both consumers and producers. Capturing and analyzing this interactive process is necessary to minimize the net benefits (both private and public) of a given technology and its contribution to total factor productivity. This aspect of IT has not however received the required attention of the researchers and policy makers. There is urgent need to study the Socio-Psychological and Economic-Environment impact of IT in a decentralized sectoral framework. The results of the two studies presented in this section highlight the scope and importance of these aspects of IT industry.

India has witnessed a BPO boom since 2000. This has opened enormous employment opportunities for the educated graduates of the country-both males and females. BPO job is round the cloak business and the employees are expected to do night shifts just like other shifts. In a tradition bound society like India how this new development of unconventional working of women in night shift is perceived? How do the parents react to this unconventional working hours of their daughters? How does the relationship between daughter and other family members is altered due to changes in work-family dynamics of women workers in call centres? Shelly Tara and P. Vigneswara Ilavarasan make a bold attempt in Chapter 15 to throw light on these issues. The study is based on the field survey conducted in National Capital Region/ Delhi and Jaipur. Contrary to the conventional ideas the study concludes that the parents are both supportive in terms of odd hours of night shifts, less family interaction, night party etc. of their daughters working in BPOs. The authors argue:

> Fathers of the women employees are 'cool' (term used by a respondent to indicate the support) about the call centre jobs. In the initial phase, mothers were little reluctant to accept the daughters' night shifts, but became adjusting and supportive later on. Parents do not consider daughters' night shifts as disgraceful. The lesser interaction between parents and daughters as an outcome of night shifts has also not been posed as a problem. Female workers accept it as part of their job and are happy to spend time with their parents on weekends. On the other side even parents understand the demands of their daughter's job and do not complain when their daughters do not spend time with them or help their mothers in household work during weekdays. The absence of daughters during festivals has also been accepted as part of their job. However, the perception of the Indian families on having such daughter-in-laws who work in call centres with night shifts is needed to be studied further as, respondents observed that life partners from the same sector are essential for continuing the job during post-marriage period.

The interaction between technology and social dynamics of organization has been subject of discussion and debate ever since the first conference on Computer Supported Collaborative Work in 1986. In this very first conference following questions were raised:

- What is the proper balance of control between human and machines?
- What are the effects of current technology on group structures?
- How will technological innovation affect the way large scale group activity is organized in the future?
- What research methods are appropriate for understanding/studying these phenomena?

M.Y. Khan in Chapter 16 analyzes this dynamic issue of interaction between technology and social dynamics of organization taking software process management as a case study. Software process is team work, cooperation, coordination and communication that occur within and among groups throughout the life of a software project. This includes those processes that occur under the broad categories of proposal writing, software engineering, development and maintenance. The problem of developing automated group support tools for software process management is interesting and critical. He concludes:

> We assert that active, intelligent, adaptive, orchestrated, groupware that manages multiple and possibly concurred accesses to shared work spaces in the desirable goal for future technology, for example, a groupware system for software process management that plays an active role (takes some initiative) suggesting

assignments for members of group, in accordance with the project's needs and schedule and based on the experience and availability of project staff. When considering groupware for software process management, however, we believe that the system must contain one important quality: flexibility. The groupware must be flexible enough to allow the dynamics of the group and the culture of the organization to influence its use. One group, for example, may need to use a tool in an entirely different manner from another group, perhaps even employing it in a manner not foreseen by the tool designer. In an authoritarian group, a manager may use an on-line group calendar facility to call meetings or even to assign particular times for employees to perform certain tasks. In a more egalitarian group, members may use the information on the group calendar facility to request for meetings with each other.

Notes and References

1. Sandeep Joshi (2009), "Software Services Exports to Reach $ 47 b in 2008- 09", The Hindu, February 5 , 2009, pp 16, Delhi Edition
2. Department of Information Technology, "Information Technology Annual Report 2007-08", Ministry of Communications and Information Technology, Government of India, 2008
3. Nasscom, "Strategic Review 2008", Executive Summary, Nasscom, New Delhi
4. Sandeep Joshi (2009), "Software Services Exports to Reach $ 47 b in 2008-09", The Hindu, February 5 , 2009, pp 16, Delhi Edition
5. Raghavan Vijaya G. and Nair V.S.M.(2001),"One of the Greatest Discoveries of our Time", Manorama Year Book, Kottayam
6. N.N. Sachitananda (2001), "Optimism at Nasscom-2001", *The Hindu*, 18 March
7. Sunder Shyam " Investing in the Future of Indian IT Industry: Prospects and Policy" Paper Presented at the international seminar on 'IT Industry in India-Prospects, Retrospects and Policies', organized by Department of Economics, Jamia Millia Islamia University, New Delhi during 1-2 November 2007
8. Nasscom, "Strategic Review 2008", Executive Summary, Nasscom, New Delhi
9. Sen Pronab (1995), "Software Export: An Assessment", EPW, February 8
10. Offshore Services: Software and Services Development on Indian land
11. Illiyan Asheref (2008), "Performance, Challenges and Opportunities of Indian Software Export", Journal of Theoretical and Applied Information Technology, Vol. 12, November
12. Dataquest, "Industry Overview" Vol. XXIII, No.13 July, 2005, New Delhi
13. Illiyan Asheref (2008), "Performance, Challenges and Opportunities of Indian Software Export", Journal of Theoretical and Applied Information Technology, Vol. 12, November
14. Joseph K.J (2009), "IT Industry in India: Leading Software and Lagging Hardware", Chapter 1 of this Volume.
15. Heeks R. (1996)," India's Software Industry: State Policy, Liberalization and Industrial Development", Sage Publication, New Delhi, Thousand Oaks, London.
16. DeLong, Bradford and Lawrence Summers (1991). "Equipment Investment and Economic Growth." *Quarterly Journal of Economics,* 106(2): 445-502.
17. Greenwood, Jeremy, Zvi Hercovitz and Per Krussell (1997). "Long-run Implications of Investment-Specific Technological Change." *American Economic Review*, 87(3): 342-362.

18. Shaw, G. K. (1992). "Policy Implications of Endogenous Growth Theory." *The Economic Journal,* 102(412): 611-621.
19. Romer, Paul M. (1986). "Increasing Returns and Long-run Growth." *Journal of Political Economy* 94(5): 1002-1037.
20. Griliches, Zvi (1979), "Issues in Assessing the Contribution of Research and Development to Productivity Growth" *The Bell Journal of Economics* 10(1): 92-116.
21. Source: http://www.expresscomputeronline.com/20020408/cover1.shtml accessed on 4 February 2009
22. See Times Ascent "Leave it to the Experts", September 26, 2007, pp iv, New Delhi edition
23. Heeks R. (2001), "Understanding E-Government for Development, i-development paper, No. 11, IDPM, University of Manchester, UK

*The terms IT Sector and IT Industry are interchangeably used in the present volume

Part I: Performance, Challenges and Opportunities of Indian IT Sector

1

IT Industry in India: Leading Software and Lagging Hardware

K. J. Joseph

Introduction

It is generally perceived that Information Technology (IT) is the General Purpose Technology of the new millennium that could be instrumental in enhancing efficiency, competitiveness and growth in all economies and help achieving the millennium development goals by the developing countries. Analytically, the contribution of IT towards development could be viewed at two, different but interrelated, levels; (a) on account of the growth of IT industries and (b) on account of the diffusion/use of IT. The former refers to the contribution in output, employment, export earning etc. on account of the production of IT goods and services and are often more visible than those from use (Kraemer and Dedrick 2001). The latter refers to IT induced development through enhanced productivity, competitiveness, growth and human welfare on account of the use of this technology by the different sectors of the economy and society. Studies on technology diffusion have shown that greater domestic availability of technology acts as a catalyst in the process of diffusion (Stoneman, 2002). As Ernst (2001) rightly remarked enhancing the diffusion of IT, however, does not imply a neglect of IT production. Both hang together and need each other. No wonder, there is hardly any developing country that has not undertaken policy measures and institutional interventions to develop IT industries and harness the new technology as a short cut to prosperity. The outcomes of these initiatives so far have been unevenly distributed; IT production capabilities, by large, are concentrated in the developed countries and sharp divides characterize the digital world order.

In this context it may be inspirational for other countries in the south that a developing country like India has acquired significant capabilities in some segments of IT like software and services and has immensely profited from its production and export. Indeed this capability has not been the handiwork of the market but has been an outcome of the innovation system built up over the years mostly at the instance of state and also other stakeholders like industry associations including the innovative efforts by the individual firms. [1] At the same time, India's performance with respect to harnessing IT for enhancing competitiveness and growth of different sectors of the Indian economy that are today exposed to international competition more than ever before, has not been highly remarkable. In this context there arise a number of issues of immense policy relevance but our understanding remains at best rudimentary. In a broad sense these issues relates India's performance with respect to production and export, the underlying factors in India's success and the limits and challenges to India's achievements. It goes without saying that an understanding of these issues is vital for informed policy making in the future. The present paper intends to shed light on some of these issues.

Growth Performance: Hardware versus Software

Performance of IT Hardware: In its broad sense the IT industries comprises of both IT hardware and IT software. The latter comprises of software products and software services that also include the IT enabled services. The former comprises not only computer hardware but also the whole lot of other electronic goods that are used in information gathering, processing and dissemination. India is one of the pioneering developing countries to make deliberate policy measures and institutional interventions as early in the mid 1960s towards developing an electronic production base in the country. The strategy during the early years was on self-reliant growth in tune with the then general industrial/technology policy framework. [2] The strategy was to build up on a deliberately derivative basis, an integrated structure so as to meet the requirements on the basis of local manufacturing (Department of Electronics 1979). Hence the Indian plans, while assigning due role for the private sector that was in its infancy, envisaged a greater role for the public sector units set up by the central government and various regional level public sector units like like Keltron (in Kerala), Uptron (in UP), Meltron (in Maharashtra) and others established by the state governments. Above all, the importance of technological self-reliance was upheld and given the disenchantment with FDI, the strategy envisaged limited role for foreign investment. To bring out the desired changes, the government used a variety of control instruments relating *inter alia* to investment and trade like the Industries Development and Regulation Act, Monopolies and Restrictive Trade Practice Act (MRTPA), Foreign Exchange Regulation Act (FERA) and other measures.

The strategy remained more or less the same in essence for a decade or so. In a general environment wherein the economy has been faced with diminishing returns to increased restrictions, the Government acted on the recommendations of the Sondhi Committee and Menon Committee (Government of India, Department of Electronics, 1981). The result was a gradual, but steady, move towards a market-oriented policy regime in the 1980s through a series of policy changes marking the second phase. These policy changes sought a liberal climate, both internally and externally, through dilution of the industrial licensing, relaxations of MRTPA and FERA provisions, liberalization of imports and greater access to foreign capital and technology. Moreover, considerable relaxations were effected in the fiscal regime including reduction in direct taxes and reduction in excise duties to provide a more propitious economic climate for private sector in the Indian industrial economy.

The new industrial policy of July 1991 set the beginning of the third phase marked by further liberalization in industrial licensing and greater outward orientation. In general, the 1990s witnessed the removal of industrial licensing for most of the products except a few products of strategic significance and further liberalization with respect to Foreign Direct Investment (FDI) and technology import along with series of fiscal and trade policy reforms to facilitate production and outward orientation. Further, as signatory to the Information Technology Agreement of WTO, import duties on all the ITA goods have been phased out by 2005. [3] The policy reforms were based on the basic premise that there is an inexorable link between trade and investment, which is apparently much stronger in electronics as compared to most other industries. Hence, the restrictive trade policies have had their adverse effect on the performance of the industry in the past. To be more specific, in the global production network, which today characterizes electronics, production of each of the component or sub assembly is made across different countries according to their comparative advantage such that the over all cost of production is minimized. This essentially means that the production in any country will call for significant imports and the output made will have to be exported to other countries rather than sold in the domestic market. Hence, if the production, and therefore, investment, in IT to take place in any country, the trade regime needs to be the one wherein the free flow of inputs into and outputs out of the economy is ensured.

Over the years the ownership profile of the industry has changed and its evolution in the computer hardware industry provides a broad indication of the way in which electronics industry evolved over the years. Broadly, one could discern following phases in its evolution. The first phase was marked by foreign domination followed by a period of where the public sector controlled the commanding heights. During the third phase the private sector firms began to dominate and in the final phase which coincided with the globalization phase, the foreign firms again had taken control over the industry (Joseph, 1997).

Today there are over 3500 firms in India's electronics industry that comprises of 11 central public sector units with 31 manufacturing establishments, 46 units in the state public sector, about 500 units in the organized private sector and more than 2900 units in the small scale sector. Over the years, with policy reforms, the share of organized private sector and the small-scale sector increased at the cost of public sector units. Today, the public sector accounts for only about 16 percent of the total output, which was as high as nearly 35 percent in 1981 (Joseph 1997). The organized private sector, that also include foreign firms with considerable share in computers and television, today accounts for about 46 percent of the total output recording an increase of over 16 percent since 1981. The increase in their share took place mostly during the last decade. Similarly, the small-scale sector also increased their share in output by about 10 percent during the last decade to reach a level of 38 percent in 2002 (Government of India, Department of Information Technology, 2004).

Broadly, the output performance of electronics industry since 1985 could be analyzed, based on policy changes, in two periods - with 1991 as the cut off. The post 1990s may be again divided into two sub-periods with 1997 as the cut off because there has been a number of institutional intervention and new policy initiatives since 1997 like the formation of a separate ministry for IT & electronics, reorganization and merger of DoE/DIT autonomous bodies, attempts at hardware promotion, fiscal incentives, etc. In a recent study that covered up to 2002 found that during the first phase the recorded output growth of total electronics output was substantially higher (28.6 percent) as compared to the second period (13.8 percent). As we move to the period since 1997, series of institutional interventions and policy reforms not withstanding, there has not been any marked increase in the recorded rate of output growth but a marginal decline (11.2 percent).

Similar to the trend in production, there has been a deceleration in exports, while the recorded export growth during 1985-90 (32.8 percent) was higher than the output growth, the subsequent period recorded not only a decline in export growth rate but the recorded export growth rate turned out to be lower than that of growth in production. The total electronic hardware exports from the country increased almost four fold during 1985-90 to reach a level of over 400 million dollars and thereafter exports declined during the first half of 1990s (Table 1.1). Notwithstanding the better performance during the mid-1990s, total exports in 1999 were less than their level in 1990. Since 2000 an upward trend appears to have set in and as a result during 1999-05 there has been more than four-fold increase in the export earning to reach a level of over US$ 1.9 billion in 2005-06.

Performance of IT Software and Services

The performance in software and IT enabled service exports, on the other hand, has been remarkable by any standard where it has been doubling in almost every alternative year. The recorded annual compound growth rate has been over 50 percent in the 1990s and 38 percent since 1997-98 and such a record has been unprecedented in independent India. As of now the software and service exports is over 20 percent of merchandise exports and even higher than one of India's principal commodity in India's exports viz. textile and textile products.

Table 1.1: Booming Software and Lagging Hardware: Trends in Exports (US$ million)

Year	Software	Share (percent)	Electronics Hardware	Share (percent)	Total
1985-86	27.79	22.01	98.51	77.99	126.30
1986-87	38.35	20.42	149.47	79.58	187.82
1987-88	54.37	22.59	186.32	77.41	240.70
1988-89	69.74	21.26	258.26	78.74	328.00
1989-90	94.28	20.02	376.65	79.98	470.93
1990-91	111.47	21.50	406.97	78.50	518.44
1991-92	134.84	36.67	232.91	63.33	367.75
1992-93	163.24	39.21	253.03	60.79	416.27
1993-94	286.94	50.08	285.98	49.92	572.92
1994-95	430.27	52.47	389.83	47.53	820.10
1995-96	702.55	53.74	604.79	46.26	1,307.33
1996-97	985.92	58.61	696.34	41.39	1,682.26
1997-98	1,641.34	67.79	780.04	32.21	2,421.38
1998-99	2,495.80	84.59	454.71	15.41	2,950.52
1999-00	3,692.36	89.08	452.55	10.92	4,144.91
2000-01	5,910.11	85.32	1,016.54	14.68	6,926.65
2001-02	7,129.09	85.43	1,216.14	14.57	8,345.23
2002-03	10,032.29	89.17	1,218.67	10.83	11,250.97
2003-04	12,961.95	88.32	1,713.72	11.68	14,675.67
2004-05	18,110.16	90.93	1,806.95	9.07	19,917.11
2005-06	23,243.24	92.39	1,914.41	7.61	25,157.66

Note: Data for the year 2005-06 are estimates

Source: Department of Electronics, Guide to Indian Electronics (different years), Nasscom Strategic Review (different years).

Needless to say the remarkable export performance has attracted the attention of researchers and well documented in the literature (Schware 1987, 1992, Heeks 1996, Kumar 2001, Arora et al. 2001, Joseph and Harilal 2001, Parthasarathi and Joseph 2002, Joseph 2002, Kumar and Joseph 2005 to list a few). The uniqueness of software export performance becomes more striking when compared with that electronic hardware exports. While software accounted for only about 22 percent of total electronics exports in 1985-86, it increased to 92 percent in 2005-06.

Contribution to GDP and Employment

The IT and ITES industry in India in value terms, according to the estimates by NASSCOM, has reached US$ 64 billion in 2007-08. This indicates an almost ten fold increase from US$ 4.8 billion in 1997. [4] As a result, the share of IT sector (defined by NASSCOM as comprising of IT hardware, software and services) in GDP increased from about 1.2 percent in 1997-98 to 5.4 percent in 2006-07. Bay way of comparison, Chandrasekhar et al. (2006) finds that IT revenues in 2004-05 were about 20 percent higher than construction sector and almost three times higher than in mining and in electricity, gas and water supply. What is more, gross revenue from IT services exceeded 12 percent of GDP generated in India's service sector that accounts for about 54 percent of the GDP. These estimates, however, appears to involve a gross underestimate because it does not include telecommunication (both equipment and services), mass communication output like Television and some of the other electronic products that are integral part of IT sector. Viewed thus the real contribution of IT to the GDP would be much higher.

According to NASSCOM (National Association of Software and Service Companies) Surveys, the software industry employed 284,000 people in 1999-00 as compared to 160,000 professionals in 1996 and reached a level of 1.63 million in 2006-07 recording an annual compound growth rate of more than 35 percent. It is shown that the industry is creating job opportunities for highly qualified (majority with an engineering degree) young graduates with a relatively short experience. It is especially notable that though the ITES accounts for only about 27 percent of the total exports, the employment generation is almost on par with the Software sector. [5]

During the last ten years the industry has been generating more than a 0.1 million employment every year and the total number of jobs created during the last decade exceeded one million. The observed performance of the software and service sector in employment generation appears highly impressive when considered against the fact that employment generation by the organized manufacturing sector has been on the decline during the last decade (Nagaraj 2004) and according to NSS statistics during 1999-00 to 2004-05 growth in employment in the rural and urban areas has been only of the order of 1.97 percent and 3.22 percent respectively (Chandrasekhar et al. 2006).

IT-enabled Services (ITES) and Broad-based Development

Having established the credibility in the export of software services, of late the Indian firms have emerged as the major players in business process outsourcing (BPO) through Internet or the so-called IT Enabled Services (ITES). In 1999-00 the total ITES exports was only of the order of US$ 565 million but as per the estimates by Nasscom (2006), the ITES exports increased from US$ 3.1 billion in 2003-04 to US US$ 4.6 billion in 2004-05 recording a growth rate of 48 percent and are estimated to reach US$ 8 billion by 2006-07. As a result, the share of ITES in total software and service exports increased from about 14 percent 2000 to 25 percent in 2006-07. Hence, but for the rise of ITES exports, the growth of IT exports would be even lower than what was observed earlier.

The recent growth in ITES exports from India has been facilitated by a number of factors. From the supply side, India with 253 universities and 13,150 colleges produces about 2.46 million graduates and about 290,000 engineering degree and diploma holders every year. English is widely used as a medium of instruction, which in turn provides an ample supply of manpower for ITES services at a much lower cost [6] as compared to other countries. In general the operating cost per employee in India is only about 20 percent of that in US and the comparable cost in Philippines is about two percent higher than in India. From the demand side, in the competitive world of today, substantial cost advantage is accrued by the firms resorting to off-shoring and outsourcing of services. Mckinsey (2003) shows that there are very large differences in the wages paid for equivalent skills between the US and developing countries like India and Philippines. For example the equivalent of a software developer who costs US$ 60 hour in the US cost only US$ 6 an hour in India. These wage differences resulting at least 45 to 55 percent savings for firms resorting to off shoring. In addition improvement in the telecom infrastructure leading to improved connectivity coupled with reduction in the cost of communication *inter alia* on account of increased competition also facilitated the ITES boom. [7]

The ITES/BPO services, experiencing a boom at present, have certain characteristics that could contribute to broad based development. While employment in the Software sector has been mainly for the highly skilled IT professionals, the ITES sector generates more broad based employment including the arts and science graduates. It is also found that ITES sector is more employment intensive with employment per million dollars of exports as high as 70, which is more than twice that of the software sector (Joseph, 2004). No wonder with 27 percent of the total exports ITES sector generates as much employment as the software sector. Thus, viewed ITES/BPO appears to have the potential of generating substantial employment for the growing number of educated youth in the country. While software industry in India is shown to have led to an enclave type development, the ITES is found geographically diffused across different regions in the country and generating more linkages with rest of the economy. Hence, for those regions, which were not successful in attracting software investment, ITES offer an alternative. Accordingly, different state governments have initiated policy measures to attract ITES activity into their states with considerable success.

Moving up the Value Chain: Select Indicators

The comparative advantage of Indian firms has been in the export of services such as customized software development, mostly on site. Indian firms were found operating mostly at the lower end of value chain by carrying out low-level design, coding and maintenance. As a result, revenue per employee in 1999 is found to be only about one-tenth of Israel and one-fourth of Ireland (Arora et al. 2001). Moreover, the net export earning has been only of the order of 50 percent of the gross FOB value of total exports of software and services (Joseph and Harilal, 2001). However, of late there are number of indications to show that the trend is changing on account of increased learning and domain expertise in a number of areas. With the MNCs increasingly looking for complementary capabilities, Indian firms are getting engaged in highly skill intensive areas like chip design and R&D and thus are moving up the value chain marked by a shift away from Business Process Outsourcing to Knowledge Process Outsourcing (Parthsarathy, 2006). Now the firms are increasingly entering into high end consulting, engineering services with the development of domain expertise. As a result the revenue per employee has recorded a six-fold increase from about US$ 6200 in 1993 to US$ 35,129 in 2000 (Athreye, 2005) and the trend continues.

Since the conventional measures of innovation like R&D intensity (measured as Research expenditure as proportion of sales) has certain limits in capturing innovation in a service sector like IT, a study by Joseph and Abraham (2005) developed an Index of Claimed Technological Competence (ITC) using firm level information on their areas of specialization. The theoretical base of the index has been drawn from the literature on technological opportunity. The estimated index revealed an upward mobility of firms. To illustrate, in 1998 over 56 percent of firms were in the low index category (less than 30 percent) where as in a short span of three years the share of such firms declined to around 44 percent. Similarly, in the higher index category (greater than 60 percent) the share of firms increased from 5.3 percent in 1998 to 8.3 percent. in 2001. The estimated index of leading IT firms like Infosys, Wipro, TCS and Satyam were found to be more than 75 percent. Thus, notwithstanding any significant increase in the R&D intensity of firms there appears to have been an upward mobility of firms in terms of the estimated ITC.

Widening Domain Expertise and Applications

There are also indications to show that the Indian companies have developed domain expertise in a wide range of domains and industries. Banking, insurance and finance has emerged as areas in which they have developed particular expertise and have even launched packaged software. An evidence of the growing ability and expertise of Indian software companies was provided by their ability to manage transition from Y2K-related projects successfully. In 1998-99, 16.5 percent of the export earnings of Indian companies were derived from Y2K related projects. Over 1996-99, a substantial part (US$ 2.5 billion) of India's software exports came from Y2K projects (Nasscom, 2000). Hence, it was widely expected that the loss of these projects with the turn of the century would lead to a decline in the growth rates of exports. However, the Y2K transition has been managed successfully on account of ability to quickly diversify into Internet and e-commerce related technologies and applications leading to a sustained increase in exports (Table 1.2).

Increasing Presence in Software Products

It is generally held that unlike countries like Ireland, India's presence in high value adding segment of software product has been limited and that led to low revenue per employee in India. The situation however, has been changing overtime. Table 1.3 provides an indicative list of firms with presence in software products. Though indicative, it is evident from the Table that a number of Indian companies have managed to enter into the area of software development. For most of these firms, domestic sales account for substantial part of their revenue. A niche market has been created in banking, financial and accounting software. These include, for instance, I-Flex that has been used by over 240 financial institutions in 69 countries.

Banking solutions from Infosys (e.g. Financle) have been adopted by 22 domestic and 16 overseas banks across 12 countries as early as in 2001. Similar to Infosys, the largest IT company from India (TCS) has launched packaged software for banking insurance, securities, accounting and health care industries. TCS also launched its branded integrated suite of software tools Mastercraft which is claimed to have been received well in the US and Europe and carries a price tag of US$ 150,000.

Table 1.2: Areas of Expertise Claimed by Indian Software Firms: 2001-2002

Area of Specialization	Number firms reporting Competence in	Percent
Antivirus/Security solutions	82	13
Business Processing consultancy/Reengineering	261	41.3
CAD/CAM/CAE	71	11.2
Call centers	114	18
CD-Rom publishing/Multimedia	86	13.6
Chip design/Microprocessor/ASIC	49	7.8
Computer Games/Computer Graphics/Animation	85	13.4
Data Processing/Data conversion	158	25
E-commerce/EDI/CRM solutions	423	66.9
ERP/MRP solutions	223	35.3
GIS/Imaging	91	14.4
ISPs/Payment Gateways	72	11.4
IT education& training	152	24.1
Localization of Software	134	21.2
Medical Transcription	27	4.3
Product distribution/Support/Implementation	188	29.7
RDBMS/Data warehousing/Data mining	345	54.6
Software maintenance and migration	345	54.6
Software Product Development	420	66.5
System integration/networking	312	49.4
Telecom solutions/Communications software	191	30.2
WAP/M-commerce	211	33.4
Web content development	229	36.2
Web technology/internet/intranet	474	75
Total number of firms	630	

Source: Joseph and Abraham (2005)

Note: Percent will not add up to 100 because the firms engage in more than one activity.

WIPRO Technologies launched a number of branded products including Teleprodigy, a billing system for ISPs, and WebSecure, an Internet security package. It is focusing on global brand building and plans to come up with a branded product every year. A number of even smaller software companies have developed packaged software, which is sold in domestic market. For example, Tally, a popular accounting package for small and medium enterprises which is being used by 50,000 companies and has been approved by the Accountants' professional bodies in India and the UK has been developed by a smaller highly specialized software company (Kumar, 2001). [8]

Table 1.3: Profile of Select Software Product Firms (2002-03)

Name of the Firms	Major Products	Revenue from Product Sales (Rs Million)	Product Sales as percent of Total Sales	Domestic Sales as percent of Product Sales
i-Flex	FLEXCUBE-caters to both corporate and retail banking	3000	64	5
Infosys	Finacle Core Banking Finacle eChannels Finacle CRM Finacle eCorporate Finacle Treasury	1810	5	47
Tally	Business accounting software-TALLY ees 6.3	NA	100	94
TCS	Products for financial banking, manufacturing and health sectors	660	7	NA
Polaris (Erstwhile Orbitech)	BANKWARE, ORBI suite of financial products	600	16.4	90
Newgen	OminFlow, OmniDocs, OmniExtract, OmniCapture, OmniReports	245	76	47
Kale	Cargo Solutions, Business Intelligence Software, Passenger Solutions	85	NA	30
ESS	Makess ERP, ebizframe	70	NA	NA
Ramco	Ramco e-applications, Ramco VirtualWorks	>500	67	NA
Aditi Technologies	Talisma –eCRM product	>500	NA	NA
Subex	Telecom sector – RevMax	2550	36	8.1

International Quality Accreditation

International orientation and the increasing professionalism of Indian software enterprises has prompted them to align their processes with global best practices and to obtain international certifications. In 2005-06 among the 401 firms that reported different international quality standards 82 had SEI CMM level 5, the highest level of quality accreditation across the globe, that accounted for more than two-thirds of such firms in the world over. As many as 123 firms had SEI CMM level 2 certification or above and 330 had ISO 9001 (Table 1.4). If the evidence presented in the table is any indication most of the Indian software enterprises have strived to attain excellence in their professionalism and best practices.

Table 1.4: Status of Quality Certification Obtained by Indian IT firms

Description	No. of Firms
ISO 9001	330
ISO 9002	23
ISO 9001/9002	345
ISO 9001:2000	72
SEI CMM Level 5	82
SEI CMM Level 2,3,4	41
CMMi Level 5	32
CMMi Level 2,3,4	14
PCMM Level 5	13
PCMM Level 2,3,4	11
Six Sigma	44
Others	41
Total	401

Note: Total refers to the total number of firms for which information available
Source: Complied by the author based on NASSCOM (2006)

Shift from Onsite to Offshore

During the early years of its development, the software and service exports from the country were carried out mostly in the form of onsite development (Heeks 1996). With the setting up of a number of Software Technology Parks, which inter alia provided access to modern telecommunication facilities, and liberalized policies towards the telecom sector, which in turn led to the entry of a number private sector telecom companies, there has been a significant shift away from onsite development. By 2005-06 nearly 70 percent of the exports take the form of offshore development.

Homegrown Success and the Emergence of IT Multinationals from India

Although Tata Consultancy Services had been exporting software since 1974, the entry for example, of Citicorp Overseas Software Ltd. (COSL) in Bombay in

1985 and of Texas Instruments (TI) in Bangalore in 1986 for software development highlighted India's potential to outside MNEs more vividly (Table 1.5).

Table 1.5: IT Software and Service Exports: By Destination

Destination	2002-03		2003-04		2004-05	
	Share (percent)	Value	Share (percent)	Value	Share (percent)	Value
Americas	69.1	6590	69.4	8884	68.4	12107
Brazil	0.0	1	0.0	1	NA	NA
Canada	1.1	101	1.1	143	0.9	159
Mexico	0.0	1	0.0	1	NA	NA
USA	67.7	6464	68.2	8725	66.5	11769
Rest of America	0.2	24	0.1	13	1.0	178
Asia, Oceania US$ Middle-East	7.9	757	7.4	943	8.0	1416
Australia	0.8	76	0.8	105	0.8	139
China	0.1	10	0.1	15	0.1	24
Hong Kong	0.2	19	0.2	26	NA	NA
Japan	2.8	269	3.0	385	2.8	500
Singapore	1.7	166	1.8	227	1.7	300
South Korea	0.2	16	0.2	23	0.0	7
Rest of Asia, Oceania US$ Middle-East	2.1	201	1.3	161	2.6	460
Europe	22.2	2123	22.6	2894	23.1	4093
UK	14.0	1336	14.5	1857	14.0	2478
Germany	2.6	248	2.8	354	2.3	400
France	0.5	46	0.5	65	0.4	72
Italy	0.1	9	0.1	12	NA	NA
Finland	0.4	40	0.5	61	NA	NA
Sweden	0.5	51	0.6	76	0.6	100
Netherlands	1.1	101	1.0	129	1.4	250
Switzerland	0.7	64	0.7	91	0.7	120
Rest of Europe	2.4	228	1.9	249	3.8	673
Others	0.8	74	0.6	81	0.5	90
GRAND TOTAL	100.0	9544	100.0	12800	100.0	17705

Note 1. Excludes hardware exports 2. Totals may not add-up due to rounding-up of individual figures

Source: NASSCOM

Subsequently, a number of other western corporations began to follow the footsteps of COSL and TI, such as HP in 1989 and followed by Novell, Oracle, among others. Seeing the potential, a number of Indian companies engaged in the manufacture of computer hardware started to spin-off their software divisions (Kumar, 2001; and Kumar and Joseph, 2005, for more details). The use of satellite links for data communication by TI's development centre in Bangalore in 1987 also served to demonstrate to the government the critical importance of providing satellite data communication links for software exports from India. Hence, the government started to provide the high-speed communication links in the Software Technology Parks (STPs).

Besides these demonstration effects, the role played by MNEs in software development in India has been quite limited. Although all the big software companies have established development base in India, their overall share in India's exports of software is rather small. The MNEs however, have been recording a faster growth rate in the recent years and they account for over 20 percent of the total exports. Yet, even in 2004-05 MNEs do not figure among the top seven software companies in India, ranked either on the basis of overall sales or the exports. Among the top twenty software companies too, no more than four are MNE affiliates or joint ventures. MNEs, however, are important clients of Indian software companies. There are hardly any Fortune *500* MNEs that do not resort to outsourcing of software from India.

Indian companies are enhancing their innovative capabilities and competitiveness through a combination of green field initiatives, cross border mergers and acquisitions and global partnerships with foreign firms. The sectoral distribution of overseas investment by Indian firms reveals that services sector, mostly led by the software, accounted for the highest share of acquisitions made. Of the 594 acquisitions during 2000-07 valued at US$51 billion, 224 (37.7 percent) were in the field of IT and ITES (Pradhan, 2007). Notably the US and UK that together accounts for over 80 percent of India's IT exports accounted substantial portion of the outward investment by the Indian firms. This tends to suggest that that the underlying strategy of the Indian firms has been to get access to the new markets and other complimentary assets needed by them from the firms developed countries.

Concluding Observations: Towards a Perspective

The experience of India in the IT sector is a clear indication of the possibility of developing countries to participate successfully in the international division of labour in knowledge intensive sectors. Though India is one among the pioneering developing countries to initiate policy measures and institutional interventions towards evolving a broad based electronics industry, the performance in the IT hardware sector has been less remarkable because of its failure the reap the linkages between trade and investment in an assembly oriented industry like electronics. While hardware lagged behind, the software sector emerged as a major source of export earning and employment generation and there are indications to show that the industry has been moving up the value chain. The existence of a vibrant sectoral innovation system built up over the years, as evident from the highly competitive production and export network led by local firms with active participation of foreign firms, network of R&D institutions and firms in a conducive policy environment, seems to have facilitated an upward movement along the value chain. This is evident from the country's increasing participation in the global division of labour in R&D, embedded software development and other high value adding activities.

While the Software and ITES exports have been emerging as a new source of income, employment and export earning in the country, a study by Forrester Research group claimed that 3.3 million white-collar jobs from the US would shift to offshore countries like India by 2015. [9] This in turn has led to the US legislations banning outsourcing of state projects and even cancellation of certain contracts offered to Indian companies. However, a study by the Information Technology Association of America, while acknowledging the fact that offshoring has led to job loss in US, shows that savings from outsourcing allowed companies to create 90,000 new jobs in 2003. The report further predicted that outsourcing will create 317, 000 new jobs by 2008. [10] Besides, Mckincey (2003) and Balli and Farrel (2004) estimate that the US economy gained US$ 1.12 to US$ 1.14 for every dollar of business offshored [11] where as the benefit to the Indian economy is only 33 cents. Balli and Farrel list out many different ways in which the US economy benefits out of outsourcing. These includes increased competitiveness of US firms, benefits to consumers on account of lower cost, additional export from the US to the offshoring country repatriation of profit etc. The overall impact has been modest because the government projects on which restrictions were imposed accounted for only 30 percent of the total market of which hardly 2 percent is outsourced. [12] The new developments exposed the weakness of India's software and service sector, which is characterized by excessive dependence on a few countries for its exports, and neglect of high value adding software products more importantly the neglect of domestic market on account of excessive export orientation (Joseph, 2004).

The other challenges that the industry needs to address include, but not limited to, lagging hardware sector that in turn cannot be de linked from the increasing competition for skilled manpower between software sector and other sectors. The adverse impact on account of the resource movement effect is not confined to the hardware sector; all the sectors that require skilled manpower like teaching, research and other activities are getting adversely affected (Joseph and Harilal, 2001; Joseph, 2006). The problem could be addressed at least partly by enhancing the supply of skilled manpower as the present rate of enrolment in higher education is at a very low level of seven percent. More importantly, India is yet to emerge as a major user of IT as the present level of IT diffusion into sectors like manufacturing remains at a very low level (Joseph and Abraham 2007). In a context wherein international competitiveness is the key to survival, IT diffusion could play a significant role in promoting international competitiveness and both by generating new opportunities and by averting potential losses. Hence it is high time for India to evaluate the social marginal product of a dollar worth of exports vis a vis domestic consumption and devise appropriate policies to promote the diffusion of IT into different sectors of the economy.

Notes

1. See for details Joseph (2006), Kumar and Joseph (2006), Balakrishnan (2006) also A. K. Sen "looking beyond the traditional domain", *The Hindu*, 16, February 2007, Chennai.
2. The resolution, which set up the Electronics Commission in 1971, stated "the government attached the highest importance to the development of an integrated and self reliant electronics industry in the country. ....an intensive promotional effort relating to both production and research & development was, therefore essential to ensure a rapid growth of self reliance" (Department of Electronics 1982, pp.14).
3. Recognizing the potential of Electronics/IT industry, the Prime Minister set up a National Task Force on Information Technology and Software Development in May, 1998. The Task Force was of the view that hardware industry and the software industry are two sides of the same coin, the success of one, whether it is export of software of US$50 billion by 2008 or IT penetration drive for realizing "IT for all" by 2008, depends on the concomitant success of the other.
4. Of this, about 32 billion is accounted from IT software and service exports and the domestic market accounts for about 15 billion. The exports have broadly three components – software services (US$18 billion) Business process outsourcing (US$8 billion) and software products and engineering services (US$5 billion).
5. This includes domestic software and as noted by Desai there are problems in estimating domestically consumed software and also the employment (see for details Desai A. V., Business Standard, 8 July 2002.
6. It has been estimated that on the average the labour cost in India in the ITES sector is only about 14 percent of that in US.
7. The cost of a one-minute telephone call from India to UK and US, for example, has fallen by more than 56 percent during 2002-03 (DoT Annual Report 2002-03) and the downward trend still continues.
8. For detailed discussion on Software Product Development experience in India see Krishnan

and Prabhu 2004).

9. As quoted in *Economist 11, December 2003.*
10. Information Technology Association of America, " The Impact of Offshore IT software and Services Outsourcing on the US economy and the IT Industry" http://www.itaa.org/news/pubs/product.cfm?EventID=1037
11. This comprises of about 58 cents from cost savings, secondly the US companies repatriate some of their profits that accounts for about 4 cents. US export to the offshore location amounts to another 5 cents. With offshoring the US workers are now freed to perform other tasks, which in turn generates another 45- 47 cents.
12. "Outsourcing Backlash unlikely to Impact in a big way", Financial Express 15, March 2004. http://www.financialexpress.com/fe_full_story.php?content_id=54731.

References

Arora, A., V. S Arunachalam, J. Asundi and F. Ronald (2001) 'The Indian Software Services Industry', *Research Policy*, 30 (8): 1267-87.

Arora, A. and S. Athreye (2002) 'The Software Industry and India's Economic Development', *Information Economics and Policy*, 14(2): 252-273.

Athreye, S. S. (2005), 'The Indian Software Industry', in Arora, A and Gambardella A. (eds) *From Underdogs to Tigers: The Rise and Growth of the Software Industry in Brazil, China, Ireland and Israel,* Oxford: Oxford University Press, 7-41.

Balakrishnan, P. (2006), Benign Neglect or Strategic Intent? Contested Lineage of Indian . Software Industry, *Economic and Political Weekly*, Vol 41, No 36 September 9, 2006.

Bally, M. N. and Farrel, l. D. (2004), "Exploding the Myths about Offshoring", Mc Kinsey Global Institute, San Francisco.

Bhagwati, J. N. (1984), "Why are services Cheaper in Poor Countries?", *Economic Journal*, 94, 279-285.

Chandrasekhar, C. P. (2006), "The Political Economy of IT-Driven Outsourcing" in Govindan Parayil (ed) *Political Economy & Information Capitalism in India; Digital Divide, Development Divide & Equity*, Palgrave Macmillan

Chandrasekhar C. P., Jayati Ghosh and Anamitra Roychowdhury (2006), "The Demographic Divident and Young India's Economic Future", Economic and Political Weekly, Vol. 41, No.49, 5055-5064.

Department of Information Technology (1996), *Guide to Electronics Industry in India* Department of Information Technology, New Delhi.

Department of Information Technology, (2003), *Guide to Electronics Industry in India,* Department of Information Technology, New Delhi.

Desai, A. V. (2002) "What is this Software?, Business Standard, July 8.

Ernst, D. (2001), "From Digital Divides to Industrial Upgrading: Information and Communication Technology and Asian Economic Development', Working Paper, Economics Series No. 36 (Honolulu: East-West center).

Heeks, R. (1996), *India's Software Industry: State Policy, Liberalization and Industrial Development,* New Delhi, Thousand Oaks, London: Sage Publications.

India, Ministry of Information Technology (2004), "Guide to Electronics Industry in India", Ministry of Information Technology, New Delhi.

Joeng, K, Jeong Huh Oh, and Ilsoon Shin (2001) " The Economic Impact of Information and

Communication Technology in Korea", in Information Technology , Productivity , and Economic Growth Edited by Matti Pohjola , UNU / WIDER Studies in Development Economics, Oxford University Press, New York.

Joseph, K. J. (1997), *Industry Under Economic Liberalization The Case of Indian Electronics,* New Delhi: Thousand Oaks; London: Sage Publications.

Joseph, K. J. (2002), "Growth of IT and IT for Development: Realities of the Myths of Indian Experience", Discussion paper No. 2002/78, Helsinki: UNU/ WIDER.

Joseph, K J. (2004), "State, FDI and Export of Software and Services from India", A Background paper for the World Investment Report: 2004, New Delhi: Research and Information System for Developing Countries.

Joseph K. J. (2005), "Transforming Digital Divide into Digital Dividend: The Role of South-South Cooperation in Its", *Cooperation south, 102-125.*

Joseph K J (2006), *Information Technology, Innovation System and Trade Regime in Developing Countries: India and the ASEAN,* Palgrave Macmillan, London.

Joseph, K. J. and Vinoj. Abraham (2005), "Moving Up or Lagging Behind? An Index of Technological Competence in India's ICT sector", in Saith, A. and M. Vijayabaskar (eds) *ICTs and Indian Economic Development*, Sage Publications, New Delhi.

Joseph, K. J. and Vinoj Abraham (2007), "Information Technology and Productivity: Evidence from India's Manufacturing Sector", Working paper No. 389, Centre for Development Studies, Trivandrum.

Joseph, K.J. (2007), "Information Technology, Innovation System and Trade regime in developing Countries: India and the ASEAN", Palgrave Macmillan, New York.

Joseph, K.J. and K. N. Harilal (2001), 'Structure and Growth of India's IT Exports: Implications of an Export-Oriented Growth Strategy', *Economic and Political Weekly*, Vol. 36 (34): 3263-70.

Kalirajan K. and S. Bhide (2005), "The Post-reform Performance of the Manufacturing Sector in India", Asian Economic Papers, 3(2):126-157.

Kraemer, K. L. and J. Dedrick (2001), "Information Technology and Economic Development: Results and Policy Implications of Cross-Country Studies", in M. Pohjola (ed.), *Information Technology, Productivity and Economic Growth,* Oxford University Press.

Krishnan, R.T. and G.N. Prabhu (2004), "Software Product Development in India: Lessons from Six Cases", in A.P. D'Costa and E. Sridharan (ed), 'India in the Global Software Industry: Innovation, Firm Strategies and Development', Palgrave Macmillan, New York.

Kumar, N. (2001), "Indian Software Industry Development: International and National Perspective", *Economic and Political Weekly*, 36(45):4278-4290.

Kumar, N. and K. J. Joseph (2005), "Export of Software and Business Process Outsourcing from Developing Countries: Lessons from India', *Asia Pacific Trade and Investment Review* 1(1): 91-108.

Kumar N. and K.J. Joseph (2006), "National Innovation System and India's IT Capability – Are there any lessons for ASEAN Newcomers?", in Bengt Ake Lundvall, Intarakumnerd P. and Jan Vang (eds), Asia's Innovation System in Transition, Edward Elgar.

Mani, S. (2005), "The Dragon vs the Elephant: Comparative Analysis of Innovation Capability in the Telecommunications Equipment Industry in India and China", Working paper No. 373, Centre for Development Studies, Trivandrum.

McKinsey (2003), "Offshoring: Is it a Win-Win Game", McKinsey Global Institute, San Francisco.

Mytelka, L. K. and John F.E. Ohiorhenuan (2000), "Knowledge-based Industrial Development and South-South Cooperation", *Cooperation South*, No.1 2000, pp 74-82.

Nagaraj, R. (2004), "Fall in organized Manufacturing Employment: A Note",, *Economic and Political Weekly*, July 24, pp. 3387-90.

NASSCOM (1999), *Indian I.T. Software and Services Directory,* National Association of Software and Service Companies, New Delhi.

NASSCOM (2000), *The I.T. Software and Services Industry in India: Strategic Review 2000,* National Association of Software and Service Companies, New Delhi.

NASSCOM (2003), *Indian IT Enabled Service Providers Directory 2002,* National Association of Software and Service Companies, New Delhi.

NASSCOM (2004), *Indian IT Software and Services Directory 2003,* New Delhi: National Association of Software and Service Companies, New Delhi.

NASSCOM (2006), *The IT Software and Services Industry in India: Strategic Review 2004*, New Delhi: National Association of Software and Service Companies.

OECD (2004), *Information Technology Outlook*, OECD Paris.

Nath, P and A Hazra (2002), "Configuration of Indian Software Industry", *Economic and Political Weekly*, 37(8): 737-743.

Parthasarathi, A. and K. J. Joseph (2004), "Innovation under Export Orientation", in A. P. D'Costa and E. Sridharan (ed.) *India in the Global Software Industry: Innovation, Firm Strategies and Development,* New York: Palgrave Macmillan.

Parthasarthy, B. (2006), "The Political Economy of Indian Software Industry" in Parayil G. (ed) Political Economy and Information Capitalism in India: Digital Divide and Equity, Palgrave Macmillan, New York.

Pohjola, M. (2001), "Information Technology and Economic Growth: A Cross country Analysis" in M. Pohjola (ed.) *Information Technology, Productivity and Economic Growth,* New York: Oxford University Press.

Reserve Bank of India (2004), *Handbook of Statistics on Indian Economy*, 2005-06, Mumbai: Reserve Bank of India.

Schware R. (1987), "Software Industry in the Third World: Policy Guidelines, Institutional Options and Constraints", *World Development*, Vol. 15, No 10/11, pp 1249-67.

Schware R (1992), "Software Entry Strategies for Developing Countries", *World Development*, Vol. 20, No. 2, pp. 143-64.

Singh, H V, Soni, A. and Kathuria, R. "Telecom Policy Reform in India", http://www1.worldbank.org/wbiep/trade/services_files/Singh.pdf

Stoneman, P. (2002) *The Economics of Technological Diffusion*, Blackwell.

TRAI (2005), "Study Paper on Indicators for Telecom Growth", Telecom Regulatory Authority of India, New Delhi) http://www.trai.gov.in/ir30june.pdf

UN IT Taskforce (2005), *Innovation and Investment: Information Communication Technologies and the Millennium Development Goals*, UN IT Task Force, New York.

Wong, Poh Kam (2001) , "The Contribution of Information Technology to the Rapid Economic Growth in Singapore", in Matti Pohjola (ed) *Information Technology , Productivity , and Economic Growth*, UNU / WIDER Studies in Development Economics, Oxford University Press, New York.

2

Economic Upgrading in India's IT/IT-Enabled and BPO Sectors [1]

Babu P. Remesh

Introduction

In India, as in the cases of many other developing countries, the recent past witnessed a massive spread of global production systems. [2] Accordingly, many new sectors and occupations came into existence, which are networked and integrated to transnational production systems. Among these sunrise sectors, the most prominent is the IT Services and ITES/BPO Sector, which currently contributes more than 5 percent of GDP, annual revenue of US$ 47.8 billion and provides direct employment to more than a million workers (NASSCOM, 2007). The boom in the employment of IT Services and ITES /BPO sector led to several transformations in the labour market and there are strikingly differing opinions regarding the performance of this sector in terms of provision of quality employment. On the one hand, industrialists and business spokesmen consider the employment in the sector superior, on account of the better work environment, higher salary structure, perks and so on. At the same time a growing body of literature (both research as well as discussions in public domain) suggests that these new generation occupations are strongly characterized by decent work deficits. [3]

In this backdrop, the present essay aims at mapping the structure and pattern of the growth of the IT Services and ITES/BPO sector in India and to understand whether and what kinds of economic upgrading strategies are being pursued by the enterprises involved in the sector, within the globally integrated supply chains. As a background for discussion, earlier part of the essay provides an overview of IT Services and ITES/BPO Sector, giving due reference to its contribution towards employment and economic growth. Towards the end of the paper, it is also attempted to briefly discuss the possible impacts of the economic upgrading on social aspects and welfare of labour engaged in the sector. As the discussion in the essay is purely based on the information gathered through some interviews (with industry representatives, entrepreneurs, managerial/executive personnel as well consultants working in the sector) and supplementary data gathered from the public domain (print and electronic media) [4], the findings need to be seen as a perspective generated from the view point of industry only. [5]

Performance and Prospects [6]

During the last two decades, the services sector in India has been witnessing the fastest growth in the economy compared to the other sectors. This higher growth has largely been accomplished by export of services in general and IT Services and ITES/BPO services export in particular. In spite of its relatively young and nascent stage, the IT and ITES/BPO industry in India had shown tremendous growth and promises and unparalleled opportunities in terms of future potentials. Over 2001-06, India's share in global sourcing is estimated to have grown from 62 percent to 65 percent for IT and 39 percent to 49 percent for ITES/BPO (Government of India, 2007).

Contributions to GDP and Export Revenues

As per the estimates of NASSCOM, the contribution of the IT and ITES sector to the GDP has gone up from 1.2 percent in 1999-2000 to 4.8 percent in 2005-06 and to 5.4 percent in 2006-07. It is also projected that the share would reach around 7 percent by 2008. The total annual revenue from the IT and ITES sector (including hardware) is estimated to be around USD 47.8 billion in 2007, with a growth rate of 28 percent.

The rate of growth of the IT and ITES sector in India is five times as fast as the global IT services industry, with a commendable compound annual growth rate (CAGR) of 28 percent since 1999-2000. Of this, software and service exports have registered a maximum CAGR of 32 percent, over the period. The share of this segment (which continues to be the mainstay of the Indian IT and ITES sector) to total revenues of the sector in 2006 was more than 63 percent. As per the projection of NASSCOM-McKinsey Report 2005, the sector would continue to grow at an annual rate greater than 25 percent and generate exports of about US$ 60 million by 2010. The report also predicts that by 2015, the total addition to the export revenue from the sector would be around US$ 15-20 billion (Government of India, 2007 a).

In view of the significance of software and services sector in the total revenues, some disaggregate and component wise analysis of growth in revenues vis-à-vis the sub-sector is also attempted (Table 2.1).

As summarized in Table 2.1, the software and ITES exports from India grew from USD 12.9 billion in the year 2003-04 to USD 23.6 billion in 2005-06. It is also expected that the total software and ITES exports from India will be about USD 31.3 billion during the year 2006-07. A component wise analysis of the data (of 2006) reveal that IT Services, accounting for 59 percent of the export revenue is the important most segment in the software and services sector, followed by ITES/BPO (24 percent); Engineering and R& D Services (17 percent).

Table 2.1: Growth in Revenues: Software and Service Sectors (US$ billion)

Sector	2004	2005	2006	2007 (*Estimates*)
IT Services				
Export	7.3	10.0	13.3	18.1
Domestic	3.1	3.5	4.5	5.6
Total	10.4	13.5	17.8	23.7
Ratio of Export to Total	0.70	0.74	0.75	0.76
ITES-BPO				
Export	3.1	4.6	6.3	8.3
Domestic	0.3	0.6	0.9	1.2
Total	3.4	5.2	7.2	9.5
Ratio of Export to Total	0.91	0.88	0.88	0.87
Engineering Services, R&D, Software				
Export	2.5	3.1	4.0	4.9
Domestic	0.4	0.8	1.3	1.6
Total	2.9	3.9	5.3	6.5
Ratio of Export to Total	0.86	0.79	0.75	0.75
Total (Export)	12.9	17.7	23.6	31.3
Total (Domestic)	3.8	4.9	6.7	8.4
Grand Total	16.7	22.6	30.3	39.7
Ratio of Total (Export) to Grand Total	0.77	0.78	0.78	0.79

Source: Compiled by author using NASSCOM Data

The pattern ratios of export revenues to total revenues across various sub-categories, indicate a strong export dominance of the sector. It is important to note that across all the service lines the shares of export revenues to total revenues are more than three-fourth. In the case of ITES/BPO, the share is more dazzling at 88 percent, in 2006. The temporal consistency of the ratios also suggests that the pattern is more or less stable for the sector as a whole, which continues to rely heavily on the offshored work from developed nations.

Employment Trends and Patterns

In terms of employment also, information based sector is prominent as it provided direct employment to around 1.3 million workers in 2006 (NASSCOM,

2007). Of the 2,35,000 additional employment created in the sector in 2006, the contribution of IT services and ITES/BPO segment were about 1,00,000 each (Table 2.2).

The available estimates suggest that, the total number of IT and ITES/BPO professionals in India had grown from 83000 in 2003 to 163000 in 2007. The estimated growth of employment in the 2007 is 3,40,000. Further to this, the sector is estimated to have created an additional 3 million employment opportunities through indirect/induced employment (Government of India, 2007 a). This indirect employment was derived in complementary and ancillary sectors such as telecom, power, construction, facility management, IT, transportation, catering and other services.

Table 2.2: Employment in the Software and Services Sector

Sector	2003	2005	2006	2007 (Estimated)
IT Services	2,15,000	2,97,000	3,98,000	5,62,000
ITES-BPO	2,16,000	3,16,000	4,15,000	5,45,000
Engineering Services and R&D and Software Products	81,000	93,000	1,15,000	1,44,000
Domestic Market (including user organisations)	3,18,000	3,52,000	3,65,000	3,78,000
Total	8,30,000	10,58,000	12,93,000	16,30,000

Note: Figures do not include employees in the hardware sector
Source: NASSCOM

The sectoral distribution of export-oriented employment [7] suggests that ITES/BPO is the major component with 44.72 percent, followed by IT services (42.88) and Engineering Services and R&D and Software Products (12.39 percent).

Key Markets for the Sector

As per the *NASSCOM Strategic Review 2006*, in 2004-05, 68 percent of India's export earnings were from U.S., followed by 23 percent to Europe (of which 14 percent from U.K). The remaining 9 percent were from the rest of the world. Thus, US and the Europe remain the key markets, which provide more than 90 percent of export earnings.

The export earnings from markets other than US and UK are also continuously on the rise, indicating a broader market base for the industry in the coming years. The Indian firms are found keenly exploring new geographies for business development, and also to strengthen their global delivery footprint. Indian firms have already invested in specialized service lines at external locations, such as China (for engineering and design) and South Africa (for insurance), along side building up competencies in near shore locations like eastern Europe and Mexico. Further, of late, more and more organizations from rest of the Europe, South-east Asia, Australia, Japan, Hong Kong, New Zealand etc. are also reaching out for Indian expertise, especially in the software sector (Government of India, 2007).

Structure and Diversification

It is widely understood that the Indian IT and ITES/BPO Industry is with a broad-based structure. While the IT segment is led by large Indian firms, the ITES/BPO segment is characterized by a mix of Indian and Multinational Corporation (MNC) third party providers and captives, which *inter alia* reflects the deep supply base of the industry. The increasing dominance of larger firms and scaling up of activities in the sector are evident in the recent years. The Indian firms have been found successful in building up the scale of operations to match the pace of increasing demands for their operations. The average size of contracts awarded to Indian firms has gone up, over the years. For instance, IT Industry Fact Sheet of NASSCOM for the year 2007 states that Indian service providers have grown their share of contracts of values in excess of US US$ 50 million dollars from 1 percent in 2002 to 7 percent in 2006.

The export orientation of the industry is evident with a high offshore component of delivery (Table 2.1). It could be seen that the growth of software and services sector exports also had a trickle down effect in the domestic market as well. The domestic IT services and ITES/BPO market is picking up, with the contribution of software and services exceeded that of hardware for the first time in 2005-06. The revenue from the domestic market (IT Software and ITES-BPO) is estimated to have grown to US$ 8.335 billion in 2006-07, as compared to US US$ 6.72 billion in 2005-06 – with an estimated growth of about 24 percent [8] (Government of India, 2007 a). The overall growth of the industry is highly related to the prospects of about 20-30 major players, who together constitute a prominent share in the industry aggregate. In terms or employment also NASSCOM reports that the top 20 companies [9] in the sector collectively employ over 500000 workers of the 1.6 million employed in the Industry. At the same time, several high performing Small and Medium Enterprises (SMEs) are also actively present in the sector, along with new entrants, in providing quality services with international standards. The broadening base of the industry in terms number of active companies is evident from the data on quality certifications. India-based centres (both Indian firms and MNC owned captives) constitute the largest number of quality certifications achieved by any single country. As of December 2006, over 440 Indian companies had acquired quality certifications with 90 companies certified at SEI CMM Level 5, higher than any other country in the world (Government of India, 2007 a).

Over the years, the firms are found expanding the gamut of their service offerings, enabling their clientele firms to deepen their offshore engagements. The sector, which started with a few subsidiaries of multi national corporations and outsourcing firms engaged in low-end tasks, has now expanded considerably in terms of variety of services provided and the types of service providers themselves. [10] The major service lines that are emerging in the sector include customer care, HR, payment services, content development, administration and finance. This also suggests that the Indian firms are simultaneously tapping the potential of both low-end as well as superior service lines, owing to the country's competitive advantages and untapped potentials in both the segments. The main verticals of the industry, accounting for 60 percent of the total, are: banking, financial services and insurance, and technology (Hi-tech/telecom). Along side this, manufacturing, retail, media, utilities, healthcare and transportation are the other growing areas (Government of India, 2007 a).

Potential for Future Growth

In spite of the impressive performance of the industry so far, it is estimated that there is still significant headroom for growth, as only less than 10 percent of the addressable market for globally sourced IT and ITES processes has been captured by India, so far. It is expected that the rate of flow of work to India through offshoring would increase considerably, in view of the mounting pressure on developed countries and global majors to resort further cost cutting. The quality standards and domain expertise attained by Indian players in niche areas are also favourable factors for attracting further business. A recent report [11] that compared the relative strengths of India vis-à-vis some of its potential competitors establishes that India's overall business climate (in terms parameters such as: government support, labour pool, infrastructure, educational system, cost advantage, quality, cultural compatibility and English proficiency is quite suitable for grabbing further business. The country's comparative advantages (in terms of cost advantage, favourable stock of educated and English speaking workers and superior quality work) far outweighs its comparative bottlenecks in terms of infrastructure and cultural compatibility (vis-à-vis near-shore competitors such as Ireland). However, with the constraints in labour supply and already unimpressive cost advantage (vis-à-vis India), Ireland is reported to be losing its grounds in IT services and moving closer to software products. It has also been pointed out in the report that India could retain its favourable labour cost differentials for the next few decades, even if the wages in the sector are growing at 9 percent per annum. Further, it also explains that it is less likely that India would considerably loose business to other destinations such as Philippines, Singapore, Australia, Mexico, Russia and China, in the next few years. As per the available projections, with a target of US$ 60 billion exports by 2010, the sector offers an estimated demand for 850000 IT professionals and 1.4 million ITES/BPO professionals. The prospective areas of future expansion of the sector include, high-end operations related to: data search, integration and management services, financial and insurance research, biotech and pharmaceutical research and computer aided simulation and engineering design. The possible impact of these transformations on India's labour markets and employment scenario would be complex and multifaceted. It is expected that massive enhancement in scale of employment in IT sector will open up considerable career opportunities as well as skill acquiring/upgradation avenues for educated youth.

Economic Upgrading: Concept and Modes

Economic upgrading is a very wide term which denotes the overall improvement in the economic capability of a firm or service provider. It also refers to the improvements in activities connected to process of production (and products themselves) as well as the evolution in the mix of activities being handled by the firm. Following the conceptualisation of Humphrey (2004), upgrading could be divided into four types, namely: process upgrading; product upgrading; functional or intra-chain upgrading; and chain or inter-chain upgrading. Process upgrading refers to the capacity upgradation of the firm in terms of: transforming inputs into outputs more efficiently; and increasing the efficiency of internal processes (both within individual links and between links of the chain). [12] In the case of product upgrading, the focus is on product related transformations of firms (such as shift towards more sophisticated product lines, introduction of new products or improvement/innovation over old products and so on). [13] The functional or intra-chain upgrading, includes: the firm level strategies towards acquiring new functions into the chain (up or down stream); and acts towards changing the mix of activities conducted within the link. [14] The final category of upgrading (i.e., chain or inter-chain upgrading) refers to the firm's act of applying the competence acquired in particular function (of a chain) to a new sector or a new chain. [15] This could also imply a movement of the firm to a new value chain.

The data gathered for the study testifies that the Indian firms have been resorting all the above types of upgradation, in the recent past. The firms are found resorting several ways of upgrading and diversifying their services to enhance their relative power and control within the supply chains. In view of this, an attempt is made in this section to delineate and discuss various specific strategies adopted by the firms in pursuit of economic upgrading.

An observant overview of the literature available and the discussions with key resource persons suggest that there are mainly seven [16] types of specific modes of economic upgradation, being followed by the service providers in Indian IT Services and ITES/BPO sector, such as: (a) Shift from voice to non-voice processes [17]; (b) Focus on high value and specialized niche markets; (c) Scaling up and strengthening competitiveness through mergers and acquisitions; (d) Improving the independence and increasing the client base; (e) Setting up R & D and innovation centres; (f) Strengthening Consultancy Services.; and (g) Extending the Outsourcing Chain.

Shift from Voice to Non-voice Processes

During the initial stage of the industry and even now, a prominent component of the ITES/BPO sector in India is voice-based services. However, the emerging trends suggest that this pattern is changing gradually, as more and more firms are focusing on more non-voice based activities. Of late, some of the major firms in the Indian BPO sector have been building their business models around the non-voice segment (e.g. WNS Global Services [18]). Many of the new entrants in the sector are even directly entering into non-voice segments and strive for creating their own niche markets. For instance, Ugam Solutions a recently entered small business providing firm has directly started its BPO operations on the high end data analytics field. Though all non-voice based processes are not necessarily superior to all voice based processes, it is widely understood that a transition towards more non-based activities is an indicator of upgradation of the firms. Non-voice services are generally more technical and thus demand greater expertise. They also fetch more revenues per unit of work.

Focus on High Value and Specialized Niche Markets

The processes handled by the Indian firms in the initial phase of offshore outsourcing were more generic and dispensable. Accordingly, the jobs in the sector demanded lower levels of domain expertise and were highly footloose [19] in nature. Contrary to this, the firms in the sector are now keener on narrowing their focus areas and ensuring that they become niche players rather than generalists. Some of the respondents highlighted that this is a way of moving closer to the core activities of the lead firm and thus shifting towards crucial positions within the supply chain. The focus on niche and superior services enable the firms to offer value-added and high margin services, which inter alia strengthens their locus positioning and bargaining strength within the value chain. An indicator of this transformation is the knowledge base of the new recruits in the sector. As against the initial trend of recruiting graduates (and even under graduates, e.g. in the case of call centres) in the initial phase, now many of the firms are looking for personnel with highly specialized and superior knowledge base. [20]

Strengthening the Resource Base through Mergers and Acquisitions

Firms, which have been traditionally handling low-end services, are also found shifting towards superior functions and processes, through acquiring other firms, which specialize in high end work. For instance, ICICI One source has acquired Pipal (a high end analytics firm) and moved to high-end service area of business research and collections. Similarly, Exl Services (a prominent Indian player in international third party business) has recently acquired Inductis, a leading consulting and analytics company serving the financial services and the insurance sector. [21] These acquisitions are found strengthening the process of upgrading of firms, as they bring broader and superior resource base to the expanding firm. In many such cases, the firms are found continuing with their traditional functions alongside focusing on the newly accessed domain strength through acquisitions. Some of the respondents opined that, this needs to be understood as a coping strategy of the firms (through straddling and product/service diversification). Some of the global majors are also found using the strategy of acquisition and mergers, by taking over Indian firms with good track record. A well discussed case of this sort is the acquisition of Daksh e-Services by IBM Corporation in 2004, to enhance the latter's competence to serve as a global hub to manage business processes for clients from across the world. [22]

Improving the Independence and Increasing the Client Base

In the early stage of Indian BPO sector, a majority of the key players in the industry were subsidiaries of MNCs [23] and Indian firms specialized in international third party operations, taking advantage of the lower cost situation in India. Most of these firms acted as captive units of their parent organization or for select clients, where the risk in business was limited (as the quantum of work was assured). As against this pattern, now, many of the MNC-captives in India are in the process of independently anchoring in the field and to offer services to other companies, while retaining their prior client contacts. (e.g. eServe International, WNS Global Services [24]). This could be viewed as a deliberate move to convert firms as profit centres as opposed to cost centres they earlier used to be, through expanding their customer base. In a similar way many of the IT and ITES companies in India, over the years, have been able to establish their own brand names and thus could retain their existing clients, along side attracting new clients as well as crucial contracts. The respondents in the interview also highlighted the desirability of having multiple clients. It was pointed out that a handful of clients, rather than one dedicated client, would be safer as single-client situations would lead to firm's dependency on a particular provider and would adversely affect its bargaining power with the client.

Setting up R&D and Innovation Centres

Besides maintaining more stable relationship with the customers, the firms are found continuously targeting new service lines through innovations, which guarantee high-value potential and better price realization. Taking advantage of India's huge front-end R&D potential and the availability of a pool of highly qualified and talented workers, many of the MNCs (like Microsoft, Google, GE, Texas Instruments, Sun, Intel etc.) have opened their development and R& D centres. Similarly, CSC, IBM and Adobe have already opened up innovation centres in India. The Indian firms are also in the same line, with several upcoming ventures in engineering design, clinical trials, biotech research outsourcing, process plant data analysis, customer analytics, market research, equity research and so on.

Strengthening Consultancy Services

The other visible shift in the recent past is a visible emergence on consultancy services, which is considered as a superior and facilitating segment of offshored business. Both the MNC players and Indian actors are attaching increased prominence to consultancy as developing this skill is crucial for the onward growth of the sector. Recently, the IBM Corporation in India has established a consulting arm (IBM Business Consulting Services). Similarly, several other firms (such as Wipro, Infosys, TCS and Cognizant are also found investing more on consultancy services in the recent past. The services offered by the consultancy firms in India include: strategy consulting, production planning, sales, product and material management, human capital and financial management solutions, e-business integration and so on. All these are obviously possible only through improving the human resource base, strengthening the frameworks and knowledge base of the company and through investing heavily on domain expertise.

Faced with a demand to lower the per seat cost, even many of the outsourcing companies located in the west (e.g. TrueDial, Teletech and Bechtel – all based in U.S) are getting into subcontracting arrangements with smaller Indian firms, through the help of Indian consultants and consulting agencies(e.g. Netspace). The concern of these firms is to lower the cost without affecting the quality of services. When smaller firms in India are ready to guarantee for both, with the mediation of established consulting agencies, these subcontracting arrangements are often found attractive to the original outsourcing companies. Here through consultancy, new actors are getting into the supply chain – as middlemen (consultants) and small players. This entry of smaller firms (with 50-100 seats) at the bottom of the chain may also affect the stakes of other actors (i.e., without moving up or down in the chain, their positions and roles may get altered, in relative terms). It was also noticed that in order to bag contracts, the small firms are also willing to offer small pilot slots of work free of charge. In many such cases, the consultants act as coordinating agencies and even as a partner to smaller firms (For instance, arranging resources and providing human capital and financial management solutions etc.), which will imply a higher share of the profit pie going towards the consultant.

Extending the Outsourcing Chains

Yet another mode, which is currently picking up its momentum, is related to the expansion of the supply chain by Indian service providers by further outsourcing a portion of the process(es) being handled by them to smaller players. It is being reported that in their run to check the rising operational costs and to circumvent the threats from growing attrition (worker turns over) in the sector, several major captive vendors operative in India are seriously contemplating to entrust smaller firms and third party service providers to manage their centres and to handle less strategic work. While resorting this extended outsourcing process, the firms are trying to retain their control over some of the key functions and to seek ways and means towards ensuring a standardised service flows from the newly identified subsidiary-service providers. This mode could be seen as a reverse of the BOT (build-operate-transfer) model. The prominent firms in India, which are currently thinking on this option seriously, include Wipro, WNS Global Services, Exl Services and Infogain India. Similarly, of late, some of the Indian firms, which have their operations spread across the US and Europe are found outsourcing some of their low-value administrative functions to US and UK-based firms, as a part of their strategy to reduce their headcount of HR professionals abroad. [25]

It is often reported that in order to retain and cope up with the increasing inflow of service work, bigger firms are found entrusting a portion of their work with smaller enterprises, thereby widening the entrepreneurial base of outsourcing. Such an extension of outsourcing chain would even enable the smaller entrepreneurs (who do not have much capital and resources) to participate in the global production chain [26] and derive some economic benefits and upgradation. This sort of extension of chains through outsourcing within outsourcing is likely to create more precariousness down the chain in terms of lowering the labour standards. Mehmood and Ghurfan (2005) provide some supporting evidence to this argument. [27]

On the whole, from the foregoing discussion, it is evident that Indian firms are making conscious efforts to move up in the supply chain, through handling superior quality services. Nevertheless, it is a fact that India still enjoys a lead and considerable competitive advantage in the so-called `low-end' jobs. While the respondents found this straddling strategy as a harmless and desirable preposition at this juncture, especially in view of the continuing employment potential of low-end operations, they highlight the desirability of putting more efforts towards upgradation, due to the inverse relationship between superiority of processes handled and the footloose nature of employment generated.

Different Ways of Understanding Economic Upgradation

The phases of upgradation in the IT and ITES/BPO sector is understood in several ways. Accordingly, the conceptualizations range from mapping of temporal evolution of the sector to a discussion of structural changes within the value chain. These conceptualizations are summarized below to gain clarity on the growth path of the sector.

Mapping the Transformation in Terms of Nature and Forms of Work

A prominent way of understanding upgradation of the IT and ITES Sector is through mapping the temporal transformation, in terms of principal nature and forms of work handled by the Indian firms, at different time periods since the advent of offshored activities in India.

Over the time period the nature of work being handled by the Indian BPO firms has changed considerably. In its first phase (i.e., from mid 1990s to late 1990s), the sector was predominantly characterized with low-end work (such as data entry, transcription services and call centre processes). However, the nature of work has been shifting since then to higher end activities. While in the early

2000 onwards Transaction processing (in non-core activities) gained prominence in the offshore basket of India, of late, the firms are also found entering into core-transaction processing jobs, which require more domain expertise and highly qualified workforce.

A similar but more detailed explanation, from 1980s onwards is provided by neoIT, a consulting and business analysis firm.

In its sprouting phase (i.e., 1980s to early 1990s), the firms in the IT and ITES sector were mostly working with out-tasking processes (which included data entry, application support, transactional data entry, technical support and maintenance. In the next phase (mid 1990s-early 2000), was the period of conventional BPO activities (including rule-based processing, IT related Customer contact/ hardware and installation support, Problem Solving/Decision Making, Customer contact in ITES and so on). Subsequent to this, the current phase is characterized with Full Process Outsourcing, R&D Services, Product Development and Testing, Research and Analytics, Design and Integration, Engineering Services and Multiple Process Outsourcing.

Upgradation as a Shift from Non-core to Core Processes

Upgradation could also be understood in terms of shift from low-end to high-end processes. Moving up in the production chain does not imply a mere increase in the rate per transaction. Moreover, it is about getting closer to the core of clients' business from the peripheral services that they outsource initially [28]

Core activities directly adding value to the end product, but are non-strategic functions, are included in square 3. Suitable strategy of such functions and processes is to spin off business or outsource through strategic alliances. For e.g. customer interface related services, product related transactions, processing services etc.

The requirement of domain skills is very high. Similarly, the value addition by the business provider is also at a higher level. The nature of relationship between the firms and client is more on the lines of a strategic partnership as compared to the client–vendor relation in the previous two cases.

In the initial phase of IT Services and ITES/BPO firms in India, the share of square 1 activities in the total service work was substantial. However, along side the process of upgrading of the industry and movement of various vendors up the production lines, the share of activities from Square 3 is gradually on the rise. There is also growing possibilities of outsourced work from Square 2, along with the growth of industry.

From Business Processing to Knowledge Processing - BPO to KPO

The discernible shift of the Indian ITES/BPO firms towards provision of superior services and to deal high-end work, in recent years, is widely understood as a movement from Business Process Outsourcing (BPO) to Knowledge Process Outsourcing (KPO). KPO, which is a recently coined term, denotes higher end knowledge based work (such as market research, competitive intelligence, engineering design, biotech/pharmaceutical research etc.). Accordingly, a KPO is considered superior to a conventional BPO due to several counts. While the work in BPOs involves mostly execution of standardized routine processes, the KPOs are focusing on processes that demand advanced information search, analytical, interpretation and technical skills as well as some judgment and decision making. From the interviews and data available in the public domain, it is evident that handling of KPO functions is on the increase in the Indian IT and ITES Sector. The trend is gaining momentum, with the expansion of a host of tele-integrated and advanced activities such as intellectual property or patent research, R&D in pharmaceuticals and biotechnology, data mining, database creation, and a range of analytical services such as equity research, competitive intelligence, industry reports, financial modelling and so on.

The growth of export revenues from the KPO segment have grown considerably in recent years. Further to this, the per worker value addition (productivity) of the outsourcing sector has gone up considerably with the advent of KPOs, leading to increased earnings and incomes. Important verticals (sub-sectors) within KPO in India are: banking, securities and industry research services; contract research organisations and bio-pharmaceutical services; data management, mining and searching analytics; and legal, paralegal and intellectual property support services.

KPO activities require considerable domain expertise (e.g. market research, business research, investment research and data mining) and specialization in niche activities. Thus, the growth of KPO segment itself necessitated a proliferation of specialized firms. As per the data provided by Evalueserve as on July 2007, there are at least 282 niche companies in India that provides third-party KPO Services. In a value chain, a typical KPO lies in between a conventional BPO and a high-end consulting firm. The relative strength of a KPO firm within the value chain is much superior to that of a low-end conventional BPO player. It is pointed out that a KPO firm and its client is more of a partnership and less of an arms-length, client-vendor relationship.

The KPO segment is still a promising area for development within the ITES/BPO Sector. As the growth of this segment warrant considerable domain expertise, it opens avenues for absorbing the favourable stock highly qualified youth in the country. The sub-sectors that are expected to `shine' within the KPO industry in the near future include data search, integration and management services, financial and insurance research, biotech and pharmaceutical research and computer aided simulation and engineering design. Also, many other sub-sectors, which are currently at their nascent stages (such as legal and para-legal support services, remote linguistic services and procurement support services, and network optimization and analytics services) are expected to gain momentum in the years to come.

Determinants and Constraints for Upgrading

Several factors are believed to have determined the evolution and growth of Indian ITES industry, along the supply chain. India's edge in terms of favourable stock of highly skilled workforce and the experience and confidence gained by Indian firms in the nascent phase of the industry are undoubtedly crucial in facilitating the growth in higher-end services, which require strong domain expertise. Apart from this, two major aspects have contributed towards the promotion of the sector in its establishing stage. These are: (a) the investment friendly and enabling environment (provided by the government); and (b) the strong presence of an entrepreneurs' association (which acted as a catalyst as well as pressure group).

Enabling Business Environment

The growth and evolution of off-shored service sector activities in India undoubtedly owe to the conducive environment provided by the Government for the investors and entrepreneurs. There have been a host of measures followed by the Government, aiming at export promotion, which included: single-window clearance systems; provision of export oriented unit (EOU) status to the firms in the BPO sector with incentives - income tax holiday, foreign exemptions in the case of exchange and excise duties and so on. All these measures facilitated the smooth evolution and growth of the sunshine industry in Indian soil. Further to this, some of the firms in the sector also derived benefits from the incentives provided to the companies located in Export Promotion Zones, Special Economic Zones (sales and service tax exemptions, advance licenses, duty free replenishment certificate and so on). The investment offers of the Government have also been attractive in the period (100 percent foreign direct investment (FDI) through the automatic approval route; zero license fee for entry to the sector; progressive reform in regulations on overseas investment and repatriation).

In the meanwhile, the country also witnessed an explosion in the telecom infrastructure, along with a planned deregulation of information technology governance. The measures towards setting up of Software Technology Parks of India (STPIs) and development of plug and play infrastructure enhanced the technological capability of the nation and enabled the Indian firms to appropriately tap the emerging business opportunities, in an age of offshoring. From the entrepreneur's viewpoint, the quality and cost-efficiency of these basic services have been much superior to conventional `Indian Standards', as these facilities were introduced along with a planned deregulation of the telecom sector, in a relatively short period facilitating healthy competition. Approvingly, it is widely acknowledged that one of the prominent drivers of Indian outsourcing sector is the marked improvement in bandwidth availability, along with a dramatic drop in both in international private lease line rates and long-distance telephonic charges. [29]

During the past one decade, the government also followed a liberal and soft labour policy to promote investment in the sector, which is expected to generate higher levels of employment to cater the growing mass of educated unemployed in the country. For instance, many of the state governments amended their labour laws or introduced special laws to facilitate a 24 x 7 working environment (by allowing flexible working hours, employment of women workers in the night shifts, removal of mandatory weekly time offs etc.). Very recently efforts have also been made to strengthen intellectual property laws in accordance with TRIPS. Similarly, the government is currently in the process of reviewing the amendment of IT Act, 2000.

The Strong Presence of an Employers' Forum - NASSCOM

A major factor that helped the growth and evolution of IT and ITES sector in India, over the past years, is the formation and active presence of a national association of producers - National Association of Software and Service Companies (NASSCOM). Over the years, NASSCOM has been facilitating the industry in the upgradation process in terms of capacity and skill upgradation, enhancing the bargaining strength within the global supply chain as well as in setting the tone for public policy for the Indian software and services industry. This association has been actively engaged on several activities which broadly fall under the head of upgradation – such as capacity and skill upgradation; corporate collective bargaining; data generation; trend analysis and so on. Over and above these aspects, NASSCOM was also found successful in terms of setting the tone for public policy for the software and services Industry in India. It is widely understood that the cumulative efforts of NASSCOM have been aimed at and to some extent were fruitful in bringing in favourable results in terms of recruiting skilled personnel, enhancing the domain expertise and charting out and planning for upcoming and potential areas of future business.

Constraints to Upgradation

There were also constraints that restricted the Indian firms from appropriately tapping the opportunities. It was widely pointed out that, quite often the companies abroad are reluctant and skeptical about outsourcing high-end services due to varied reasons such as data security, worries about quality and professionalism in a remote location and so on. Sharing of data to firms located abroad is often considered risky by many of the potential service providers, who include data management itself as a core and non-separable function of the lead firm. The growing political attention in developed countries on the adverse consequences of offshoring of services also brought in certain regulations in the free flow of opportunities to countries like India in the recent past. For instance, the growing protests in US highlighting the possible adverse consequences of offshoring even led to certain legislations and changes in the policy framework (e.g. restriction on government departments apropos offshore outsourcing). The pressures from international trade unions as well as social activists also bring in certain restrictions to offshoring of service work to countries like India. In the context of growing discontent among the international trade union movement regarding offshoring (especially due to its inherent potential to shift jobs from bargainable categories –and regions – to non-bargainable categories and regions), many of the clients are now insisting on certain level of labour and social standards to be maintained by the service providers. An HR manager, during a discussion, explained that the style of organisation of modern service sector firms in India and many of the worker caring strategies followed by them have been closely linked to the preference of customer firms – which are keen to have certain employee benefits and standardized facilities at workplace. To him, in some cases, there are even mandatory corporate social responsibility (CSR) allocation insisted by clients. All these things affect the profitability of firms, besides warranting several efforts (and resources) to obtain quality certifications and client approval vis-à-vis standards.

Despite the wide recognition regarding higher and improving quality of services being provided by Indian firms, still there are discomforts for the clients and developed nations which include: the uneasiness regarding outsourcing strategic work to remote locations; [30] the concern over data security; [31] the fear of losing control over core processes; and the possible threat of lagging behind the developing countries and their players in the technology and R&D fields, in future.

There are also restrictions paused by the mismatch regarding the skill base of workforce. With the improvements in the quality and type of works (high-end) being handled in India, the firms are finding it difficult to find appropriate workers with required skill base. As detailed training and skill imparting to recruited workers are found to be highly resource demanding, a major concern of the employers is to identify resourceful workers and retain them. Worker retention is also yet another issue, as the sector is characterized by 40-60 percent of attrition.

A large chunk of workforce in the sector is with an average service of less than two years with their current employer. As a dedicated pool of talented workforce is a prerequisite for the industry, a major challenge of human resource departments is to attract and retain skillful workers. The substantial levels of employee turnover in the sector also adversely affects the quality of services (and thus the pace of upgrading), as employees are unable to gain sufficient domain expertise and knowledge, during their short spell with the firm, before they move on to their next job. Yet another issue is poaching (or inter-firm competition in attracting resourceful workers), which is gradually being resolved between the firms through bilateral agreements and mediations by employer collectives and so on.

Perceived Social Gains of Upgrading

The above discussion, based on a firm level perspective, indicates that the process of upgrading or moving up in the supply chain may bring several benefits to the economy as well as the labour market. These include: increase in the revenues/ export earnings; benefits in terms of enhanced productivity, quality benefits and cost savings; increase in employment and quality of employment; favourable impacts of technological upgradation on job quality and so on.

A closer look at the perceived advantages in the labour market suggests that the most important among these is the stable potential of the offshoring sector to open up huge employment opportunities to the pool of technically educated youth in India. India, which is currently in the beginning phase of reaping its demographic dividend (with more than 50 percent of the population in the younger age categories), is viewing this sector as the most promising and estimates suggests that within few years 1-2 percent of the total workforce would find employment in the sector. The available information suggests that, over the years, most of the firms in the sector could increase their workforce size through the process of upgrading (both horizontal as well as vertical expansion). Along with the progress of the IT services and ITES/BPO sector in the supply chain, it is also viewed that, the quality of employment has enhanced, besides making it more stable and secure. While the employment in low-end offshored work (e.g. call centres) is highly vulnerable, with its footloose nature, the emerging job categories in high-end operations are more superior (with certain level of domain expertise and skill base), which makes quick shifting of these jobs to other competing regions a difficult affair to the clients. Further, the emerging higher-end processes in the sector are relatively free from the threat of further labour saving automation and standardization of work. [32] The positive correlation between the process's position in the supply chain and the complexity of interaction, imply a higher level of stability of employment and job security to workers, when the industry is moving up in the supply chain. Confirming this, the firms, which are engaged in high-end processes in India, do not currently entertain any fears regarding an immediate erosion of jobs to other destinations.

Another perceived advantage is in the field of training and skill upgradation. The industry experts and respondents opine that with the movement of Indian firms in the supply chain, more and more resources and efforts are being diverted to the area of training and skill upgradation of workforce. The discernible improvements in the grade of offshored work being handled by Indian firms, *inter alia*, implies more investment in human resources vis-à-vis training and skill creation/upgradation. [33] It is also pointed out some of the entrepreneurs and executives that (to appease their clients) firms are more and more inclined to improve the workplace standards and spending on social issues (CSR), when they move to higher layers in the supply chain.

Conclusion

From the foregoing analysis, it is evident that the firms in IT Services and ITES/BPO Sector in India are currently in an intensive phase of economic upgrading. From the perspective of the entrepreneurs and employer bodies in the industry, this process of economic upgrading is expected to bring certain favourable results in the economy as well as labour market, which include: enhanced contributions to GDP and export earnings, advancements in technology and economic infrastructure, gains in employment levels and quality of employment, increase in labour standards (in terms of better work environments, improved working conditions, higher salaries, perks etc.), increased allocation of funds towards training and skill upgradation of workforce, greater earmarking of resources towards social spending and so on. In this context, it is important to have more detailed and socially/worker oriented enquiries that confirm whether the perceived advantages of economic upgrading under global production chains are ultimately realized by the workers in the sector as well as the society at large.

Notes

1. The paper is based on the findings of a research study supported by International Labour Organisation/Japan Institute for Labour Policy and Training (ILO/JILPT) and was carried out by the author at the V.V.Giri National Labour Institute, NOIDA (Remesh, 2008).
2. For a detailed discussion on the concept of global production systems refer Barrientos, 2007.
3. To get clarity on the concepts of decent work and decent work deficits, refer ILO (2002; 2004).
4. These supplementary data included: published interviews of industry representatives and entrepreneurs; views of business houses and employer organizations on various aspects of upgrading; analyses prepared by various business research/ consultancy organizations and so on.
5. While this industry oriented perspective provides a useful overview of the upgrading in the sector and briefly explains the anticipated benefits in the social sector (and in labour standards), there is a need for cross-checking whether these expected benefits are really happening. Accordingly, an enquiry was done as an extension of this analysis, though it does not form part of the present essay. For the findings of the overall study see (Remesh, 2008).
6. The discussion in this section is largely based on the information available from NASSCOM, the prominent association of software and service companies in India (www.nasscom.org). Though the authenticity of NASSCOM data can be debated, in the absence of any other reliable source of data, it is considered as the single most prominent source that could at least provide a rough understanding of various aspects of the sector.
7. Export-oriented employment is calculated by deducting the contribution of domestic sector from total employment. Accordingly, as given in Table 2.2, the export-oriented employment in the sector for the year 2006 is 928000 (1293000 – 365000).
8. The growth forecast for the domestic market, however, is relatively lower as compared to exports. Notwithstanding this, with progressive deregulation and rising competition, companies are increasing their focus on the domestic market; especially on emerging sectors such as retail, logistics, telecommunications and SMEs (Government of India, 2007 a).
9. As per the latest list of NASSCOM, the top 20 IT-ITES employers in India in 2006-07 (in descending order) are: Tata Consultancy Services; Infosys Technologies; Wipro Technologies; HCL Technologies; Cognizant; Satyam Computer Services; Hewlett Packard; Genpact; Oracle; Intelenet Global Services; WNS Global Services; Mphasis; Patni Computer Systems; Convergys; Firstsource Solutions; Sutherland Global Services; Citigroup Global Services; Aegis BPO Services; ExlService Holdings; and L&T Infotech.
10. Of late, the sector has been witnessing a visible shift from low-end business processing to higher value operations (for instance, knowledge based processing).
11. HCL BPO (2007): "BPO India: A White Paper", HCL Technologies BPO Services Ltd.
12. Process upgrading could be understood as the increase in total factor productivity, where same amount of factors of productions can produce relatively more output. Accordingly, in the context of IT Services and ITES/BPO industry in India, process upgrading can be understood as a firm's capability to offer more quantum of services with the same cost and same amount of resource deployment.

13. This conceptualisation, thus, helps in situating the shift of firms from low-end to high-end or (low value added to high value added) products.
14. The efforts of firms apropos moving the locus of activities to different links in the chain are central to this conceptualisation. Accordingly, the strategies of service providers in the Indian IT Services and ITES/BPO industry to enter into the new niche markets (especially through merger and acquisitions), could be seen as acts towards functional or intra-chain upgrading.
15. For instance, a firm that has acquired certain level of competence in banking services can apply its gained experience towards building up its service lines in insurance sector.
16. This categorization is only indicative and not exhaustive. Also, many of the categories are interlinked. The purpose of categorization, thus, is mainly for delineating and discussing some of the discernible changes in the nature of functioning of firms.
17. Here, it is to be noted that all non-voice based services are not perceived as superior to voice based activities. Yet, it is widely understood that a movement from voice-based activities (mostly telesales, voice based customer care etc.) to non-voiced activities (which transaction processes, data analytics etc.) is an indicator of upgradation of the Industry.
18. WNS Global Services operates in the knowledge services business segment and offers high end services such as market, investment and business research.
19. For instance, a low end and voice based customer care operation may require only limited domain knowledge. Similarly, the skill involved in the work (English speaking etc.) is minimal, which makes the shifting of the process to any other low cost destination an easy affair to the lead firm.
20. For example, Wipro Spectramind was found hiring Ph.Ds in molecular biology to maintain a database for a genomics knowledge management company in U.S.
21. In similar way, WNS Global Services acquired : (a) ClaimsBPO (Nashik), of the US-based Green Snow Inc.; Town and Country Assistance (Ipswich, U.K) and entered in healthcare business; (b) Trinity Partners (US), to strengthen its business in the mortgage and financial services; and (c) Ipswich (UK; Town and Country Assistance), to enter the insurance claims processing market.
22. The new enterprise known as IBM-Daksh is one of the major players in offshored business in India, with its business centres in Delhi, Mumbai, Bangalore, Pune, Kolkata and Chandigarh. It has also a centre at Manila.
23. The pioneering firms in the Indian IT Services and ITES/ BPO industry were GE, American Express and British Airways.
24. WNS, which started as a captive and Indian arm of British Airways, had gradually established as WNS Global Services and became one of the major Independent third party providers in India.
25. For instance, it is reported that Hanna Global Solutions, a US based full-service human resource management firm, is currently serving over 30,000 employees belonging to prominent Indian companies such as Satyam, Infosys, Wipro and so on.
26. During the filed survey of this study, it is found that there are even vendors who provide infrastructure packages (such as work area and device support - which includes: seating space, computer, telephone and internet connectivities) to smaller entrepreneurs on a monthly basis – e.g., @Rs.7000 per seat per month.
27. To quote from Mehmood and Ghurfan (2005): "A few months before, a middle level manager of an elite international call centre contacted Guddu [owner of small internet café located in a Delhi suburb]. The preposition was to share some load of that call centre in

return for some forty thousand rupees a moth. Guddu agreed. He hired some out-of-work and some working agents who had prior experience of this industry on `payment on performance' basis, from 2:00 A.M till 10:00 A.M. With no frills of dinner and snacks, pick-up-or drop, and no facilities like toilets, cash coupons or tickets to multiplexes, Guddu operates an `international call centre'. He manages a complete gamut of campaigns and products, just like any other Business Process Outsourcing unit in Noida and Gurgaon. For the agents of this international call centre, there is no accent training, minimal process training and a more informal relationship with the boss. Unlike many other premium call centres, less monitoring, supervision and surveillance takes place here, even as the work contract is more fragile."

28. The discussion in this section is developed upon the conceptualisation of Sha, Jignesh (undated): Business Process Outsourcing: A View from India, Business Consulting Group, Mumbai.
29. In just two years (2000-2002), there was a 110 percent drop in the international private lease line rates with an equally dramatic fall in long-distance telephonic charges.
30. To address this concern, the Indian companies, of late, are found setting up their centres in nearby locations (to Europe and US) or acquiring other companies located abroad, as part of their efforts to provide `near-shore comfort' to their clients. The concern over data security acts as a major deterrent in the case of banking and financial institutions. Certain controversies and growing discussions in the international media also found adversely affecting the flow of such work to destinations like India.
31. For instance, of late, the low-end operations such as call centres are losing their headcount considerably, on account of introduction of interactive voice response systems (IVR), which reduces the requirement of human component in the process to the minimum. As against this, high-end operations are relatively safer from the threat of intrusion of IVRs, as they imply more complexities in interaction, which is not easily amenable to automation. It is generally understood that higher the position in the value ladder, lesser is the scope for getting replaced by IVR.
32. While the training required for a call centre job is of a few weeks' duration, the training of workers in fields such as market research, data analytics, R & D, intellectual property and legal processing may take several months or even a year or more.
33. As mentioned elsewhere, one such enquiry as extension of the analysis provided in this essay was attempted by the author, the findings of which are provided in Remesh (2008).

References

Barrientos, Stephanie (2007), "Global Production Systems and Decent Work", Working Paper No.77, Policy Integration Department, International Labour Office, Geneva.

Government of India (2007), "India: E-readiness Assessment Report 2005", Department of Information Technology, Ministry of Communications & Information Technology, New Delhi.

Government of India (2007 a), "Information Technology Annual Report, 2006-07", Department of Information Technology, Ministry of Communications & Information Technology, New Delhi.

Humphrey, J. (2004), "Upgrading in Global Value Chains", Working Paper No.28, Policy Integration Department, International Labour Office, Geneva.

ILO (2002), "Decent Work and the Informal Economy", International Conference, 90th Session, International Labour Organisation, Geneva.

ILO (2004), "A Fair Globalisation: Creating Opportunities for All", Report of the World Commission on the Social Dimension of Globalisation, Geneva.

Mehmood, Taha and Ghurfan, Iram (2005), "Chronicling the Remote Agent: Reflections on Mobility and Social Security of Call Centre Agents in New Delhi", Paper presented at the International Conference on New Global Workforces and Virtual Workplaces: Connections, Culture, and Control, August 12-13, National Institute of Advanced Studies, Bangalore.

NASSCOM (2007), "Indian IT Industry: NASSCOM Analysis", www.nasscom.org

Remesh, Babu P. (2008), "Decent Work in Global Production Systems: A Study of India's IT Services and ITES/BPO Sector", Research Report Submitted to International Labour Organisation/ Japan Institute of Labour Policy and Training (ILO/JILPT), International Labour Organisation, Bangkok.

3

Productivity Growth, Firm Dynamics and Knowledge Spillovers: Evidence from the Indian Electronics Industry

Shishir Saxena

Introduction

Discussion of the positive effects of investment in plant & machinery and R&D is reasonably extensive in economic theory. Greenwood *et al.* (1997) among others, argue that machinery of a relatively new vintage embodies the latest technology and therefore, fosters higher labour productivity growth. Meanwhile, DeLong and Summers (1991) argue that investment in machinery generates significant learning effects, which spill over to other firms (Shaw 1992). Furthermore, Romer (1986) stresses investment in research and development as an engine of long-run growth, while Griliches (1979) highlights the favourable impact of spillovers from such investments. The empirical literature using firm-level data from developed countries has established that a firm's equipment embodies the latest technology (Sakellaris and Wilson, 2004), and that R&D and its spillovers, promote growth (Wieser, 2005). However, evidence on these claims from the developing world, is either lacking or contentious. Moreover, no studies, either from developed or developing countries, have considered the effect of spillovers from investment in machinery on output.

The Indian Information Technology ("IT" hereafter) industry is an appropriate case to examine in this context. Several authors including Arora *et al.* (2001) and Lal (1999) among others have highlighted the exemplary growth performance of the Electronics industry in India. However, as Arora *et al.* (2001) highlight, Indian Electronics firms export resource-intensive goods and adopt new technologies only slowly. No study till date has tried to examine whether a firm's own R&D has led to substantial growth for these firms and whether knowledge spillovers, could explain such exemplary growth. It is vital to examine these issues to verify if future growth of the industry could be sustained even if cheap labour costs does not remain the sole reason for success of these firms.

This paper provides recent evidence on the impact of investment in plant & machinery and R&D, and their associated spillovers, on growth of output of Indian electronics firms. More importantly, it is the first piece of research, to simultaneously consider two different sources of spillovers for a firm; one each, from the industry-wide stock of equipment and R&D. [1] The industry-wide stock of R&D measures the available industry-wide knowledge generated from advances in technology, which increases the productivity of human capital associated with the research sector of a firm (Romer, 1990). On the other hand, the industry-wide stock of equipment is meant to capture either or both, of the following effects. The first is based on an extension of the "learning by watching" hypothesis, which argues that investment in machinery by one firm has a "demonstration effect" onto production workers of other firms (King and Robson 1993), as it creates an intangible capital stock (Hammond and Rodriguez-Clare 1993). The second, and the less known, is the level of development effects, dating back at least to Frankel (1962). According to this argument, each firm's output is an increasing function of the level of development achieved by that industry, an idea, similar in essence to that of "balanced growth" advanced in Rosenstein-Rodan (1943). Whereas Frankel (1962) chooses capital intensity to represent the level of development, this research uses the industry-wide stock of equipment as its measure. These two spillovers for a firm are independent of each other. Learning from investments/ level of development effects, impact(s) output of all firms, regardless of advances in knowledge, and vice versa. Furthermore, the industry-wide equipment stock appropriately measures the level of development of an industry, which, the industry-wide R&D stock may not, as not all firms invest in R&D.

Performance of Indian Electronics Firms; 1994-2006

This study uses data from over the period 1994 to 2006 taken from Prowess, an electronic database, maintained by the Centre for Monitoring Indian Economy (CMIE), Mumbai, which has built the largest database on Indian firms. Prowess compiles information from the annual reports of large- and medium-sized Indian firms including government undertakings, whose shares are regularly traded on major Indian stock exchanges. Data on Electronics firms covers between 143 and 324 firms for different years. The dataset is an unbalanced panel of 3,655 firm-year observations.

This section provides insights into the productivity of Electronics firms and the industry over the period 1994 to 2006. Productivity at the firm-level is estimated using the specification given in Aw *et al.* (2001), who measure productivity of each firm against that of a hypothetical firm measured at time period 0 as follows,

$$LnTFP_{ft} = \left(LnY_{ft} - \overline{LnY_t}\right) - \frac{1}{2}\sum_{i=1}^{n}(S_{fit} + \overline{S_{it}})\left(LnX_{fit} - \overline{LnX_{it}}\right)$$

$$+\sum_{s=1}^{t}(\overline{LnY_s} - \overline{LnY_{s-1}}) - \sum_{s=1}^{t}\frac{1}{2}\sum_{i=1}^{n}(\overline{S_{is}} + \overline{S_{i,s-1}})(\overline{LnX_{is}} - \overline{LnX_{i,s-1}})$$

where i indexes inputs; S_{fit} and X_{fit} measure the expenditure share of the ith input of firm f at time period t in nominal prices and the real quantity of that input respectively. An over-bar indicates the mean value over a relevant quantity. For example, $\overline{S_{it}}$ measures the average of the expenditure share of the ith input over all firms at time period t. Similarly, $\overline{LnX_{is}}$ measures in logarithm, the mean value of the real quantity of the ith input over all firms at time period s. The first two terms on the right hand side of this equation measure the relative gap in TFP of firm f to that of the representative firm in that year. The next two terms measure the difference in level of TFP of the representative firm in year t and the base year 0. Figure 1 highlights that electronics firms in India have had a systematic rightward shift in productivity over the entire period from 1994 to 2006, as given by both the constant returns to scale ("CRS" hereafter) and non-constant returns to scale ("NCRS" hereafter) specifications.

The summary distributions undoubtedly provide a deep insight into the several productivity episodes which these firms have been exposed to since 1991. However, it must be noted that productivity of firms at each of the percentiles could be rising synchronous with rising productivity of all firms, or because of the exit of the least productive firms. These distributions fail to highlight the inter-temporal movement in productivity of individual firms, which would provide deeper insight into a firm's productivity dynamics. On the other hand, a transactions matrix can reveal important information about the dynamics of firm turnover as well as the inter-temporal movement in productivity of firms at different relative productivity levels. The purpose of this matrix is to ascertain whether individual firms retain their productivity from the beginning of the period under consideration to the end-period, or do they become relatively more or less productive over time? Table 3.1 displays the transition matrix of the inter-temporal movement of TFP of all firms, segregated by quintiles over two time periods. The top panel considers the period 1994 to 2000, while the bottom panel considers the period 2001 to 2006. However, it must be noted that the data does not permit distinguishing between firms which exit the industry because of low past productivity, from those which get merged or acquired. Although, an intuitively appealing argument would be that the high productivity firms get merged or acquired (and therefore, appear as firms which exit), whereas the low productivity firms shut down and exit the industry. All results are based on the CRS specification. Row values in these tables denote the movement in productivity of firms from the quintile category corresponding to that row in the initial time period, to other quintile categories in the end time period. For instance, it can be gathered that of all the 29 top quintile firms in 1994, 17.24 percent remained in the top quintile in 2000, 6.9 percent had dropped down to the second quintile, 13.79 percent had dropped down to the third quintile, and so on, while almost 48.28 percent of these firms had exit the industry.

Table 3.1: Transition Matrix of Firms (values in percent)

TFP Level 2000 TFP Level 1994	No. of Firms	Highest TFP Quintile	2nd TFP Quintile	3rd TFP Quintile	4th TFP Quintile	Lowest TFP Quintile	Exit
			1994 to 2000				
Highest TFP Quintile	29	17.24	6.90	13.79	6.90	6.90	48.28
2nd TFP Quintile	29	13.79	24.14	13.79	13.79	17.24	17.24
3rd TFP Quintile	29	3.45	13.79	17.24	24.14	10.34	31.03
4th TFP Quintile	29	13.79	6.90	10.34	13.79	24.14	31.03
Lowest TFP Quintile	29	3.45	3.45	3.45	10.34	31.03	48.28
Entry	117	23.93	21.37	23.08	19.66	11.97	
TFP Level 2000 TFP Level 1994	No. of Firms	Highest TFP Quintile	2nd TFP Quintile	3rd TFP Quintile	4th TFP Quintile	Lowest TFP Quintile	Exit
			2001 to 2006				
Highest TFP Quintile	43	18.60	13.95	6.98	0.00	0.00	60.47
2nd TFP Quintile	43	11.63	20.93	9.30	6.98	6.98	44.19
3rd TFP Quintile	43	2.33	2.33	13.95	18.60	2.33	60.47
4th TFP Quintile	43	4.65	0.00	9.30	18.60	13.95	53.49
Lowest TFP Quintile	43	2.33	0.00	0.00	9.30	34.88	53.49
Entry	91	25.27	24.18	21.98	15.38	13.19	

Note: The number of firms in each quintile is not the same because not an equal number of firms in each quintile survive from one period to the next.

It must be noted though, that firms in a particular quintile in the two panels of the table may not be the same unless a firm remains in the same quintile from

1994 to 2000. This is because, a firm for instance, which moves from the 2nd to the 1st quintile from 1994 to 2000, will appear in the 2nd quintile in the left panel and in the 1st quintile in the right panel. This table highlights several noteworthy comments.

There is a noticeable degree of persistence in a firm's relative TFP level within the industry. A majority of firms in the top two quintiles in most industries, and over both time periods, maintain their position in those same productivity cohorts. The same applies to firms in the bottom two quintiles. These proportions for the least productive firms are quite high, suggesting that over this time-period under consideration, firms chose to persist than exit. One possible reason for persistence of low productivity firms is that this industry is capital-intensive, making a quick shut-down process difficult. Another reason is, as argued in NMCC (2006) and Panagariya (2002), that India has traditionally had highly stringent exit laws even for inefficient firms and the most recent reform process through the 1990s have failed to completely address this problem. This situation needs to be addressed soon, the urgency of which is adequately highlighted in NMCC (2006), failing which industrial productivity would be held down.

Turning to the exit of firms over this former time-period, it can be noticed that these rates suggest a story of flux. Exit is confined to both the highest and least productivity firms alike. This finding partially supports the claim above, that whereas the most productive firms exit because they get merged or acquired, the least productive firms shut down. The distribution of entrants, dispersed across the productivity distribution, reveal an interesting pattern. It can be observed that, nearly 68 percent of the entrants (those firms which entered the industry between 1995 and 2000 inclusive) occupy the top three TFP quintiles in 2000. A similar observations is also encountered in Shiferaw (2007) for African manufacturing firms and Jensen *et al.* (2001) for U.S manufacturing plants. The first study attributes this finding to the size effect, which states that some entrants are large in size and therefore, can derive economies of scale. On the other hand, Jensen *et al.* (2001) puts forth the vintage effect, which argues that new firms bring in new technology, and therefore, are more productive.

Turning over to the second period under consideration, 2001 to 2006, the results presented in the bottom panel of Table 3.1 reveal that although the observed persistence of firms appears to be slightly lower over this period relative to the former, primarily because the exit rate is much higher, close observation reveals that the switch over of firms from highest to lowest productivity quintiles and the reverse is much less over this period. It can be observed that none of the top TFP firms in 2001 end up in the bottom two quintiles in 2006. Similarly, only 2.33 percent of the bottom two quintile firms climb the productivity ladder to reach the top two quintiles from 2001 by the end of 2006. Therefore, it appears that firms over 2001 and 2006, have either maintained their relative position in that productivity cohort, or have exit the industry. One can gather from this evidence that the first wave of economic reforms initiated in 1991, led to a large churning in the productivity of firms. A sizeable number of firms in the bottom-most productivity quintiles had managed to increase their relative standing by the end of that decade. However, by 2000, the more efficient firms had climbed the productivity ladder and the least efficient had either quit or receded in productivity, after which, there is no major jump observable across productivity quintiles over the period 2001 to 2006. What is striking to note is that the exit rate is much higher over this latter period compared to the former, for both the high and low productivity firms. Note that this was a period of intense mergers/ acquisitions in India, and it is not unusual therefore, to find a high exit rate among high productivity firms. Analysis not reported here reveals that as compared to the period 1994-2000, a higher proportion of quitting firms over the period 2001-2006 were larger in size than the 75th percentile firm in that distribution, which gives more credence to the argument on mergers and acquisitions. Statistics on entry reveal a similar story as that for the earlier time-period; a sizeable proportion of entrants occupy the top three TFP quintiles over this period.

Given the high exit rate observed in the transition matrix, this analysis was repeated using data only on survivor firms. Results not reported here underscore the main findings presented above. The surviving firms seemed to have consolidated their position during 2001 and 2006.

Next, the measures of industry-level productivity and productivity growth are summarized. The former is measured using the following equation,

$$LnTFP_t = \sum_f \theta_{ft} LnTFP_{ft}$$

where the left-hand side of Equation is the level of TFP of an industry; and θ is the share of firm f's sales relative to total industry sales in that year. Once the level of TFP has been determined for an industry, it is possible to calculate TFPG for the industry between two time periods, $LnTFP_t - LnTFP_{t-1}$. Foster *et al.* (2001) have shown that TFPG for an industry can be calculated as the aggregate of the following five effects,

where S is the set of survivor firms which remain in the industry over the years t-1 and t; E is the set of new entrant firms which enter in time period t; and X comprises those firms which were in the market in time period t-1 but exit in time period t. All other variables are as defined before. Table 3.2 provides the average level of productivity of the Electronics industry. As with firm-level estimates of TFP, the level of TFP for the industry as an aggregate has continuously increased from 1994 to 2006.

Within effect

$$\sum_{f\epsilon S} \theta_{f,t-1} \cdot (LnTFP_{ft} - LnTFP_{f,t-1})$$

Between effect

$$\sum_{f\epsilon S} (\theta_{ft} - \theta_{f,t-1})(LnTFP_{f,t-1} - LnTFP_{,t-1})$$

Covariance effect

$$\sum_{f\epsilon S} (\theta_{ft} - \theta_{f,t-1})(LnTFP_{ft} - LnTFP_{f,t-1})$$

Entry effect

$$\sum_{f\epsilon E} \theta_{ft}(LnTFP_{ft} - LnTFP_{t-1})$$

Exit effect

$$\sum_{f\epsilon X} \theta_{f,t-1}(LnTFP_{t-1} - LnTFP_{f,t-1})$$

Table 3.2: Industry-level TFP Measures

	Average 1994-2000	Average 2001-2006	Average of all years
CRS	0.479	1.264	0.841
NCRS	1.244	2.222	1.771

As regards TFP growth, based on different methods of calculating TFP growth, Electronics industry has grown at between 13.1 percent and 20.2 percent per annum over all years, as reported in Table 3.3. The production function ("PF" hereafter) estimates of TFP growth are based on the production-function approach of Levinsohn and Petrin (2003).

The following section delineates a production-function based approach to verify if the phenomenal growth of this industry is a result of knowledge creation efforts of firms and whether spillovers have affected their growth?

Table 3.3: TFPG Estimates

	Average 1994-2000	Average 2001-2006	Average of all years
CRS	0.140	0.121	0.131
NCRS	0.167	0.236	0.202
PF	0.248	0.061	0.155

Conceptual Framework

The econometric model is based on an extended Cobb-Douglas production function framework. It is postulated that output (*Y*) of the ith firm at time period *t,* is a function of its private inputs, including the stock of capital (*K*), labour (*L*), materials (*M*), and R&D (*R*). The share of recently purchased machinery in the total capital stock approximates the effect of technology embodied in a firm's machinery (R*PM/K*). The spillover effects are captured by the industry-wide stocks of equipment (*EQI*) and R&D (*RDI*), net of a firm's own stocks of these variables. It is assumed, as in Los and Verspagen (2000), that technical progress is endogenised in a firm's own R&D and equipment, and therefore, a time trend treating such progress as exogenous is not included in the production equation.[2]

$$Y_{it} = A\ K_{it}^{\alpha}\ L_{it}^{\beta}\ M_{it}^{\gamma}\ R_{it}^{\mu}\ (RPM/K)_{it}^{\rho}\ EQI_{it}^{\theta}\ RDI_{it}^{\omega}$$

The superscripts in the above equation measure the elasticity of the concerned inputs. A logarithmic transformation of this equation is used to measure the elasticity of the concerned inputs,

$$y_{it} = a + \alpha k_{it} + \beta l_{it} + \gamma m_{it} + \mu r_{it} + \rho(rpm/k)_{it} + \theta eqi_{it} + \omega rdi_{it} + \epsilon_{it} \quad \text{Eq.1}$$

where lower-case letters denote natural logarithm of the corresponding variable. In this analysis, R&D is treated as a stock variable, and not as a flow. Griliches (1979) argues in favour of conversion of R&D expenditures into a stock variable, as the contribution of R&D extends beyond the time period in which it is conducted.

To estimate the parameters in Eq.1, it is essential to have deflated measures of output and inputs. Nominal values of these variables drawn from Prowess have been deflated using industry-wide deflators (up to 5-digits when available) accessed electronically, from the Annual Survey of Industries (ASI), which is the principal source of industrial statistics in India; and various publications of the Central Statistical Organization (CSO), Government of India. The methods used, are outlined below.

Output (*Y*): The current value of output is deflated using the 5-digit commodity-wise Wholesale Price Index of India (WPI), with 1993-94 as the base year, available electronically from the Office of the Economic Adviser (2000). [3]

Physical capital (*K*): The aggregate of the net fixed capital stock ("NFCS" hereafter) of equipment (plant & machinery, transport and communication equipment) and structures (land & buildings), at constant 1994 prices, is chosen as the measure. The perpetual inventory method ("PIM" hereafter) is used to convert expenditure flow on capital of both types into a stock variable. The depreciation rate chosen is 10 percent for equipment and 3 percent for structures. To apply the PIM, a base year capital stock of each type of capital needs to be determined. This is determined using the methodology adopted by Basant and Fikkert (1996).

The gross value of each type of asset is reported at historical cost. To derive NFCS in 1994, the average age ("AA" hereafter) of each type of capital needs to be determined. This is done by deducting the reported net value of the total capital stock from its gross value, and assuming that equipment and structures at that time, took 20 and 50 years respectively to depreciate fully, as given in CSO (1989). One drawback of this assumption is that both equipment and structures are assumed to be of the same vintage, however, no other feasible alternative is available. Using AA, the reported value of the asset in "1994-AA" prices is deflated to 1994 prices using a price index constructed for each type of asset with the help of the constant and current price series of gross fixed capital formation on "construction" and "machinery & equipment", available electronically from CSO (2006). [4] NFCS of equipment (K^{equip}) and structures (K^{Str}) in 1994 prices, are then constructed as follows:

$$K_{i,1994}^{Equip} = \frac{Equip_{i,1994}}{CD_{1994-AA_i}^{equip}} * (1-0.10)^{AAi}$$

$$K_{i,1994}^{Str} = \frac{Str_{i,1994}}{CD_{1994-AA_i}^{str}} * (1-0.03)^{AAi}$$

where *Equip* and *Str* are the book values of equipment and structures respectively, reported in 1994; $CD_{1994-AA_i}^{Equip}$ and $CD_{1994-AA_i}^{Str}$ are the deflators for equipment and structures respectively for "1994-AA$_i$". After having determined the initial values of NFCS for each type of asset, the PIM is used to convert subsequent investments into a stock variable.

Labour (*L*): The total number of employees is used as the measure of labour. It is not reported in Prowess and has been calculated by dividing each firm's consolidated wage bill (wages and salaries paid to all types of workers including managers and owners and contract workers), by the 4-digit industry average for compensation per employee per year. The latter is calculated by dividing "total emoluments to employees" with "total persons engaged", both of which are available from the ASI, and have been accessed electronically from www.indiastat.com (2006). [5]

Raw materials (*M*): The real value of all intermediate inputs (raw materials, power and fuel, stores and services) at 1994 prices is chosen as the measure. The Input-Output Transactions Table of India 1998-99, which has information on raw material use segregated by 4-digit industry code, available electronically from CSO (2006), is used to deflate nominal values. The price of each raw material is deflated using the 5-digit commodity-wise WPI series. The industry-specific materials deflator is calculated as the weighted average of the deflated values of all raw-materials, with composition of raw material used, acting as weights.

R&D capital stock (*R*): Net real R&D stock at constant 1994 prices is the measure used. PIM is used to convert a firm's expenditure on R&D into a stock variable. The chosen rate of depreciation is 15 percent, and it is assumed that R&D affects output with a one-year lag (Coe and Helpman 1995), and that R&D ceases to have an impact on output after five years (Griliches 1979). The initial stock in 1994 is determined as the cumulative past real R&D expenditures, including technical fees and royalty paid, as follows:

$$r_{i,1994} = \sum_{\tau=0}^{5} rdex_{i,1993-\tau}(1-\delta)^{\tau}$$

where r_{1994} is the R&D stock in 1994, and $rdex_{1993}$ is the R&D expenditure inclusive of technical fees paid by the firm in 1993. Raut (1995) argues, that in a developing country like India, disembodied technology purchases by a firm (technical fees and royalty), also form a source of its private knowledge. Hence, these are also included in R&D expenditures. Analysis was also carried out by excluding expenditures on imports of such disembodied technology and results were found to be robust to this change. The series used to deflate R&D expenditure is the average of the deflator series for equipment capital and wages for industrial workers. The latter is obtained from the Ministry of Labour (2006). The R&D stock for subsequent years is generated using the PIM as follows

$$r_{i,t+1} = r_{i,t}(1-\delta) + rdex_{i,t}$$

Since the final sample contains many firms which do not expend on R&D (65 percent of total observations), following Hasan (2002), 1 has been added to all observations on the final R&D stock variable for these firms, to avoid taking logarithms over a zero R&D value.

Recent stock of plant & machinery (*RPM*): As adopted in Hasan (2002), the stock of RPM is constructed as the cumulative past real expenditures on plant & machinery, as follows:

$$RPM_{it} = \sum_{s=0}^{4} Inv_{i,t-s}(1-\delta)^s$$

In this equation, *Inv* is the real expenditure on plant & machinery, calculated as the difference between their book values of two consecutive years. It is deflated using the equipment capital deflator. [6] To avoid collinearity, the stock of recent investments is considered as a ratio of total physical capital (*RPM/K*)

Equipment spillover (*EQI*): This is calculated as the aggregate of the equipment capital stock of all firms in the industry, net of the capital stock of the firm under consideration:

$$EQI_{it} = \sum_i K_{it}^{equip} - K_{it}^{equip} = \sum_{q \neq i}^{N} K_{qt}^{equip}$$

where N is the number of firms in the same industry as the ith firm.

R&D spillover (*RDI*): This is measured as the aggregate R&D capital stock of all firms in the same industry as the ith firm, net of its own R&D stock.

$$\sum_i R_{it} - R_{it} = \sum_{q \neq i}^{N} R_{qt}.$$

Table 3.4 reports descriptive statistics of the variables. To examine stationarity of production variables, this study conducts panel unit root tests based on the Im, Pesaran and Shin (IPS) and Maddala and Wu (MW) tests. Results reported in columns 4 to 5 alongside the descriptive statistics in Table 3.4, plausibly if not convincingly reject the presence of a unit root for every variable.

However, it was found, that the two spillover variables were highly correlated. The degree of collinearity was observed to be 0.98. Including both these variables in a single regression consistently yielded insignificant and negative estimates of these variables. Therefore, several proxies were considered for these variables. For instance, the industry-wide investments in equipment and industry-wide stock of total physical capital were tried instead of industry-wide stock of equipment; and industry-wide R&D expenditures were chosen instead of industry-wide stock of R&D. However, this did not mitigate the problem. There was no single proxy, which could be used without leading to collinearity. It was considered appropriate therefore, to use two separate regressions, for analysing the impact of each of these spillover variables, without making use of any proxies. Two different specifications of Eq.1 were used. Specification 1 estimates the impact of conventional inputs, the recent machinery stock, and industry-wide equipment, whereas, Specification 2 assesses the impact of conventional inputs, R&D and industry-wide R&D.

Table 3.4: Descriptive Statistics and Results of Panel Unit-Root Tests

Variable	Mean	S.D	Median	IPS	MW
Output	2,136,929,833	9,883,174,196	222,954,903	(-) 5.40657 (0.000)	839.090 (0.000)
Physical Capital	203,961,067	1,112,962,669	2,172,461	(-) 4.10274 (0.000)	891.964 (0.000)
Labour	2,187	13,416	169	(-) 2.15992(0.015)	720.746 (0.040)
Materials	335,920,930	1,285,451,031	24,200,001	(-) 6.52913 (0.000)	645.410 (0.215)
Recent Equipment Stock	90,142,302	671,413,034	0	(-) 6.96928 (0.000)	622.557 (0.253)
Equipment Spillover	76,750,760,008	35,628,632,289	66,875,700,208	(-) 26.4166 (0.000)	2851.46 (0.000)
R&D	21,377,108	172,896,809	1	(-) 73.4237 (0.000)	632.167 (0.000)
R&D Spillover	10,684,687,017	6,233,083,018	8,824,153,159	(-) 84.406 (0.000)	(-) 6462.84 (0.000)

Note: Descriptive statistics are based on actual numbers. S.D stands for standard deviation. Unit root tests are performed on the logs of each of the variables, which are used in the final analysis. IPS is the panel unit-root test based on Im, Pesaran and Shin and MW is the panel-unit root test based on Maddala and Wu. The null hypothesis for each of these tests is presence of a panel unit root. Values in brackets are p-values. All tests include individual effects. Lag selection is based on the Schwartz Information Criterion

Estimation of Eq.1 is based on both the static fixed effects panel data technique and the dynamic panel data system GMM technique advanced by Arellano and Bond (1991) and later developed by Blundell and Bond (1998). The dynamic panel data technique should take care of the endogeneity problem encountered in production function estimations. In both types of estimations, the residuals are corrected for both heteroskedastic and serially autocorrelation of the first order. For this purpose, the static estimator proposed by Baltagi and Wu (1999) is used. This estimator corrects for AR(1) and ensures homoskedasticity, and is the only suitable estimator to handle unbalanced panel data when the residuals do not follow the standard normal assumptions.

Empirical Findings

This section presents production-function estimates of Eq.1, based on both the static and dynamic panel data estimates reported in Table 3.5. The left panel considers results from the static estimator whereas the right panel considers results from the dynamic estimator. The top panel reports results from Specification 1 and the bottom panel for Specification 2. Since two different regressions are estimated for each industry, it should be noted that the estimates of conventional inputs and the recent machinery stock in these two regressions differ in magnitude, although are same in sign and significance.

Table 3.5: Empirical Findings

Specification 1: $Y_{it} = A\ K_{it}^{\alpha}\ L_{it}^{\beta}\ M_{it}^{\gamma}\ RPM_{it}^{\rho}\ EQI_{it}^{\theta}$

	Fixed Effects Estimator	**Dynamic Panel Data Estimator**
Lagged Y	-	1.05***
Physical Capital	.178***	.504***
Labour	.404***	.057**
Materials	.304***	.003
Recent Equipment Stock	.0369***	.071**
Equipment Spillover	.369***	2.05*
No. of Observations	2,415	2,303
R^2	0.79	-
m2	-	0.922
Hansen	-	0.534

Specification 2: $Y_{it} = A\ K_{it}^{\alpha}\ L_{it}^{\beta}\ M_{it}^{\gamma} R_{it}^{\mu}\ RDI_{it}^{\omega}$

	Fixed Effects Estimator	**Dynamic Panel Data Estimator**
Lagged Y	-	.932***
Physical Capital	.225***	.146**
Labour	.397***	.272***
Materials	.301***	.436***
R&D	.004*	.003
R&D Spillover	.361***	.512**
No. of Observations	2,233	2,121
R^2	0.79	-
m2	-	0.001
Hansen	-	0.433

Note: *** significant at 1 percent level; ** significant at 5 percent level; * significant at 10 percent level. Residuals corrected for heteroskedasticity and autocorrelation. Time dummies included in all regressions. m(2) is the test statistic which tests the null hypothesis of serial correlation of residuals in first differences. The Hansen test has the null hypothesis that all instruments used in the system GMM equation are valid and pass the over identifying restrictions. R^2 not reported for the GMM module, "xtabond2" in Stata.

Results from both the static and dynamic estimators concur with each other. All coefficients on conventional inputs are found to be of an acceptable magnitude. If we focus on Specification 1, it can be observed that the recent stock of machinery has a considerable positive impact on output, according to both estimators. Findings suggest that a 1 percent increase in this stock has between a 0.36 percent and 0.94 percent increase in output, which concurs with estimates from aggregate samples.

Results for the presence of the equipment spillover are highly convincing. Both the static and dynamic panel data estimators support the view. The elasticity of the spillover ranges from 0.36 to 2.05. One of the plausible reasons for this high elasticity might be because the proportion of skilled workers employed in the Electronics industry is very high. Greiner and Semmler (2002) point out, that the spillover from physical capital has ameliorated effects for those firms, whose workforce has a high proportion of skilled workers. This is because, such workers are better able to assimilate information generated from their contemporaries. Preliminary investigation of the data shows that the ratio of skilled to unskilled workers in this industry is 0.55, which is higher relative to the average of 0.33 of all industries combined. This result gives strong support to the argument of the "level of development" effects of the industry. As more and more firms enter the industry, each firm gains from the experience of the other and grows.

Specification 2 presents results of regressions using data on conventional inputs, own R&D and its spillover. These results give partial support to a firm's own R&D proving to be a significant contributor to its output. Both the static and dynamic estimators suggest that a 1 percent increase in R&D leads to 0.04 percent increase in a firm's output, albeit the coefficient using the dynamic estimator is insignificant. This is an interesting finding as previous firm-level studies on Indian firms, including Hasan (2002) and Raut (1995) had not found R&D's role to be significant for most industries.

Turning to the effects of the R&D spillover, it can be seen that both estimators point to a highly significant and a coefficient of similar magnitude for the output of a firm. It is worth noting, that the estimated elasticity of the spillover is much larger, compared to the R&D coefficient which is not alarming. It suggests, as in the results reported in Wieser (2005), that a firm learns much more from the available industry-wide knowledge relative to the knowledge it creates itself.

Implications of Findings and Concluding Remarks

This research finds that, not only do knowledge generated by a firm's internal efforts, but also, intra-industry spillovers generated through all other firms' investment efforts, favourably affect the output of Indian Electronics firm. As noted before, these findings add considerable value to the literature, as previous studies on India, had not found a firm's own R&D's contribution to be significant. Moreover, no studies till date, have evaluated the impact of spillovers from the industry-wide equipment stock. Since entrepreneurship is considered a vital conduit to ameliorate the effects of spillovers for firms (Acs and Szerb 2007; Parker 2006), it is important that institutions promote rather than hinder entrepreneurship.

The analysis suggests that the recently purchased equipment stock of a firm favourably affects output. This stock may comprise domestically purchased and/or imported machinery. Since the latest machinery embodies contemporary technology, it is suggested that firms, including India's large number of labour-intensive firms, should ensure that their equipment stock incorporates the latest design and technology. However, a recent change in the Indian Government's Foreign Trade Policy, defeats this claim, at least with respect to the import of machinery. As of August, 2004, firms were once again allowed to import second-hand machinery of any vintage. The importation of second-hand capital goods, which may incorporate outdated technology, may retard the competitiveness of Indian firms in the world market and dampen exports.

There is strong evidence linking a firm's productivity to the "learning by watching" effects from investments in equipment, made by all other firms. This further underscores the need for using equipment of a new vintage, which would positively affect the output of all firms in that industry. Furthermore, this spillover suggests that a firm enjoys the "level of development" effects of the industry to which it belongs. All firms gain from this spillover. A firm need not be large in size or investing heavily in either equipment or R&D to benefit from this spillover. However, its effect is found to be higher for a firm which employs more skilled workers. In a labour-abundant country such as India, the need for training the workforce therefore, assumes even more significance.

This research also finds, albeit only partially, that a firm's own R&D, makes a significant contribution to output. Furthermore, not only does R&D help a firm build its own stock of knowledge, it also exerts a significant positive externality onto other firms within the same industry. It appears from these results, that India's Electronics industry's output is being increasingly knowledge driven. Even if small- and medium-sized firms cannot afford independent research, the possibility to organise cooperative R&D facilities is worth considering. Kumar and Aggarwal (2005) find that R&D intensity in India, when considering only small-sized firms, declines with firm-size. Therefore, such firms can consider cooperative R&D, which would benefit all of them as well as others in the industry. Medda *et al.* (2006) have gathered evidence from Italy that such collaborative research yields high returns to firms.

The findings of this research suggest that output of an Indian manufacturing firm is being increasingly driven by knowledge stocks and spillovers, over the period 1994 to 2006. Therefore, to sustain growth, it is imperative that equipment be modernised and that R&D opportunities explored. Such investments can bolster the performance of Indian manufacturing, and can help it become, yet again, one of the engines of growth for the entire Indian economy.

Notes

1. The stock of equipment comprises the stocks of plant & machinery and transportation & communication equipment. It is larger in magnitude compared to the stock of plant & machinery alone.
2. Regressions inclusive of a time trend were also tried. Results were found to be of the same sign and significance for all inputs. However, the coefficient on R&D and its spillover were found to be relatively lower when such a trend is included. This gives credence to the assumption, that technology embodied in R&D captures technological progress.
3. The financial year in India runs from 1st April to 31st March. For expositional convenience, 1993-94 is written as 1994 hereafter, and the practice is followed throughout this paper
4. Data on capital deflators with base 1994 is available only up to 2004. Thereafter, the new series with base 2000 has been introduced. For 2005 and 2006 therefore, this study has generated deflators using the growth rates of deflators provided by the new series.
5. ASI data for 2005 and 2006 is still not available electronically. For these two years, data has been extrapolated based on the mean growth of wages over the period 1994 to 2004.
6. The recent stock of plant & machinery is chosen to comprise of investments in the past four years. Different values of two and three years were also tried, but results remained robust to these changes.

References

Acs, Zoltan J. and Laslo Szerb (2007), “Entrepreneurship, Economic Growth and Public Policy.” *Small Business Economics* 28(2): 109-122.

Arellano, Manuel and Stephen Roy Bond (1991), "Some Tests of Specification for Panel Data: Monte Carlo Evidence and an Application to Employment Equations." *Review of Economic Studies* 58(2): 277-297.

Arora, Ashish, V.S Arunachalam, Jai Asundi and Ronald Fernandes (2001), "The Indian Software Services Industry " *Research Policy* 30(8): 1267-1287

Aw, Bee Yan, Xiaomin Chen and Mark J Roberts (2001), "Firm-level Evidence on Productivity Differentials and Turnover in Taiwanese Manufacturing." *Journal of Development Economics* 66(1): 51-86.

Baltagi, Badi H and Ping X Wu (1999), "Unequally Spaced Panel Data Regressions with AR(1) Disturbances." *Econometric Theory* 15: 814-823.

Basant, Rakesh and Brian Fikkert (1996), "The Effects of R&D, Foreign Technology Purchase, and Domestic and International Spillovers on Productivity in Indian Firms." *Review of Economics and Statistics* 78(2): 187-200.

Blundell, Richard and Stephen Roy Bond (1998), "Initial Conditions and Moment Restrictions in

Dynamic Panel Data Models." *Journal of Econometrics* 87(1): 115-143.

Coe, David T and Elhanan Helpman (1995), "International R&D Spillovers." *European Economic Review* 39(5): 859-897.

CSO, Central Statistical Organization (1989), "Sources and Methods, Chapter 22: Capital Stock and Consumption of Fixed Capital, National Accounts Statistics, Government of India.

CSO, Central Statistical Organization (2006), "National Accounts statistics", Ministry of Statistics and programme implementation, Government of India.

DeLong, Bradford and Lawrence Summers (1991), "Equipment Investment and Economic Growth." *Quarterly Journal of Economics* 106(2): 445-502.

Foster, Lucia, John Haltiwanger and C J Krizan (2001), Aggregate Productivity Growth: Lessons from Microeconomic Evidence. New contributions to productivity analysis. C. R. Hulten, E. R. Dean and M. J. Harper. Chicago, The University of Chicago Press: 303-372.

Frankel, Marvin (1962), "The Production Function in Allocation and Growth: a Synthesis." *American Economic Review* 52(5): 995-1022.

Greenwood, Jeremy, Zvi Hercovitz and Per Krussell (1997), "Long-run Implications of Investment-specific Technological Change." *American Economic Review* 87(3): 342-362.

Greiner, Alfred and Willi Semmler (2002), "Externalities of Investment, Education and Economic Growth." *Economic Modeling* 19(5): 709-724.

Griliches, Zvi (1979), "Issues in Assessing the Contribution of Research and Development to Productivity gGowth." *The Bell Journal of Economics* 10(1): 92-116.

Griliches, Zvi (1979), "Issues in Assessing the Contribution of Research and Development to Productivity Growth." *The Bell Journal of Economics* 10(1): 92-116.

Hammond, Peter J and Andres Rodriguez-Clare (1993), "On Endogenizing Long-run Growth." *Scandinavian Journal of Economics* 95(4): 391-425.

Hasan, Rana (2002), "The Impact of Imported and Domestic Technologies on the Productivity of Firms: Panel Data Evidence from Indian Manufacturing Firms." *Journal of Development Economics* 69(1): 23-49.

Im, Kyung So, Hashem M. Pesaran and Yongcheol Shin (2003), "Testing for Unit Roots in Heterogeneous Panels." *Journal of Econometrics* 115(1): 53-74.

Jensen, J. Bradford, Robert H. McGuckin and Kevin J. Stiroh (2001), "The Impact of Vintage and Survival on Productivity: Evidence from Cohorts of U.S Manufacturing Plants." *Review of Economics and Statistics* 83(2): 323-332.

King, Mervyn A and Mark H Robson (1993), "A Dynamic Model of Investment and Endogenous Growth." *Scandinavian Journal of Economics* 95(4): 445-466.

Kumar, Nagesh and Aradhna Aggarwal (2005), "Liberalization, Outward Orientation and In-House R&D Activity of Multinational and Local Firms: A Quantitative Exploration for Indian Manufacturing." *Research Policy* 34(4): 441-460.

Lal, K. (1999), "Determinants of the Adoption of Information Technology: a case Study of Electrical and Electronic goods Manufacturing Firms in India " *Research Policy* 28(7): 667-680

Levinsohn, James and Amil Petrin (2003), "Estimating Production Functions Using Inputs to Control for Unobservables." *Review of Economic Studies* 70(2): 317-341.

Los, Bart and Bart Verspagen (2000), "R&D Spillovers and Productivity: Evidence from U.S Manufacturing Micro-data." *Empirical Economics* 25(1): 127-148.

Maddala, G S and Shaowen Wu (1999), "A Comparative Study of Unit Root Tests with Panel Data and a New Simple Test." *Oxford Bulletin of Economics and Statistics* 61(4): 631-52.

Medda, Giuseppe, Claudio Piga and Donald S Siegel (2006), "Assessing the Returns to Collaborative Research: Firm-Level Evidence from Italy." *Economics of Innovation and New Technology*, 15(1): 37-50.

Ministry of Labour (2006), Government of India.

NMCC, National Manufacturing Competitiveness Council (2006), The National Strategy for Manufacturing. Government of India.

Panagariya, Arvind (2002), India's Economic Reforms: What has been Accomplished? What Remains to be done? ERD policy brief series, Economics and Research Department, Asian Development Bank.

Parker, Simon C. (2006), "Learning about the Unknown. How Fast do Entrepreneurs Adjust their Beliefs? ." *Journal of Business Venturing* 21(1): 1-26.

Raut, Lakshmi K (1995), "R&D Spillover and Productivity Growth: Evidence from Indian Private Firms." *Journal of Development Economics* 48(1): 1-23.

Romer, Paul M (1986), "Increasing Returns and Long-run Growth." *Journal of political economy* 94(5): 1002-1037.

Romer, Paul M (1990), "Endogenous Eechnological Change." *Journal of Political Economy* 98(5 part 2): 71-102.

Rosenstein-Rodan, P N (1943), "Problems of Industrialization of Eastern and South-eastern Europe." *The Economic Journal* 53(210/211): 202-211.

Sakellaris, Plutarchos and Daniel J. Wilson (2004), "Quantifying Embodied Technological Change." *Review of Economic Dynamics* 7(1): 1-26.

Shaw, G. K. (1992), "Policy Implications of Endogenous Growth Theory." *The Economic Journal* 102(412): 611-621.

Shiferaw, Admasu (2007), "Firm-heterogeneity and Market Selection in Sub-Saharan Africa: Does it Spur Industrial Progress?" *Economic Development and Cultural Change* 55(2): 393-423.

Wieser, Robert (2005), "Research and Development Productivity and Spillovers: Empirical Evidence at the Firm Level." *Journal of Economic Surveys* 19(4): 587-621.

4

Quality Aspect of Indian Software Industry

Santhosh Rebello

Introduction

The Indian software exports, as is well known, have grown from less than US$ 100 million before 1990 to over US$ 5 billion in 2000. What is not well known is that revenue per person per year has grown from less than US$ 20,000 to over US$ 50,000 in most large companies in the last five years. This is not simply due to inflation? Though the industry is still focused on tapping the huge software services market, most of the big players have moved from lower value services to higher value services. In the early days, the service was primarily providing technical manpower, which later moved to providing low-value services like coding and testing. Now Indian companies are operating even at the top end of the spectrum in terms of technology (Corba, Java, E-commerce, etc.) or the services (complete business solutions, consultancy, strategy, etc.). The high rates they command is an indication of the perceived value in the eyes of the customer of the services they provide.

The Indian software industry truly symbolizes India's strength in the knowledge based economy. Highly skilled human resources coupled with low wage structure and world class quality have transformed India into a global powerhouse in the Information Technology (IT) software services and solutions sectors.

Strengths of Indian Software Industry

India has moved rapidly on the **quality** front. Indian software organizations adopted the ISO model soon after it came and then when the Capability Maturity Model (CMM) started becoming more important, rapidly transitioned to the CMM. Though no doubt the success of Indian companies with these frameworks is due to the exam taking abilities that our people have (what with exam taking being part of any career), still these models have been by-and-large absorbed quite well. This has given the Indian companies solid project and process management strength these models are supposed to bring.

This is a tremendous strength that is not easy to emulate. For example, China, despite the fact that it has about a 1000 software companies in Bejing area alone, has probably none at level 4 or 5 and is now actively looking at CMM as part of the country-level strategy to tap the global market. The situation in the rest of Asia (not counting the developed countries) is similar. The situation in Latin America is no different. In fact, outside of the developed countries, it is only India that has companies that have been able to successfully implement these global quality models.

The difficulty of achieving this can be illustrated by a small anecdote. Based on experience of Indian companies transitioning from ISO to CMM, most Indian companies moved from ISO to Level 4 or above with a matter of a year or two. Overall, the project and process management abilities of Indian companies are very advanced and compare well with the best in the world. In fact, many Indian companies are now getting requests from their customers seeking help in setting up quality systems in their organizations.

The second strength is the business expertise that now exists in the industry. With time and experience with projects and dealing with clients across the world, many managers have developed good understanding of business practices in other countries and have acquired domain knowledge and an understanding of the cultural context of the customer. These are the factors that are essential for moving from low level technical services to higher level services, i.e. for going up the value chain. They are also essential for customer satisfaction, which, in turn, is essential for business. Many Indian software companies boast of repeat business of over 70 percent - a huge asset when trying to grow at rapid pace. The past customers remain for growth at least loss of past customers does not need to be handled. At the same time, the senior management seems to have transformed from thinking like India-based supplier to being a global player. This has brought in organization management and business practices that are in line with the developed world.

Finally, Indian software companies have developed the ability to change rapidly with technology. This has been amply demonstrated by the quick movement to Java, E-commerce related technologies, etc., which are now major portion of the business for many companies. In today's technology world, if a service provider does not have the ability to reprogram its people and processes to adapt to new technologies then it will not remain in service for too long.

These three solid strengths: processes, business skills and reprogammability are not easy to acquire. There is a lot that goes into these. And how did India Software Pvt. Ltd. acquire these? Obviously there can be no simple explanation to this. Let me point out a few reasons.

Perhaps the biggest factor has been the export orientation of the industry. This has helped the industry being receptive to technology and has brought about the quality orientation. There is nothing like fighting for customers in the global market for improving an industry. This has been shown earlier by the auto industry in Japan, the toys industry in China, semiconductor industry in Korea, etc. In all of these, like the Indian software industry, the domestic market was not the main market; the industry grew, and kept its focus on the global market.

The second reason is that the software industry grew without government help. Perhaps the only thing that the government did was to get out of the way by liberalising imports and exports in the early 1990s. The ministry of IT has been formed only recently, and then too it had to go out of its way to explain that it will try to aid and not hinder the industry and focus on internal consumption of IT. Contrast this with the situation in other developing countries almost all countries in Asia and Latin America are looking for government funding and help for developing the IT industry. Government help and funds can perhaps help for domestic market, but for global market this just cannot work. The government money and policies, particularly in developing countries, are too slow, while the IT business is very fast. Latin America has many examples of lost opportunities because the government provided funds and subsidies to some industry sector, thereby encouraging it to remain inefficient and uncompetitive in the global market.

The business leaders have played a key role. The leaders in the major software organizations were quick to tap the opportunity when it became available and set high standards for the rest of the industry. After the standards set by TCS, Infosys, Wipro, HCL, etc., no company today aspires to be just a few-crore companies. The aspirations have all gone up. And with increased aspirations came drive and innovation, which lead to increased success, particularly since the demand in the global market was available and the outsourcing model for software had matured.

Education must have played a role. Though we complain a lot about our education system, it is also true that no country outside of developed countries has been able to develop an education system like that of IITs. And though IITs have directly contributed little manpower to the software industry, it has indeed provided many success icons and many leaders across the world. And the IITs have set a high standard in education something that indirectly aids education across the country. Even the fact that students frequently have to self educate because many colleges are not doing their job might have helped after all, this probably helps develop the learnability skills, which are essential for re-program ability.

All this does not mean that the industry or the country can rest on its laurels. The lead can, at best be viewed as a few years lead. There are many areas where the limits are reaching, particularly on the trained manpower front, and the use of IT with the country itself. These can start becoming the bottleneck. But for now, we can justifiably feel proud of what India Software Pvt. Ltd. has achieved and feel proud that many other countries want to emulate it but are finding it tough. After all, how many such shining examples are there in India today which can make one feel proud?

Profitability: Mantra of Indian Software Industry

The Indian software industry started in Mumbai in the late 1970s but the real turning point for the industry was the establishment of a US company, Texas Instruments' (TI) design centre in Bangalore in 1986 which created a snowball effect and Bangalore became an IT software hub with Indian and international companies clustering in and around the city. Large Indian companies such as Infosys and Wipro are headquartered in Bangalore. Opening up the Indian market to foreign players and the economic reforms of the early 1990s acted as a catalyst for the Indian software industry. Today, most top international computer and software companies (HP, Digital, Compaq, IBM, and Intel etc.) have wholly owned subsidiaries or joint ventures with Indian companies based in Bangalore and other cities. This impressive rise and growth of the software industry has spawned "Silicon" metros such as Bangalore, Hyderabad, Mumbai, Pune and Chennai. In addition, smaller cities like, Bhubaneshwar in the East, Pondicherry, Thiruvananthapuram and Mysore in the South, Gurgaon and Chandigarh in the North are in a strong position on the map of Indian software today.

Balance sheets of Indian software companies indicate that they are going through a metamorphosis. The three behemoths in the market — TCS, Infosys and Wipro — as well as others have shown impressive growth potential. In financial year 2005-06 TCS made a net income of Rs. 11,282.81 crore a sharp rise of 38.90 percent from the previous year, Infosys secured a net income of Rs. 9,028 crore with a 31.60 percent rise from last year. Wipro's net income was Rs. 10,625 crore with a rise of 30 percent over the previous year .The most fascinating thing is that top five players (Tata, Wipro, Infosys, HP, HCL) in the Indian IT market accumulated a gross total of Rs. 49836 crore and are moving towards the remarkable landmark of US$ 5 billion. [1]

The Indian software industry increased their headcounts in India and abroad last year either by recruiting or through mergers and acquisitions. In the previous year 2005-06, TCS had a net addition of 21,140 employees which brought its total strength to 62,832 employees. Infosys added 12,500 new employees increasing its organizational strength to 44,658 (excluding Infosys' BPO subsidiary previously known as Progeon). Wipro had 55,000 people in the last financial year. [2]

The Indian software industry uses two different models for revenue recognition. The first is the Time and Material (T&M) Contracts model in which customer are billed on the basis of hours worked by the employees of supplier software companies. Hourly rates are agreed on by both parties and are applied to the total hours worked to arrive at the revenue that is to be recognized. The second is the Fixed Price Project Model (FPP). Under the Fixed Price Project Model, the total contract price is agreed upon between the parties. Billing may be done either at the end of the contract or over the period of the contract on the basis of the agreed milestone for billing.

India's competitiveness in the software industry is of world class quality, cost-effectiveness, high reliability and rapid delivery to the customer. The software industry in India has matured as compared to earlier days. The industry has moved beyond 'Body Shopping' to 'End to End delivery of business solutions'. It is easy to interpret from historical data that the Indian software industry depends heavily on its human resources for revenue generation. The greater number of employees provides the capability to fetch more work thus generating more profit and growth. In other words, a company's headcount and revenue are in direct proportion. Availability of a large pool of skilled labor in the Indian market makes it easy for software companies to leverage the demand for IT in the global market. This business model is effective and successful to a certain extent.

The biggest question for the Indian software industry is how far a company can scale only in terms of headcount. The linear model of growth is not justifiable in the current global market and top global players are already demonstrating this. For example, IBM's headcount is currently approximately 330,000. In the terms of manpower it is nearly six times larger than any of the top three Indian players but in terms of revenue the gap is 45 times. The total revenue of IBM in 2006 was US$91.4 billion. In context of a product development company like Microsoft which is operating currently with around 71,172 employees, the revenue is 22 times higher than the top Indian players. The combination of Offshore-Onshore development is working as a panacea for the Indian software industry but companies like IBM or Accenture are moving fast in the same direction and the day is not far when the competitive advantage of the global delivery model will start eroding for Indian software companies. Indian companies need a new strategy for growth which should not be completely based on linear growth patterns. [3]

Global Profitability on the Service Front

India is the undisputed leader of the software services market at the global level .The magic mantra of offshore-onshore development is the key factor behind the success of Indian players. According to a fact sheet currently published by NASSCOM on the Indian IT industry the total combined Indian IT services and IT enabled services export market in fiscal 2005 was nearly US$ 18 billion. A report published by NASSCOM-KPMG in 2004 [4] indicated that total Indian IT and IT enabled services market export is expected to grow to US$ 49 billion by 2009. Significant cost benefits to customers, availability of low cost skilled professionals and labour, ability to effectively integrate onsite and offshore execution capabilities to deliver seamless and scalable services and increase in depth and breadth of the services offered to customer are some of the key factors which are responsible for the sustained growth of Indian software.

Top Indian players are moving up the value chain and providing a range of services to customers. Indian companies also understand that they can not scale up only in terms of adding manpower and they will have to develop models that will not be based on the linear model of people, revenue and profit. Top players are coming out with new business models in the current scenario. One such business model is the 'Risk Reward Model', in which companies take a joint stake in the development of a product with technology companies. Software companies are combining efforts and resources with technology companies and sharing the benefits thereof. They are partnering with companies operating in the field of telecommunications, data communications, semiconductors etc. Wipro is the leader in this area making an average of three times the normal service business in the segment. More than 10,000 employees are working in the product development stream in Wipro. The limitation in this kind of business model is that Indian software companies do not own the product i.e. they do not have the intellectual property rights for the product. They are only getting incentives from the owner company to make a successful product.

Indian companies are also implementing intellectual property (IP) components. They are not end products but are a vital part of the end solution. These components are basically related to networking and telecommunications for example wireless LAN components which will be used by most telecom-vendors. The limitation with this model is that companies are getting a one time service fee only not the licensing fees as they would if they owned the product. Another common predicament with these models is the concept of 'Branding'. The customer is seeing the partner technology company's name as the brand and not the Indian company's name; in the long run this may hinder the overall growth of Indian companies as they lack name recognition for putting effort and resources as far as the brand awareness with the end customer is concerned.

'Do the work where it makes the most sense' is the success formula for the Indian software industry and to effectively enable the onshore-offshore development model, Indian companies are increasing their local presence in the global market. This provides a competitive advantage over others in cost. Moreover, it helps companies widen their domain expertise. Mergers or acquisitions will certainly expand a company's knowledge pool and expertise in different verticals. Furthermore, acquiring a well managed company with less employees and great business can undoubtedly help tackle the problem of linear growth.

Over time big players in the Indian market have understood the problem of linear scalability and every one of them is trying to get out of the situation. Every one is building capacity, a business model and expertise. For example, Wipro is building expertise in Product Engineering, Technology Infrastructure, and Enterprise Platform implementation. Infosys is concentrating on the global market of consulting on the lines of bigger players like IBM and Accenture. TCS which has a larger global footprint and is the biggest among the three is taking the middle path by concentrating on a technology-centric approach. In 2005, TCS announced a relationship with CollabNet, the leading provider of on-demand (SaaS) distributed software development solutions. TCS has adopted CollabNet Enterprise Edition 4.0, the company's flagship solution for distributed application lifecycle management and software development, for its internal projects. TCS also earned three patents in the US and one in India during fiscal 2005-06 and applied for 13 new patents in various technology areas during the same period. Infosys alone generated over 82 invention disclosures and filed over 20 patents last year.

Another model dubbed the 'Long Term relationship Model' is one of the best ways to reduce cost and increase profitability. It substantially reduces the cost of marketing and sales and provides enhanced revenue visibility. Indian companies are recognising the benefit of long-term relationships with customers but still the model is in a nascent stage. If we will leave the big players and concentrate on the mid-sized and small players then it can be seen that a significant part of revenue is coming from small projects (US$ 1- US$ 10 million). Not only small and mid-sized but even bigger players in India find it uncomfortable to compete with the global players in the case of billion dollars projects. Indian companies are on the verge of developing 'Partnership Models' which are helpful in dealing with bigger projects. Competition is intense in the global arena and everybody is bidding for the same project. Instead of losing out to bigger players, companies are coming with joint ventures with other Indian or foreign companies. Once a contract is won, work is divided on the basis of area of expertise. It is like a symbiotic relationship between two or more players that helps grow revenue without raising headcount.

Domestic Profitability on the Services Front

Changes in economic policy and deregulation of key sectors have driven IT adoption in the country. The amount of global competition is the largest in banking and telecom due to presence of global competitors in the market and that is the reason these segments account for 35-40 percent of Indian domestic IT spending. Similar competition can be seen in the Insurance and Airlines sector and the Retail sector is booming. Several e-governance plans launched by Indian government under the National e-governance Plan (NEGP) are expected to provide sustainable growth in the domestic IT market. According to a forecast by IDC, the Indian domestic IT market is projected to be worth Rs. 238 billion in 2009 as compared to Rs. 103 billion in 2004 (IDC-2005). [5]

Some key challenges for the Indian software industry in the domestic market are low government spending and stiff tender mechanism. In 2004-05, Indian spending was 0.13 percent of GDP in IT investments as compared to 0.24 percent in Russia and China, 0.34 percent in the US and 0.49 percent in the UK. Domestic projects are fixed price projects (FPP) without clear specifications. Tenders are awarded to companies on the basis of proposed cost of projects not on the basis of proposed solution. Even in the private sector IT is not high on the agenda. SMEs (Small and Medium Enterprises) are still in early stages of adopting full fledged IT solutions. To tap the market, some Indian software companies are now coming with customised solutions for the public and private sectors. India is a big country with 15 different official languages and 500 dialects. Indian companies are providing multilingual solutions to domestic clients at an affordable price.

According to recent research, done by Gartner, [6] the Indian economy will surpass the US economy by 2050. This statement may be a forward looking statement but it is sufficient enough to express the opportunity in the domestic market for the Indian software industry. Currently software is 'made in India' but it is not 'made for India'. The software industry will have to recognize the difference between the Indian market and the global market and that will be a big leap for revenue recognition. The verticals will be the same even in the Indian market (Banking/ Finance, Retail etc.). The only difference is going to be the cultural difference and Indian companies are in much better position to study and implement it. That said foreign players are far ahead in understanding the Indian market and catering to a large portion of market when compared to Indian players.

The combination of Offshore-Onshore development is working as a panacea for the Indian software industry but companies like IBM or Accenture are moving fast in the same direction and the day is not far when the competitive advantage of the global delivery model will start eroding for Indian software companies. Indian companies need a new strategy for growth which should not be completely based on linear growth patterns.

Global Profitability on the Product Front

Product development is different from project based development. It requires proper research, funding, branding and intelligent marketing communication. The global packaged software market has been projected to be worth US$350 billion in 2007. Indian software companies can leverage the market but the Achilles heel for an Indian Software Company is that it does not have many recognised products in the market. Indian software products, which are global brands, can be counted on your fingers and the major reason is investment. Any software company needs to invest a huge amount of resources in brand building and marketing and the top management is unwilling to do that. Take for instance, behemoths like TCS or Infosys. Even as they gather growing global clout, their top management publicly announces their desire to emulate an IBM Global Services, EDS or Accenture but never do they ever dream of trying to become the next Microsoft, Oracle or SAP. That is the Indian dichotomy, on the one hand we are the leader of global services and on other hand we stand relatively nowhere when it comes to products.

The software industry in India contributes to about 5 percent of the country's GDP and nearly 20 percent of its exports but it still account for only a meagre 0.2 percent of the US$180 billion global market. [7] It shows that without software products the dream to become an IT powerhouse is likely to remain just that, a dream. According to a study conducted by IIM-Bangalore in 2007, [8] the revenue potential of India's software products is as high as US$7 billion by 2010 — a near ten-fold rise from the current levels. However, this is still a drop in the ocean.

Many small Indian software companies have tried to make an entry in the global market but they could not survive due to extreme competition or lack of funds. The most important reason was lack of knowledge of the culture abroad. When Indian software companies enter the global market they face an alien culture without having any insights into it resulting in a less than successful entry. Global leaders are using extensive market research data and analytics to get an in-depth understanding of what the customer wants and what he is willing to pay for a product. Successful products can change the fate of an Indian software company. The biggest advantage is that they need not concentrate on the whole market to grab the business opportunity. Make a competitive product for the market and concentrate on branding and customer communication. .Once the product is successful then the same product or product line can be a major revenue earner for the company without large expansion in the team size. It takes years to create a product that is a successful brand and Indian companies must have that much of patience if they wish to step away from the linear growth mentality.

The biggest challenge for Indian companies in raising their market share is to understand the difference between project and product development. Projects require understanding of only one customer whereas products require understanding of the whole market. If the project does not meet the requirement of the client it can easily be redesigned according to what suits the customers but in case of a product the customer either selects or rejects it. In many cases it is possible that product companies may face failures but there lies the art of survival of any product based software company and it hugely depends on the funding as well as the research and development. The speed with which a company can eliminate causes of failures once the product is launched in the market is of utmost importance. The most successful word processing software in the market Microsoft Word also faced stiff competition in its early days in 1981 from Word Perfect. Customers were reluctant to buy Word at that point as they were unwilling to shift to a new product from the existing one. Microsoft learned from the market and Word is the most popular word processing product today.

Domestic Profitability on the Product Front

It is not only in the global market, opportunities exist in the domestic market as well and only a handful of companies have tapped this so far. Tally is a case in point of a product that has understood the needs of Indian Small and Medium Businesses (SMBs) and enjoys the status of being a preferred brand for accounting in India. 94 percent of the product revenues of Tally Solutions are generated from India itself which is similar in the case of Polaris/Orbitech (90 percent domestic) or Infosys BBU (Finacle: 47 percent domestic). Most other companies like i-flex Solutions (5 percent domestic) still heavily depend on the global market.

The biggest hurdle in the Indian market is the fact that most solutions are not affordable for SMBs. So either they turn to open source solutions or to copyright infringement. Still the way business is done in India is very traditional and not only large businesses but SMBs are also family run with a thrifty mentality. Huge investments in IT are considered unnecessary. When IT is perceived as a necessity, the business relies heavily on pirated software. Software as a service (SaaS) is a potential solution. So far the SaaS market has dominated by the Salesforce.com and Right Now Technologies; companies that have stressed or focused on SFA (Sales Force Automation) and CRM (Customer Relation Management) .However with success of this concept, companies hopping on to the SaaS bandwagon are offering everything from ERP to collaboration tools from SCM to HRMS (Human Resource Management System). The approval of the success of this model is evident from the fact that even a traditional software company like Oracle that gets most of its revenue from licensing is talking about the SaaS model of business. Gartner predicts [9] that by 2010, around 30 percent of new license purchases in APAC excluding Japan will be in the form of SaaS or delivered through this Model. If Indian companies come with less expensive solutions to cater to the domestic market it can dramatically boost the revenue of software companies because India is heavily dependent on foreign players as far as software products are concerned.

Future Trends and Tactics

The last decade was the decade of Indian software. India has been the leading offshore destination during the period and now accounts for 65 percent of the global industry in offshore development. Benefits of the global delivery model will induce more companies to outsource to India in the new decade. According to a McKinsey–NASSCOM report in 2005, the addressable market for global offshoring is US$300 billion and so far we have realised only 10 percent of this market. While the addressable market is very big the realisation of this market depends on two things—global and domestic demand and the supply of resources (productivity and quality).

Indian companies will have to understand the changing demand patterns of the global market and build core competencies by any and all means necessary. In the next decade we will probably see Indian companies working with global players in joint ventures and negotiating terms as per their needs. The supply of competent resources primarily depends upon the Industry-Academic partnership and diversified training of personnel. Till now Indian companies have gained an advantage in the latter but the relationship with academic institutes is still in a nascent stages. Only the bigger players are connected with academic research and development projects and that also in a partial way. In the future we can expect a stronger relationship between the industry and academia, not only at the domestic level but also at the global level. The line between software and services continues to blur. The notion of SaaS and Hosted Application Management (HAM) services are well known and are proving to be quite disruptive to the software industry. These new models of software development and distribution have gathered momentum and customer mindshare in previous years and are expected to continue in future. The Indian software industry has a bright opportunity to tap the market with its low cost, better quality solutions in this area. They should develop competencies in these areas to tap the market as soon as possible.

Tier I cities in India (Mumbai, Bangalore etc) are already running out of resources. Companies will have to shift their operations to Tier II cities to gain a competitive advantage. Infrastructure is the biggest issue in Tier II cities. It requires huge investments from government and private bodies. China has developed a huge amount of infrastructure to lure foreign players and it is currently getting more Foreign Direct Investment (FDI) than India. Public-Private partnership becomes the most important priority to tackle the growing competition from China and other countries. In the future we will see a number of Special Economic Zones (SEZs) and software parks in India but the concept of Tax Holidays for software companies is the most disputed topic in the political corridor nowadays. The software industry must deal with political and government bodies to enjoy significant tax benefits that are a major benefit to the industry over others. The involvement of the government in the development of software companies is becoming more essential over time.

Indian software companies are in a confident position in terms of going out to the US and Europe to acquire companies. Still a lot can be done in the European market. Developing countries should also be there in the list of Indian companies. If the market cannot be captured directly then the partnership model can be used to make a local presence in a new or underdeveloped market. In the future Indian software companies will have to compete on several fronts. Global IT services and consulting companies are expanding their operations in India. Accenture has built human capital of almost 50,000 employees in India alone to gain a significant cost advantage. More than 80000 employees of IBM global services are operating in the Asia-Pacific region. It is an alarming sign for the Indian software companies. They will have to move up the value chain and form joint ventures among themselves to deal with global giants.

Till now India is the preferred offshore destination due to its cost advantage. NASSCOM has predicted that Philippines, China, Malaysia, Mexico, Brazil, Poland, Hungary, Russia etc. are in a good position to compete with India. The Philippines is at par now with India as far as the labour cost in concerned (US$13 per hour as compared to US$12 per hour for India). Improvement in productivity is a must for the Indian software industry to sustain strategic benefits. Indian software companies will have to be one step ahead by increasing their presence in other low cost countries as the salaries in India are surging due to lack of skilled professionals and the entry of MNCs (Multi National Companies) in the Indian market.

The first decade of IT services was dominated by labour arbitrage, the second decade by process improvement and this decade will be dominated by the creation of IP assets. The bigger players have a head start but the smaller players are far behind. The creation of huge IP assets will be the greatest challenge for the Indian software industry. IP assets will drive the future of the industry in the new decade but it requires a large amount of investment from any company. Even bigger players are hesitant to invest a large amount of money in building IP assets and this could turn out to be harmful to the Indian software industry in the long run. Another important area is consulting which is one of the most profitable areas and Indian companies stand far behind in comparison with global players here. In the future Indian companies must increase their revenue from consulting front to deal with the problem of linear growth.

Indian companies will have to do both: not demolish the business of development and maintenance which is still a cash cow and build an IP-driven business for the future. Leading Indian companies are realising the necessity to re-engineer themselves. They are investing enough now in terms of non-linear models though not as much as they should. Many of them are building infrastructure so that they can drive 8 to 10 percent productivity growth a year, and also 8 to 10 percent growth. That means that they need not scale up in terms of headcount. Successful implementation of this strategy depends on the increase of the number of people vis-à-vis revenue. Problem of growth without scalability is solved if a company raises its productivity by X percent without increasing its headcount or if the company is increasing its headcount by Y percent then certainly productivity should increase by more than Y percent to maintain sustainable growth. Indian companies are headed in the right direction with an average growth rate of 30 percent as compared to global players which are growing in the range of 10 to 12 percent annually. But the problem of scalability can be solved only when their will be a limit to the linear growth pattern. All the development which is visible to us is the effort of a couple of decades and the industry will require more time to mature. It all depends on the visionary leadership of the industry and where they want to land in the future — a successful industry which caters to the nation and its people effectively or an industry which is struggling to manage its internal infrastructure and resources to become more productive and effective to survive in ruthless competition.

Conclusion

Overall, the transition of India to becoming a knowledge-based economy is giving all the education companies a platform to display competencies. But like all growing economies, the weightage on services is high, and hence the margin of error in this business is getting narrow. Players in the unorganized sector shall find it difficult to sustain themselves, and this shall result in the IT training space being dominated by a few knowledgeable performers. As the famous management law states, the top few companies shall control around 80 percent of the market which is a good news for the IT industry, corporate world, and its stakeholders i.e students, franchisees, potential industry entrants and all of us. So it is the best time to be part of this industry as a new entrant or as a new centre of an established brand.

Notes

1. The last statistics available till 2006.
   Source: http://www.wipro.com/investors/pdf_files/Wipro_annual percent20report_2005-06.pdf
2. The statistics have been taken from
   http://www.nasscom.in/Nasscom/templates/LandingPage.aspx?id=51336
3. "Flooding ERP (SAP, PS, ORACLE Apps)" by Sriram N A, Jayanth IBM India 2003.
4. NASSCOM-KPMG Study, 2004, "Choosing a location for offshore operations in India", 2004 accessed from
   http://www.citigroupglobalservices.com/pdfs/Executive_Summary_KPMG_Report.pdf
5. "The Next wave of Business intelligence", Praveen Sengar, IDC (India) Ltd. 2005.
6. "Indian IT growth outpaces global biggies: Gartner Report", Business Standard, 2005.
7. "IT's contribution to GDP up: Study", NASSCOM India Leadership Forum 2007.
8. "Key highlights of IT and ITES performance", IIM Bangalore, 2007 accessed from
   http://www.mdi-delphique.com/information_management.html
9. "Business needs will force a refresh of IT Change management policies", Gartner RAS Core Research Note, UK, 2006.

References

Akella, R. and Dossani, R. (2001), "A Report on the Software Value Chain: the Indian Suppliers During the Downturn", Working Paper, Asia-Pacific Research Center, Stanford University.

Arora, A. and Athreye, S. (2002), 'The Software Industry and India's Economics Development", *Information Economics and Policy,* 14(2), 253-273.

Arora, A., Gambardella, A. and Torrisi, S. (2001), "In the Footsteps of Silicon Valley: Indian and Irish Software in the International Division of Labor", Working Paper, Stanford Institute for Economic Policy Research.

Athreye, S. (2005), "The Indian Software Sndustry", in Arora and Gambardella (ed.), *From Underdogs to Tigers*, Oxford University Press: Oxford.

Athreye, S. (2003), "The Indian Software Industry and its Evolving Service Capability", Working Paper, The Open University, Walton Hall, Milton Keynes, UK.

Balasubramanyam, V. and Balasubramanyam, A. (2000), "The Software Cluster in Bangalore", in Dunning (ed.), *Regions, Globalization and the knowledge-based economy*, Oxford University Press: Oxford.

Correa, C. (1996), "Strategies for Software Exports from Developing Countries", *World Development,* 24(1), 171-82.

D'Costa, A. (2000), "Technology Leapfrogging: the Software Challenge in India", in Conceicao et al. (eds), *Knowledge for Inclusive Development*, Quorum Books: Westport.

D'Costa, A. (2002), "Software Outsourcing and Development Policy Implications: an Indian Perspective", *International Journal of Technology Management,* 24(7/8), 705-23.

D'Costa, A. (2002), "Export Growth and Path Dependence: the Locking-in of Innovations in the Software Industry", *Science, Technology and Society*, 7(1), 51-87.

Dedrick, J. and Kraemer, K. (1993), "Information Technology in India: the Quest for Self Reliance", *Asian Survey*, 33(5)

Desai, A. (2003), "The Dynamics of the Indian Information Technology Industry", Working Paper, London Business School.

Dossani, R. and Kenney, M. (2002), "Creating an Environment for Venture Capital in India", *World Development,* 30(2), 227-253.

Dunning, J. (1992), "The Competitive Advantages of Countries and the Activities of Transnational Corporations", Economist, Innovative India, April 3 2004, 65-66.

Gereffi, G., Humphrey, J., Kaplinsky, R. and Sturgeon, T. (2001), "Introduction: Globalisation, Value Chains and Development", *IDS Bulletin*, 32(3), 1-8.

Gadrey, J. and Gallouj, F. (1998), "The Provider-customer Interface in Business and Professional Services", *The Services Industries Journal,* 18(2), 1-15.

Ghemawat, P. (2000), "The Indian Software Industry at the Millennium", *Harvard Business School,* Case 9-700-036.

Hobday, M. (1995), *Innovation in East Asia: The Challenge to Japan*, Blackwell Publishers Edward Elgar: Cheltenham.

Heitzman, J. (1999), "Corporate Strategy and Planning in the Science City: Bangalore as Silicon Valley", *Economic and Political Weekly*, January 30.

Joshi, V. (1999), "*India's Economic Reforms 1991-2001*", South Asia Books: New Delhi.

Lateef, A. (1997), "Linking up with the Global Economy: a Case Study of Bangalore's Software Industry", www.ilo.org, downloaded May 20, 2002.

Kelkar, V., Chaturvedi, D. and Dar, M. (1991), "India's Information Economy: Role, Size and Scope", *Economic and Political Weekly*, September 14 1991 2153-60.

Kripalani, M. and Engardio, P. (2003), "The Rise of India", *Business Week* December 8 2003.

Ministry of Human Resource Development (2001), "Technical Education Quality Improvement Project of the Government of India", Ministry of HRD, New Delhi.

Mowery, D. (ed.) (1996), "The International Computer Software Industry: a Comparative Study of Industry Evolution and Structure", Oxford University Press, Oxford and New York.

Naidu, B. (2006), "Growth of IT Export Industry in Karnataka", STPI: New Delhi.

Nair, J. (2002), "Singapore is not Bangalore's Destiny", *Economic and Political Weekly*, April 29, 2002 available at www.epw.org.in.

Nasscom. (2002, 2003, 2004, 2005), *The IT industry in India*, Nasscom: New Delhi.

North, D. (1990), *Institutions, Institutional Change, and Economic Performance,* Cambridge University Press: New York.

Ohlin, B. (1933) *Inter-regional and International Trade,* Harvard University Press: Cambridge.

Park, A. (2003), "EDS: What Went Wrong", *Business Week,* April 7 2003, 60-63.

Patibandla, M. and Petersen, B. (2002), "Role of Transnational Corporations in the Evolution of a High-tech Industry: the Case of India's Software industry", *World Development,* 30(9), 1561-77.

Parthasarathi, A., and Joseph, K. (2002), "Limits to Innovation in India's ICT Sector", *Science, Technology and Society*, 7(1), 13-50.

Parathasarathy, B. (2000), "Globalization and Agglomeration in Newly Industrializing Countries: the State and the Information Technology Industry in Bangalore, India", Ph.D. thesis, University of California, Berkeley.

Porter, E. (2004), "Indian Techies lack Creative Ability, feel US firms", *New York Times*, April 29 2004.

Porter, M. (1990), "The Competitive Advantage of Nations", The Free Press: New York.

Porter, M. (1998), "Clusters and the New Economics of Competition", *Harvard Business Review*, Nov-Dec 1998, 77-90.

Sen, S. and Frankel, F. (eds.) (2005) "India's Strategy of IT-led Growth." Center for the Advanced Study of India: Philadelphia, PA.

Siwek, S. and Furchtgott-Roth, H. (1993), *International Trade in Computer Software*, Quorum Books: Westport, Connecticut.

Torrisi, S. (2002), "Software Clusters in Emerging Regions", Working Paper, University of Camerino.

Tschang, T., Amsden, A. and Sadagopan, S. (2003), "Measuring Technological Upgrading in the Indian Software Industry: a Framework of R&D Capabilities and Business Models", Working Paper, Asian Development Bank Institute,Tokyo, Japan.

Whitman, J. (1990), "Key Factors for Software Success", mimeo, Oakland Group: Cambridge, MA, December 26, 1990.

5

Tardy Growth of IT Hardware Industry in India

D.N. Rao

Introduction

Over the last decade the Indian software, IT-Enabled Services (ITES) have demonstrated their competitive edge in the international market. This sector has grown steadily with Indian companies expanding their service offerings and shifting from low end business processes to high value ones. But the computer hardware industry has not been able to keep pace with the software industry. The two industries are complementary and their demand has been growing together in the global markets. The software and ITES exports grew from Rs. 10,940 crore in the year 1998-99 to Rs. 1,41,800 crore in 2006-07 whereas the electronics components(mainly computer hardware) industry exports increased over the same period from Rs. 1,800 crore to Rs. 11,500 crore (Table 5.1).

Table 5.1: Electronics Exports from India

Year	Computer software	Electronics components
1989-99	10,940	1800
1999-00	17,150	1400
2000-01	28,350	4788
2001-02	36,500	5,800
2002-03	46,100	5,600
2003-04	58,240	7,700
2004-05	80,180	8,000
2005-06	104,100	9,625
2006-07	141,800	11,500

Source: Department of Information Technology, Government of India

Growth in the IT and ITES sector has had a salutary effect on the employment scenario with the total number of professionals employed in this sector growing from an estimated 2,84,000 in 1999-2000 to 1,287,000 in 2005-06 (*Government of India, Economic Survey, 2006-07, page 147*). In addition, Indian IT and ITES sector is estimated to have helped create an additional 30 lakh job opportunities through indirect and induced employment in telecom, power, construction, facility management, IT transportation, catering, and other services. All this has led to a sense of euphoria about Indian success in this sector. But will this euphoria last? What are the perceived bottlenecks in sustained growth in the IT sector?

Dependence on Imported Hardware

Computer hardware is an essential requirement for smooth production of IT and ITES output. The hardware not only has to be available in sufficient volumes at cheap rates but it has also got to be the latest in technical innovation. India's IT sector is becoming increasingly reliant on imported hardware and components Only about 25 percent of India's commercial IT hardware is actually made of domestic manufacture and the share is moving towards 20 percent as India moved towards a zero duty regime for component imports in line with WTO's information technology agreement signed in 1995.

Indian IT sector is projected to export software worth US US$ 60 billion by 2010 which, in turn, is expected to generate demand for IT hardware to the tune of US US$ 77 billion (ASSOCHAM Report). Growth in the Indian economy in 2005-15 is expected to increase electronics demand to US US$ 159 billion by 2015. Thus, it will not be long before Indian hardware imports exceed software exports. IT and ITES could become the biggest drain on India's foreign exchange reserves.

Pattern of Computer Hardware Production in the World

The US is world's largest producer of electronics accounting for more than 20 percent of world total Asia has emerged as the fastest growing region for electronics production in the world (Table 5.2).

Table 5.2: Electronics Production as Percent of GDP in Some Asian Countries

Country	Electronics production as percent of GDP
Taiwan	17.0
South Korea	15.1
China	9.6
India	1.7

World PCB production and Laminate Report 2005 presents estimates of PCB production and Laminate market by country and product type. The report indicated a market expansion for the third successive year. Asia has grown to nearly 80 percent of world's PCB production. US occupied the fifth place with 10 percent of world production(www.ipc.org). Thus it can be seen that the world market for computer hardware is an oligopoly with about four or five players and those too with a highly skewed distributed. It will not be surprising if the prices of computer hardware rise faster than software and render the IT and ITES business in India loss making.

What Needs to be Done?

Something needs to be done on war footing to stimulate innovation and domestic production in the computer hardware industry so as to make it globally competitive. The union ministry of information technology (IT) has to design and pursue policies for stimulating growth in this sector. The IT Ministry was expected to set up a separate task force for this purpose but no information is available about any such task force.

In September 2007 the government issued the guidelines for the operation of semiconductor manufacturing policy. The guidelines provide definitions of fab (a factory that takes raw silicon wafers and creates chips with them. A special incentive package scheme was announced in February 2007 under which Centre or its agents would provide 20 percent of the capital expenditure for the units in Special Economic Zone (SEZ) and 25 percent of the capital for non-SEZ units during first ten years. Capital expenditure of non-SEZ units would be exempt from countervailing duty. Reports indicate that the Union government is in the process of the fine tuning the new hardware policy.

6

TRIPS, R&D in Microsoft and Reforming Copyright Laws for IT Competitiveness

Neha Aggrawal and Ravinder Goel

Introduction

It is generally understood that a strong Information Technology (IT) sector can make substantial contributions to any economy. In countries where software, hardware or IT services are generated in abundance, their contribution to gross domestic product is above 5 percent. [1] In many countries, the IT sector has also been a major source of labor productivity growth. No wonder, then, that many governments have tried to promote the growth of local IT producing industries. In their quest to promote local IT industry, they have over the last two decade been working on the understanding that the IT software and hardware industries have special legal requirements and need strong IPRs (Intellectual Property Rights).

Based on the belief that a vigorous framework to protect IP (Intellectual Property) encourages both inward investment and homegrown innovation, each country strengthened its IP protection laws and consequently IP protection rules were also introduced in the multilateral trading system of the World Trade Organization (WTO) through the TRIPs (Trade Related Intellectual Property Rights) Agreement (1995). India also provides extensive protection to software under the Copyright Act of 1957 (as amended in1983, 1994 and 1999). The Indian Act protects copyright in computer programs (software) as a 'Literary Work'. The act confers upon the owner of copyright in the relevant software program, all rights available to the owner of copyright in a literary work plus the right to sell, or give on commercial rental, or offer for sale, any copy of a computer program.

The rights of a software owner are protected for a period of 60 years and the 60-year period is counted from the year following the death of the author/ owner of the software. Under Indian Copyright Law, there are stringent penalties for infringement of copy rights of the software owner. Significantly, 'knowing use of an infringing copy of a computer program' upon a computer (Section 63B) attracts a minimum mandatory sentence of 7 days' imprisonment, which may extend to three years and with fine ranging from a minimum of Rs. 50,000, to a maximum of Rs. 2,00,000. In the case of offences by joint stock companies, not just the company but also all those persons in charge of the affairs of the company at the time such offence took place, shall be subject to criminal liability.

It is time that emulation of the US approach, of one of the strongest IP laws in the world, and an active approach to their enforcement, of promoting IT sector growth is put to critical analysis both theoretically and in the light of empirical data and suitable lessons drawn from it for development of IT sector in developing countries of the world. Following sections explore the theoretical rationale and other related issues associated with grant of exclusive rights to copyright owners. We also analyse the data (accounting information) from two big software companies of the world - Microsoft Corporation of USA and Infosys Technologies of India to understand how monopoly protection to software owners promotes research and development. We also examine the question of strong IP protection from the perspective of developing economies and discuss the results of the debate at WTO forum on availability of medicines at affordable costs in developing countries and its outcome in terms of provisions for patent protection. Finally, conclusion with suggestions for reforming the copyright laws to promote growth of IT sector in developing countries is presented.

Nature and Impact of Copyright Laws

Essentially, copyright confers monopoly on its owners. This has been long recognized and the negative impact of monopoly on consumers too has been recognized. Elementary economic theory also tells us that monopoly leads to inefficiency, profit maximizing behavior of the monopolist keeps the prices above marginal cost and the total production by the monopolist is below the level which is compatible with least cost and thus monopoly is a waster of economic resources. [2] At a social level the profit maximizing behavior of the monopolist copyright holder leads them to pricing of their products above reasonable levels. Though 'reasonable level' is a vague term, an idea can be had in terms of what it alternatively costs to duplicate any IPR protected work. This question has become relevant in the current context where advances in technology have made duplicating technology widely available and fairly cheap. If a written page can be photocopied for 50 paisa, a publisher's pricing of the same at Rs. 3 or 4 a page hampers the spread of information and knowledge. Similarly, a software programme which can be copied on a CD available in the open market for a price below Rs.10 cannot be priced at a few hundred or a few thousand rupees. Sustaining this monopoly leads to socially unjustified demands for legal protection as incorporated in the World Intellectual Property Organization (WIPO), Copyright and Phonogram Treaties and the US Digital Millennium Act of 1998. [3] It is also an invitation to users to break the law and irrespective of what the law contains, it cannot be enforced.

Most interestingly, many times, because of the mismatch between the declared price and the marginal cost of a product, monopoly copyright owners might themselves promote piracy. An example from the Indian music industry makes this amply clear. Peter Manuel quotes an employee of T-series as saying "... the big ghazal singers on HMV were privately coming to us and asking us to market pirate versions of their cassettes, for their own publicity since HMV wasn't nearly able to keep up with demand." [4] Further piracy has its advantages to the singers. The singer reaches to more people, becomes more popular and consequently has more demand for live concerts where the earnings are much more than the earnings from the cassette sales and they are exclusive owners of the income. It is a fair guess that overwhelming dominance of the desk top market by windows is also to a certain extent built on willful promotion of pirated windows software by Microsoft Corporation.

Beyond monopoly, copyright, as a legal and philosophical concept, is a product of western capitalist development at a particular historical moment. There was no idea of copyright in the ancient world. Books could be copied by anyone who could read and write and could get some writing material. Later on with the industrial revolution, printing was developed and printing required specialized, fairly expensive equipment that an ordinary reader would not have. So in effect it created a situation in which copies could only feasibly be made by specialized businesses. Thus, copyright in content was initially an industrial regulation aimed at curbing competition from other printers to the exclusive right of copyright owners/publishers. It did not encroach upon the right of readers to do what they wanted to do with the purchased work. The question of readers' rights has become relevant now only when technology has immensely facilitated duplication of any work.

But, strengthening of IP is proposed on the ground that in the long run it is beneficial to the society. Even though in the short-run, it reduces competition and increases price in the market in which the IP protected product is sold, it is claimed that in the long-run, by providing rents or monopoly profits, it increases the incentive to undertake research and development, by allowing the fixed costs of R&D to be recouped. This in turn confers long-run dynamic gains in terms of improved technology and better products.

Monopoly and Microsoft Corporation: An Analysis

In the light of the above analysis it may be instructive to see how far facts support the proposition that the monopoly to software owners promotes R&D and offers improved technology and products to users. To understand this we analyzed the income statements of Microsoft Corporation since FY (fiscal year) 2001 (1 January 2000 to 30 December 2000), the results of which are given in Tables 6.1 and 6.2.

Table 6.1: Yearly Income Statement of Microsoft Corporation (US$ million)

S.N	Particulars	FY2001	FY2002	FY2003	FY2004	FY2005	FY2006	FY2007
1	Revenue	25,296	28,365	32,187	36,835	39,788	44,282	51,122
2	Cost of Revenue*	3,455	5,699	6,059	6,596	6,031	7,650	10,693
3	R & D**	4,379	6,299	6,595	7,735	6,097	6,584	7,121
4	Sales and Marketing***	4,885	6,252	7,562	8,195	8,563	9,818	11,455
5	General and Administrative Expenses	857	1,843	2,426	5,275	4,536	3,758	3,329
6	Operating Income	11,720	8,272	9,545	9,034	14,561	16,472	18,524

Note: *Cost of revenue includes manufacturing and distribution costs for products sold and programs licensed, operating costs, costs incurred to support and maintain Internet-based products and services, warranty costs, inventory write-downs, costs associated with the delivery of consulting services, and the amortization of capitalized research and development costs associated with software products that have reached technological feasibility.

**Research and development expenses include payroll, employee benefits, stock-based compensation, and other headcount-related costs associated with product development.

***Sales and marketing expenses include payroll, employee benefits, stock-based compensation, and other headcount-related costs associated with sales and marketing personnel and advertising, promotions, tradeshows, seminars, and other marketing-related programs.

(The above definitions are in accordance with the USGAAP (Generally Accepted Accounting Principles)

Source: Microsoft Corporation, Yearly Income Statement, accessed from their website: http://www.microsoft.com/en/us/default.aspx

It turns out that since FY 2001, the biggest cost component for Microsoft Corporation is sales and general expenses and not the cost of revenue or the so called R&D expenses which are supposed to have been promoted by strong IP protection. Only in FY 2002 were R&D expenses equal to sales and marketing expenses. The impact of monopoly is, however, apparent when we see the cost of revenue of the firm over the years. It is true that it has increased but still it fluctuates around 20 percent of the total revenue.

Table 6.2: Yearly Income Statement of Microsoft Corporation (in percent)

Particulars	FY2001	FY2002	FY2003	FY2004	FY2005	FY2006	FY2007
Revenue (US$)	25,296	28,365	32,187	36,835	39,788	44,282	51,122
Cost of Revenue as a percent of revenue	14	20	19	18	15	17	21
R&D as a percent of revenue	17	22	20	21	15	15	14
Sales and Marketing as a percent of revenue	19	22	23	22	22	22	22
General and Administrative as a percent of revenue	3.38	6.5	7.5	14	11.4	8.5	6.5

Source: Microsoft Corporation, Yearly Income Statement, accessed from the website: http://www.microsoft.com/en/us/default.aspx

The role of monopoly in profits of Microsoft Corporation is apparent when the different cost components of the company for the FY 2007 are compared with another leading software company, Infosys Technologies of India. Table 6.3 shows that the absolute values for two companies are substantially different but the comparisons are instructive. The cost of revenue for Infosys Corporation is 58 percent of its revenues, while for Microsoft it is only 21 percent. On the other hand the selling and marketing expenses or the expenses related with delivering copies of software- the system of enclosing physical disks or tapes in attractive packaging, shipping large numbers of them around the world and storing them for sale along with advertisements is 22 percent of revenues for Microsoft Corporation as against only 7 percent for Infosys Technologies. This cost is presented as an expense of doing business. In truth, it is part of the waste caused by having to develop a system of selling a product much above its marginal cost. Interestingly, despite a much higher cost of sales and marketing and also spending some money on R&D, the operating income of Microsoft Corporation is 36 percent of revenues, while it is only 28 percent for Infosys Technologies.

IP Protection and Developing Countries

For developing countries, the case of strong IPR protection has always been debatable. Firstly, they are net users of research and development-intensive products and they do not benefit from the monopoly profits that are created by IP protection but their consumers suffer from the higher prices. Second, because their markets are small in relation to global demand actions by them to strengthen IP protection has little impact on the incentive to undertake additional R&D. Thus, a combination of higher costs in the short-run and the likely absence of dynamic gains over time mean that raising levels of protection would not be beneficial to them. Further, as suggested by the comparative analysis of Microsoft and Infosys, the claims of strong IP laws promoting R&D can also be questioned. In fact, in the case of software, at least it seems that guarding or maintaining monopoly, which is technically not feasible, is leading to heavy wasteful expenditure on sales and marketing only.

Table 6.3: Costs and Revenue of Microsoft Corporation and Infosys Technologies: A Comparison

[in US$ million and percent] (*F.Y. 2007*)

S.N	Particulars	Microsoft	Infosys
1	Revenue	51,122	3,090
2	Cost of Revenue	10,693	1,777
3	2 as a percent of 1	21	58
4	R&D	7121	NA
5	4 as a percent of 1	14	NA
6	Sales and Marketing	11,455	209
7	6 as a percent of 1	22	7
8	General& Administrative	3,329	249
9	8 as a percent of 1	7	8
10	Operating Income	18,524	852
11	Operating income as a percent of revenue	36	28

Source: Microsoft Corporation, Yearly Income Statement, accessed from the website: http://www.microsoft.com/en/us/default.aspx, and Infosys Technologies (2007) USGAAP Datasheet accessed from the website: http://www.infosys.com

The authors are not aware of any studies which quantify the economic welfare losses of strong IP protection in the field of IT sector. But, to the extent it raises costs and hence restricts affordability it will impact the spread and penetration of IT in society. A number of studies have also shown that the net economic welfare losses of higher patent protection for pharmaceuticals to developing countries could be substantial. [5] An indirect idea of the perceived higher costs by software users can be had from high rates of software piracy, defined as the percentage of unauthorized copying, reproduction, usage, or manufacturing of packaged software to total software installed in a country (Table 6.4). Interestingly, software piracy rates of 22 percent and 35 percent in North America and European Union point to the legal limits of enforcing software monopoly.

Table 6.4: Software Piracy across the World (2004)

Region	Piracy Rate (in percent)
Asia Pacific	53
European Union	35
Latin America	66
Middle East & Africa	58
Non EU Europe	44
North America	22
World	35

Source: BSA White Paper (December 2005) Expanding the Frontier of Our Digital Future, pp. 8, available at www.bsa.org

Thus, initially developing countries were hesitant in negotiating multilateral IP protection rules under the Uruguay Round of WTO negotiations. This does not mean that the developing countries had no IP protection laws. Instead, they maintained a lot of flexibilities in these matters and generally their laws were much flexible than the IP protection laws of developed world. However, they were made to agree to uniform and higher standards of protection for IP in exchange for better access to developed country markets for textiles & clothing and agricultural goods.

Affordable Medicine and IP Protection at WTO Forum

The relevance and need for strong IPR laws were indeed questioned very soon. The issue of flexibility in TRIPS was first raised by USA to deal with cross border epidemics like the Anthrax scare in the background of US war in the Afghanistan. The issue was widened to cover other killer situations also. The AIDS crisis also focused attention on the very high costs of AIDS treatments. IP protection was identified as a cause of these high costs and issues were raised whether such protection—enforced around the world by TRIPS—was defensible either economically or ethically. Pushed to a corner, the developed countries conceded that the balance in the protection of intellectual property between the short-term interests in maximizing access and the long-term interests in promoting creativity and innovation needs to be more flexible than mandated in

TRIPS. WTO members, on 30 August 2003, agreed to a deal which made it easier for countries, with no manufacturing facilities, to import cheaper drugs for fighting diseases like "HIV/AIDS , tuberculosis , malaria and other epidemics" under compulsory licensing. The decision allows WTO member countries to digress from their obligations under the TRIPs agreement relating to patent rights in pharmaceuticals, if the drugs are required for meeting national health emergencies in one's own country or other countries.

One important contribution of this decision is that it shattered the myth that WTO provisions are sacrosanct and not subject to change. A related fall out of the debate has also been that the consumer and civil society groups in developed and developing countries are also asking whether IP protection laws are too restrictive and whether the resulting level of prices of medicines are high?.

Beyond medicines, the issue has made a section of people to seriously think as to whether the current system of IP protection is the best way of ensuring the optimal creation and dissemination of knowledge and R&D. Currently the approach is simple. IP system grants monopoly power to the creator (and that too) for a fixed number of years irrespective of the type of creation, sector of creation or other characteristics of creation.

An ideal system would, however, have to be more creative in not only ensuring the best incentives to creating knowledge and recovering the large fixed costs involved in this process, but also ensuring that once created, the invention should be made available at reasonable costs to maximize the benefits to society from diffusion and dissemination. Debate over the issue will prevail and will be influenced by technological developments and values and politics of different proponents, but to the extent the drugs decision has focused discussion on the pricing aspect of IP protection, it will be helpful in the long run.

Conclusion

IT sector is a proven engine for economic growth. In the last decade IT investments have unleashed unprecedented new innovations that have transformed almost every aspects of our lives. And the IT sector's greatest untapped potential lies not only in what it enables individuals to do today but in its ability to fundamentally transform economies for tomorrow. The question before policy makers today is to design policies that can maximize the IT sector's potential and harness rapid growth to benefit different sections of society.

Such a policy design will need to find an appropriate balance between the short-term interests in maximizing access and the long-term interests in promoting creativity and innovation, beyond the current 'one size fits all' approach propagated by the TRIPS agreement. It needs to be appreciated that it is not an Agreement about simply maximizing the level of protection for IP. TRIPS Article 7 on "Objectives" recognizes that the protection of IP should contribute 'to the promotion of technological innovation and to the transfer and dissemination of technology', to the mutual advantage of users and producers of technological knowledge and in a manner conducive to social and economic welfare and to a balance of rights and obligations. TRIPS Article 8 (2) clearly states "Appropriate measures, provided that they are consistent with the provisions of this agreement, may be needed to prevent the abuse of intellectual property rights by right holders or the resort to practices which unreasonably restrain trade or adversely affect the international transfer of technology."

In the light of the above discussion, some suggestions for reforming the copyright law and increasing IT competitiveness of developing countries can be:

- From the long term perspective it is important that **duration of protection under copyright law may (or should) be reduced.** Copyright used to be much shorter in its original incarnation. It has been extended over and over again in the past and in fact it appears that the increase in term will keep on extending so that they will never expire. The term of protection should also be in tune with the reality which prevails in the market. Today most books that are published are out of print within three years. Even successful works are usually out of print well before ten years. So there is no need for copyright to last for a long period of 60 years (as currently prevailing) after the death of the author/owner of copyright. Though arbitrary, a term of ten year copyright may be sufficient. Once things settle down, this period can be modified. In any case the term can be statistically determined on the basis of empirical data of sales of published work. Statistical studies can be undertaken to determine the necessary length of protection for different kinds of work. Computer software may have less; films and other capital-intensive projects can be longer. But in today's world with shortening shelf lives of all types of products, there appears to be no case for long copyright protection.
- Another dimension of copyright policy is the **extent of fair use**: some ways of reproducing all or part of a published work that are legally permitted even though it is copyrighted. The digital era is a revolution in duplicating knowledge and information. Technologically we are in a situation more like the ancient world, where anybody who could read could also make a copy of it that was essentially as good as the best copies anyone could make.

Attempts at fencing access to the information/ digital material by passwords, encryption technologies and stringent IPR laws will only create hindrances in the realization of the promise of the digital era and are bound to fail. Laws of the past are now harmful because they were in a different context. This development in technology changes the situation in which copyright law operates. In essence copyright law which fitted with the technology of the printing press now fit badly with the era of digital information technology.

It has also to be recognized that there are various forms of copying. There is commercial sale of copies passing them as originals in the stores at one extreme and at the other there is privately making a copy for yourself once in a while. In between there are things like selling a cheaper copied version to those who cannot afford or want to pay the price of the original and are even satisfied with aesthetically inferior versions of the product. Different situations have to be treated differently in copyright. One way can be to place the limits of copyright somewhere in the middle so that the activities that are most directly relevant to peoples' private lives or pockets become free. When governments prevent such activities they have to start intruding into everyone's lives by using harsh punishments. The only way basically to stop people in their private lives from sharing information is to give license for a police state. There is no place for provisions like section 63 B of Indian Copyright Act which mandate that knowing use of an infringing copy of a computer program upon a computer will attract a minimum mandatory sentence of 7 days' imprisonment, which may extend to three years and with fine ranging from a minimum of Rs. 50,000, to a maximum of Rs. 2,00,000.

- In most countries, government is the biggest purchaser of IT services. Government buying decisions can have a big effect on the growth of domestic IT industry and IT literacy. The **government can use its purchasing decision to extract concessions** in terms of free of cost/ concessional price use of functional computer programs (to get jobs done) for educational purposes or for personal non commercial use by residents of a country. Or it may be negotiated that it will be universally available after a delay of two or three years from the program's publication.
- In the area of pharmaceuticals, there are suggestions by eminent economists like Jeffrey Sachs and Michael Kremer to create a fund to reward the discovery of cures for rampant diseases like malaria and AIDS. This is a welcome step to **finding the most efficient ways to fund and deliver the supply of global goods of particular and specific importance to the poorest countries.** [6] Similar efforts can be promoted to develop and propagate functional computer programs like Linux which will go a long way in promoting IT sector growth in developing countries and reduce the scar of piracy in these economies.

To sum up, technological innovations update equipments and items of copyright relevance at an unprecedented pace these days and their contribution to economic value addition exercises and practices require an altogether new approach on the part of policy makers, law creators, enforcement agencies and all other stake holders.

Notes

1. Economist Intelligence Unit (EIU) estimates based on 2006 data from each country's national accounts quoted in EIU (July 2007) *The means to compete: Benchmarking IT industry competitiveness*, pp. 2.
2. Samuelson & Nordhaus *Economics, Seventeenth Edition* (Tata McGraw Hill Edition), pp.-193-194.
3. US Digital Millennium Act, 1998, allows copyright owners or publishers to impose any restrictions on the use of a work, and these restrictions take the force of law. Based on this law, the movie industry in USA has imposed censorship on free software for reading and playing DVDs, and even on the information about how to read them. A Russian programmer, Dmitry Sklyarov, was arrested in 2001 while visiting the U.S. to speak at a conference, because he had written such a programme in Russia, where it was not unlawful to do so. The US law follows two WIPO treaties-- the WIPO Copyrights Treaty and the WIPO Performances and Phonograms Treaty .The two treaties are not yet part of TRIPs agreement and thus not binding on India. The treaties contain provisions and obligations which are TRIPs plus.
4. Peter Manuel (Oxford University Press 2001), *Cassette Culture: Popular Music and Technology in North India,* pp. 68-69.
5. See Arvind Subramanian *TRIPS, Medicines, and Patents* accessed from http://www.petersoninstitute.org/publications/author_bio.cfm?author_id=488
6. The suggestion is referred to in the paper of Arvind Subramanian -*TRIPS, Medicines, and Patents- see details at 5 above.*

7

Information Technology, Globalization and International Migration: The Indian Experience

Zakir Hussain [1]

Prelude

Globalization and international migration is Siamese twins. The latter is a natural consequence of the former. The advent of cheap information and communication technology (ICT) and its 'wiring up' effect during the second half of the twentieth century has considerably been effective and affective in setting the contemporary trends and patterns of international economic migrants. It is believed that if globalization has liberalized rules of national boundaries, information technology has made the borders porous through leakages of information and reduction of the cost of transportation and telecommunication across the globe. Today, every 35th person on the earth is an international migrant. According to Philip Martin (2004) 'if all the international migrants, approximately 200 million i.e. 3 percent of world demography, are clubbed together, it would form the fifth largest nation, the 'Migrant State', after China, India, USA and Russia'.

The present paper is an attempt to discern and delineate the long standing phenomenon of labour migration across the national boundaries which is 'nothing new' as Nayyar Deepak (1994) says, from the perspective of modern day globalization and the expansive role assumed by the information technology [2] in modern world, including India. It is also pertinent here to explain that the role of globalization and technology (IT) in electrifying international migration has been taken from two angles: one, how the first two forces, globalization and ICT, have reinforced and sustained upon each other and helped accelerate and assist, and streamlined the overall flow of international migration; and second, the role of globalization and ICT has been assessed from the point of view of inducing and abetting specific kind of labour outflows, i.e. IT workforce, particularly from India.

In order to understand and explicate the broad linkages and implications of the three interacting and something complementary forces – IT, globalization and international migration – we have schematized the present paper into three sections. Section I is committed to elicit the nexus and role of globalization and technology (IT) on international migration. Section II is specifically devoted to examine the role and implication of globalization and technology (IT) on international migration from India. It is pertinent here to underline that although we find that globalization and technology has deeply been associated with Indian migration system, they have hardly acted in synergy as before as the 1990s. The role of 'globalization' on Indian migration system has been all pervasively noticed since the ancient period but it has never been so affective until the advent of British rule in India. Likewise, the role of 'technology' has predominantly been effective on Indian migration only after 1990. Therefore, Section II has been parted into two sub-phases purposively: one, migration during the British period or colonial rule; and second, migration of skilled workers to the developed nations. In fact, migration of skilled workforce which has variously been termed, nowadays, as "talent workers" "knowledge workers", etc. invokes as well as embodies the role of technology which we widely notice flows towards the developed nations, particularly during the second half of the twentieth century. The Middle East phase, although accounts a landmark era in Indian migration system, is an extension of 'First Phase of Migration' which marks the influence of globalization, so this phase is skipped off. Section III carries a limited deliberation upon the implications of the skilled emigration for India.

Section I

In this section, the nexus between IT, globalization and international migration has been explored with brief historical discussions and considerable emphasis has also been laid down to examine its contemporary role in abetting and assisting international migration, especially since 1990.

Without going into the debate which 'process' started first, there is universal agreement that globalization and international migration has reinforced each other. [3] According to Spencer Wells, Oxford scientist, "all of us are literally Africans...brothers and sisters separated by 2,000 generations". The first human migration occurred 50-70,000 years ago when we [Indians] moved out of Africa and into India through sea and land routes. [4] The rise of 'colonial' powers and 'capitalism' during the eighteenth century laid down the foundation of the modern day 'economic' nexus between globalization and international migration. Slave trade, which transported more than 15 million Negros from Africa most to sugar plantations to various parts of the world, indentured labour system, which created the modern diaspora community, accounts between no less than 12 to 39 million or more [5] or 59 million Europeans migrated between 1847 and 1939, the largest migration known to the history of mankind, are historically linked to the same process. [6]

During the second half of the twentieth century, appearance of some new forces such as economic liberalism, emergence of global production system and global supply chain, ICT revolution and demographic imbalances accompanied with shortages of specific labour force in some parts of the world evolved a situation where '*techno-eco-human*' alliances emerged as rudimentary requisites for sustaining globalization, (economic) growth and international migration. During the last quarter of the 20th century, the synergy between globalization and human mobility, in fact, became more rapid, robust and visible when it was uniquely associated with and catalyzed by modern high-speed communications technology networks (satellite, phone links, the spread of cellular phones digitalized data transfer, video conferencing, live international news coverage, etc.) and fast transportation developments such as airplane travel for passengers which has led to "compression" of time and space. (Simmons, 2005; Castells, 1985, 1996). In fact, the expanding international travel and international migration are *de facto* tightly linked elements of contemporary globalization (Simmons, 2005). Thus, under such historical settings we may reach at a consensus that although the nexus between globalization and international migration existed since back to the origin of humanity itself, the development of 'technology', especially ICT, supplemented with global economic dynamism, has been added as a 'third' significant force in the whole process of globalization and international migration, especially pronounced during the last quarter of the 20th century.

This is also evident by the fact that during this period the rate of international migration has relatively increased at a greater pace than earlier. Undoubtedly, this acceleration in the volume of international migration is attributable to the growing momentum of globalization and the expansive role assumed by the information technology. Statistics speak out that during the last quarter of the 20th century magnitude of international migration has risen by more than 2.5 times, i.e. from 75 million in 1965 to around 200 million in 2005. This pace has further risen during 1990-2005. The annual average growth rate of the migrant stock has been accelerating, increasing from 1.4 percent during 1990-1995 to 1.9 percent in 2000-2005. Between 1965 and 1985, the average annual flow of international migrants was 1.5 million people, which more than trebled in the next 20 years (1985-2005) to nearly 5 million per year (UN, 2006; UNFPA, 2006). This imbalances are disproportionately all the more pronounced when we decompose the developed and developing countries accounts of international migration. We find that out of total international migration, 115 million, or 60 percent, are concentrated in the developed countries. This rise in scale and magnitude of international migration as well as predominant presence in developed countries, especially in the high income countries (48 percent) and as a single country USA alone attracted increasing number from 23.3 million in 1990 to 38.4 million in 2005 obviously foretells the increasing role of globalization and ICT revolution in accelerating and speeding up the flow of international migration during this period. The high-income countries and the USA as well as other advanced nations are the throbbing centres of both activities, viz. globalization as well as development of ICT industries.

In case of international economic migrants, the role of globalization and ICT seems relatively more pronounced and optimistic. ILO (2004) estimates approximately 86 million migrant workers worldwide, not including family members who move them to join them. Indeed, subtract global refuge figure from 175 million people living outside their country of birth in 2000 and were left with approximately 163 million individuals who have "voluntary" chosen to cross borders, in search of better economic and/or social opportunities. The IMP report (2006) interestingly considers them as the fastest growing category of migrants. During the last five years it is estimated to have increase by 36 million or 61 million during the last decade. This expanding stock of economic migrant is sustained by rising inflows and outflows, with more people emigrating and going back home, enabled by the rising forces of globalization and ICT revolution (IMP, 2006).

Perhaps one of the most significant and crucial implications of the ICT revolution on international migration, particularly since the last quarter on the twentieth century 1990s, which induced and fuelled the mobility of specific kind of labour forces was the huge demand for the IT workforce all across the globe in general and developed countries in particuiar. The sweep of digital technologies and the information to a knowledge-based economy has created a robust demand for workers highly skilled in the use of information technology. "Indeed, the information, communications, and technology (ICT) revolution sparked a worldwide effort by governments to court ICT workers...the ICT revolution is not the first, but it is perhaps the most startling example of how an information-age industry can quickly generate global demand for labour" (Lowell, 2001). ILO (2004) in one of its studies has also reached at similar conclusion that ICT and globalization has enormously fuelled the demand for IT workforce across the globe. "Since the early 1990s, the international mobility of high skilled workers has been increasing with rising global demand for skills, alongside advances in globalization and phenomenal growth in information and communication technologies [ICT] (ILO, 2004). In fact, the host countries encourage the immigration of the IT specialists because they perceive the demand-supply gap is only temporary as a result of current economic boom. So importing of the IT specialists can allow that industry to expand as fast as market allow, generating additional economic output via multipliers or by augmenting strategic economic sectors (Martin, 2003). In fact, the demand for workers can create, apply and use information technology goes beyond these industries, cutting across manufacturing and services, transportation, health care, education and government (Berman, et al. 1998). It is responsible for new products and services; creating new companies and industries; revitalizing existing products, services, and industries; providing new venues for commerce; enhancing our ability to manage information and to innovate; and improving our productivity, quality of life, and national standard of living. IT is changing the way we live and work, and transforming the economy at a fundamental level.

In the past ten years alone, employment in the US computer and software industries has almost tripled (Berman, et al. 1998). This growth has been more pronounced in highly trained computer scientists' jobs which are projected to grow by 99 percent while the somewhat lesser skilled programming jobs are forecast to grow 30 percent. Both of these core IT occupations are set to outstrip the forecast 14 percent growth for the labour force overall (Lowell, 2000). According to Braddock (1999) this is strongly expected to continue, especially in case of computer scientists and engineers, through 2008. Table 7.1 depicts the projected growth rate of occupations in the US during 1996-2006. The fastest among them are computer scientists (108 percent), followed by computer engineers (109 percent) and system analysts (103 percent). Growth rates in personal and home care aids and all occupations are holding the lowest position, 58 percent and 14 percent respectively, in one decade.

Table 7.1: Fastest Growing Occupation in the US, 1996-2006

S. N.	Occupation	Projected Growth Rate (in percent)
1.	Computer Scientists	118
2.	Computer Engineers	109
3.	System Analysts	103
4.	Personal and Home care Aides	58
5.	All Occupations	14

Source: US DOL Statistics, 1996 Industry-Occupation on Employment Matrix.

It is noted that during the 1990s, like the United States, employment in the information technology industry grew rapidly in other advanced countries as well, and some IT employers in the industrial countries asserted that there were too few workers to fill IT jobs, and they needed easy access to foreign IT workers to fill what they estimated to be 10 to 15 percent vacancies. [7] Thus, the US and other governments like Canada, Australia, UK, Germany, etc. responded to the plea with easy-entry programmes for foreign IT workers, many from India. (Martin, 2003) We notice that during the 1990s, most of the developed and high income countries have introduced new temporary entry programmes to attract and invite high skilled workers, especially the IT professionals. Among others, the H-1B visa programme which targets the 'specialty workers' launched in 1990 in the US is the most successful temporary programme in the world. Estimates suggest that during the 1990s, the H-1Bs played significant role in supplying IT workers to the United States. In 1989, H-1B made up 4 to 7 percent of IT labour force which grew to 54 to 60 percent in 1999. Today H-1Bs comprise about 10 percent of the total IT labour force in the US, which increased in absolute number from 974,000 in 1989 to 2,450,000 in 2000 (Lowell, 2000).

Another major country where ICT-induced migration has been noticed is the United Kingdom. Although the share of IT work permits in total work permits has declined over the period from 23.5 percent in 2001 to 15.5 in 2004, still it is one of the largest sectors, perhaps next to health sector, where foreigners are invited in largest number in the country.

Another significant effect of globalization and ICT revolution has been felt on the migration of the student community, whom Khadria (2001) terms as 'semi-finished human capital', 'to get polished and finished' particularly from the developing countries to the developed countries. For instance, in the OECD (Organisation for Economic Cooperation and Development) countries foreign students are concentrated mainly in US (34 percent), UK (16 percent), Germany (13 percent), France (11 percent) and Australia (8 percent) (OECD, 2002). These students opt more to the subjects that are more transferable internationally, e.g. science and engineering, IT rather than law and languages. Other significant reason, apart from high quality education, these countries provide lucrative job opportunities to these students during and after the completion of their education. Among them, IT specialists are more appreciated and acclaimed in advanced countries.

Another indicator of the rising role of technology and globalization in international migration is the growth of temporarisation, especially of PTKs (professionals, technical and kindred). In recent decades, out of shortages of specific labour force, most of the developed countries have introduced temporary entry programmes to attract skilled workers especially IT professionals and computer-related occupation workforce. The H-1B visa attracted approximately 200,000 specialty workers just a year before the dot com bubble and thereafter it registered a steep decline to half, i.e. 100,000 in 2002. This indicates the intensity of the H-1B visa used for attracting the ICT workers in the US. The situation in UK and New Zealand do not appear so fluid. In UK the number of Work Permits issued to the workers before 2001 increased substantially, but it tapered off with 2001 recession and turn down slightly last year, while New Zealand demonstrates a steady upward climb. This is likely due to transition to permanent status (Lowell, 2007).

The cheap, easy and widely accessible telecommunication has also enabled international migrants to stay longer in destination countries as they are now able to establish direct and frequent links with their families. With the help of Internet and webcam devices they are now able to see and directly talk with their wives, children, parents and friends left behind. They are now even swiftly able to intervene in running their houses at home. This has helped in reducing homesickness and nostalgia not only in the migrant workers working in far away countries but also the newly married wives left behind. During 1960s and 1970s the newly married wives, left over by their husbands were suffering from this psychological trauma which doctors in Pakistan termed, as 'Dubai Syndrome' was very common among the Gulf migrant families in South Asia. [8] The IT, in fact, has built a 'human bond' and helped overcome the problems of loneliness, especially among the international migrant workers. This was not when telecommunication systems and cellular phones were neither economically affordable nor easily accessible. The direct and swift links possible by telecommunication system has also helped in reducing problems of the international migrant workers, especially the international women migrant workers, in the host countries. Today, during pre-departure training programmes every international migrant has been reminded to keep telecommunication links with their family members, embassies and other helpline, so they could easily link up with if they face any crisis-like situation.

The ICT has also assisted international migration by enabling swift, safe, and cheaper and hassle free transmission of remittances. Nowadays, some international units such as Western Union, MoneyGram have expanded their networks to far and wide. In India Western Union claims to deliver money within 24 hours; it has linked up with the Indian Postal department and using extensive networks of post offices and has covered those remote areas where formal banking system has yet not opened up. Philippines and US are developing 'wire system' on the similar pattern of the US-Mexico, to help transmit remittances. This would allow the Filipinos immigrants in the US to transmit their money within hours and at much cheaper rates. The emergence of such trans-national remittance-transmitting agencies has considerably reduced the informal operation of money transferring systems such as *Hundi*, *Hawala*, *fi-Chin, etc.*, hence, reduced the chances of misuse of money by terrorist outfits. This aspect of international migration is distinguished and discussed after the 9/11 events all over the world when terrorists funded were funded through these means.

The above in-depth role of globalization, information technology on international migration elicits that in the contemporary era the three forces, international migration, globalization and information technology, appears to be integral components and sustain and reinforce each other in a 'cyclic' manner. The movement of the 'cyclical flow' can be visualized in the likewise manner: in the *first round* globalization initiates the process or paves the way for trans-national mobility by reducing the legal barriers; this trans-national mobility gets accelerated by technology (ICT revolution) by making the borders porous to information leakages and this leads to greater flows of knowledge, hence, ensures greater mobility of people across the globe. In *second round*, the cycle moves on from the other way round, that is 'technology' takes the lead role in sustaining as well as accelerating the momentum of international migration which further accelerates the pace of world integration. In the second round technology plays, in fact, two vital roles: one, weaving networks among the migrant community; and second, creating scope for further migration through innovations and transformation in the production system. In this manner each force reinforces each component, thus, the cycle spins off.

In another way, the whole nexus of the three forces can also be curiously seen through the interplay of classical 'push', 'pull' and 'network' point of view. Necessity to survive 'pushed' the human being to be a transnational wanderer which can be labeled as *Human Survival Phase.* The search in anticipation for safer, prosperous and lucrative destinations 'pulled' transnational communities to particular regions. These pull factors created *Economic Expansion Phase.* The evolution and use of technology gave rise to the development of 'forward' and 'backward' networks among the moving stocks. This facilitated as well as speeded up the process international migration and globalization. This can be termed as *Technology Catalyst Phase* (diaspora networks, PKT migrants' network, recruitment agencies network, etc) of international migration. [9]

Section II

As discussed in the beginning, Part II is devoted to explicate the role of ICT and globalization on intentional migration from India. It is categorized into three broad sub-sections dealing with the beginning of international migration from India as result of rising wave of globalization which is associated with the advent of British rule in the 18th century in the sub-continent; followed by the role of technology on Indian migration system, which has started since the second half of the twentieth century, mainly directed towards the developed nations which were in need of skilled labour forces as a result of their transforming economies; and the last section which examines the pattern of migration since 1990 onward with particular emphasis on delineating the impact of ICT revolution and response of the Indian IT professionals.

Currently, according to the estimates of the Ministry of Overseas Indian Affairs (MOIA), 'approximately more than 25 million strong Indians are dwelling in more than 130 countries. This is the third largest diaspora community, next to Britain and China, in the world. Similarly, as estimated, around 7 to 8 million stocks of the Indian workers with an annual flow of more than half a million, are engaged as migrant workforce in different parts of the world. This is the second largest workforce, next to Philippines, working on *contractual* basis outside the country of birth in the world. [10] According to Stalker (1993), the sheer size of the modern-day Indian communities [and workforce] dwelling and [working] in more than 130 countries is the effect of the long and diversified pattern of migration that took place from all spectrums of the sub-continent way back to the annals of history and more particularly since the advent of the Colonial Rule in India in the 18th century. Motwani (1993) has classified migration of Indians into three categories:

- Classical wave starting in 5th century B.C. includes cultural colonization of Hinduism and Buddhism and early trade adventures of Indian traders.
- Colonial wave emphasizing on indentured labour (*coolies*) to European colonies, specially the British. The human cargo was exported to Africa including Mauritius, Caribbean region, Fiji and South East Asia.
- Contemporary wave is the most recent one which started in the 1960s of the 20th Century. This is also known as the "Brain Drain" wave because most of the migrants have been professionals.

Migration to the Developed Nations: First Phase of Globalization and Technology on Indian Migration System

After independence international migration from India entered into a new phase. We notice an obvious role of technology and globalization on the Indian migration system. This is primarily actuated by the transforming technological base as well as liberal immigration policies of the industrialized countries, mainly the US, the UK and Canada. These countries, in fact, felt an increasing demand for workers enarmed with high skills and professional expertise such as engineers, scientists, management personnel which their economies were in need of at that time. In order to meet their requirements, these countries opened up their borders by liberalizing their immigration policies and attracted the skilled workforce from all across the world. India, owing to vast reserves of skilled workforce at its disposal emerged as a frontline labour sending State to these countries, especially since 1950s. According to Nayyar (1994) "The *first phase* of the Indian migration during post-colonial period was characterized by a movement of persons with technical skills and professional expertise to the industrialized counties mostly on a permanent basis. The major destinations were the United Kingdom, the United States, Canada and, on a much lower scale, Western Europe and Australia". According to Chakravartty (2006) "Between 1950 and the mid-1970s, the majority of Indian emigrants were "skilled" professionals moving permanently to the UK, US and Canada".

Although migration from India to the three traditional industrialized countries, the US, the UK and Canada started during the same period, we notice variations in terms of magnitude as well as composition.

Migration to Developed Nations during the Post-1990: Second Phase of Globalization and Technology on Indian Migration

Coming to the post-1990 period, we notice that technological progress and globalization of the economy have increased the importance of human skills. This is diametrically discernible in the field of information technology, which resulted into an explosive growth in the demand for IT professionals. (Burman, et al. 1998) In fact, these changes were deeply entrenched with some significant developments occurring at global level such as greater openness of the world economy, rise and spread of the MNCs (Multinational Corporations), demographic imbalances, ageing of the work force in developed countries, etc. In such circumstances, the Indian labour forces both "talent workers" and "service workers" due to their high quality services and sincerity of the purposes, as many view, established their trustworthiness in the world as an "alternate community" to the West. [11]

Another noteworthy development impressed upon the trends of Indian labour migration, particularly the highly skilled one, was the transformation of the immigration policies of almost all the developed labour-receiving countries. During this period, we notice two major trends: one, rise of the temoprarisation in international migration; and second, special short term entry programmes and schemes introduced by the developed countries to attract 'specialty occupation' workers, with strong emphasis on IT professionals. We find a number of countries introduced temporary entry programmes such as EU's Blue Card Scheme, H-1B Visa programmes of the US, HSMP (High Skilled Migration Programme) of the UK, German Green Card, Job Card Scheme of Denmark and Pilot Project "Active Selection of Qualified Workforce" of Czech Republic, e-

Japan, etc., with the intention to first invite them for temporary stay but after meeting certain conditions granting them permanent residence.

Migration of Professionals and High Skilled Workers

Still we don't have our own data set, we have to rely on the information of the receiving countries which show the unabated flow of Indian labour force to both, traditionally industrialised countries such as the US, the UK, Canada and other newly joined countries like Australasia, Continental Europe and South East and East Asia.

The available evidence on trends on permanent immigration from India to the selected industrialised countries, the USA, the UK, Canada and Australia, [12] during the period 1991 to 2005 is presented in Table 7.2.

Table 7.2: Indians Immigrating on a Permanent Basis to Major Industrialized Countries: 1991-2005

Year	US	UK	Canada	Australia	Year	US	UK	Canada	Australia
1991	45064	5680	12850	5608	1999	30157	6295	17458	2557
1992	36755	5500	12675	3553	2000	41903	8045	26126	4631
1993	40121	4890	20445	2643	2001	70032	7280	27904	6336
1994	34921	4780	16637	3908	2002	70823	8005	28838	5091
1995	34748	4680	18262	3908	2003	50228	11460	24593	5783
1996	44838	4620	21291	3700	2004	70151	11870	25576	8135
1997	38048	4645	19615	2681	2005	84681	N.A	33146	9414
1998	36414	5430	15376	2786	--	--	--	--	--
				Percentages					
1991	65.1	8.2	18.6	8.1	1999	53.4	11.1	30.9	4.5
1992	62.8	9.4	21.7	6.1	2000	51.9	10	32.4	5.7
1993	58.9	7.2	30	3.9	2001	62.8	6.5	25	5.7
1994	58.0	7.9	27.6	6.5	2002	62.8	7.1	25.6	4.5
1995	56.4	7.6	29.6	6.3	2003	54.6	12.4	26.7	6.3
1996	60.2	6.2	28.6	5.0	2004	60.6	10.3	22.1	7.0
1997	58.5	7.1	30.2	4.1	2005	66.6	N.A	26	7.4
1998	60.7	9	25.6	4.6	--	--	--	--	--

Source: OECD, 2006.

The data clearly shows that considerable numbers of Indians are immigrating on a permanent basis to these industrialized nations. The average annual inflows of the Indian immigrants to all these countries recorded substantial growth since the 1990s as compared to the earlier decades. In the case of the United States, the average annual inflow of immigrants recorded at 26,184 persons during the 1980s (Nayyar, 1994) almost doubled to reach 48,592 during 1991 to 2005. As regards Canada, the average annual inflow which was 7,930 persons in the 1980s more than tripled to reach 21,386 during the period 1991-2005. Similarly, the average inflow of immigrants from India to the United Kingdom increased from 5,400 persons during the 1980s to 6,212 during 1991-2005. Australia has recently emerged another major destination of Indian professionals and high skilled workers. Number of Indians immigrating on a permanent basis to Australia has recorded significant increases since the 1990s and especially after the turn of the 21st century. The average inflow of Indian immigrants to Australia has almost doubled in the recent years with the numbers increasing from 3482 during 1991-2000 to 6565 during 2000-2005 (Table 7.2).

Composition

Table 7.3 provides the skill composition of the Indian emigrants to the US during 1991-2001.

Table 7.3: Immigration from India to United States by Major Occupation Group: 1991-2001

(number of persons/percentage)

Occupation Group	1991-93	1994-96	1997-98	1999-00	2001
Professional & Technical	20,395 (16.7)	19,603 (17.1)	12755 (17)	12124 (16.8)	19935 (28.4)
Executive, Administrative & Managerial	6174 (5.1)	6246 (5.5)	3463 (5)	2756 (3.8)	3062 (4.4)
Clerical & Administrative Support	2719 (2.2)	2390 (2.1)	1339 (2)	1149 (1.6)	643 (0.9)
Sales	975 (0.8)	1489 (1.3)	1247 (2)	1337 (1.8)	842 (1.2)
Services	2965 (2.4)	3487 (3.0)	1817 (2)	1357 (1.9)	N.A
Farming, Forestry & Fishing	18875 (15.8)	3567 (3.1)	3005 (4)	2408 (3.3)	N.A
Skilled Workers	1263 (1.0)	1613 (1.4)	1128 (2)	609 (0.8)	N.A
Total Above with Occupation	53366 (44.0)	38395 (33.5)	24754 (33)	21740 (30.1)	27073 (38.5)
No-occupation or Occupation not reported	68574 (56.0)	76133 (66.5)	49799 (67)	50543 (69.9)	43217 (61.5)
Total Immigration	121940	114528	74553	72283	70290

Source: Figure 1991-93 & 1994-96 from Khadria, 1999; rest from Annual Reports, INS (US)

It is noted that overall the picture of almost all the categories of the skilled workers remained stable or experienced slight decline, except the professional and technical category of the immigrants which registered a jump of 12 percent from 16.8 percent in 1999-2000 to 28.4 percent in 2001. This indicates the rising demand for the skilled and professional category of the workforce in the US, especially after 2000 onward. Other categories such as skilled, clerical, farming, forestry and fishing witnessed a decline. This is likely due to the boom in IT sector which absorbed the highest percentage of the foreign nationals, particularly since 2000 onwards.

In terms of the share of Indian immigration to the total immigration of the US, we notice that it has increased only 1 percent, from 5.5 to 6.6 percent during 1997-2001, whereas the share of skilled workforce in total skilled immigrants admitted to the United States from all countries of the world, Indians contributed around a quarter during 1997-2001. It rose from 11 percent in 1997 to 24 percent in 2001 (Table 7.4). This shows that during the 1990s, the skilled Indian workforce constitutes a significant proportion of the total foreign-born workforce in the US.

Table 7.4: Proportion of Indian Immigrants in Total and Professional Immigration Category of the US Immigration: 1997-2001

Year	Indian Share in Total US Immigration	Indian Share in Total Professional Immigration
1997	5.5	11
1998	4.7	13
1999	4.9	9
2000	6.6	15
2001	6.6	24

Source: INS Annual Reports (US)

IT and Indian Migration

One of the major characteristics of immigration from India during the post-1990 was the temporary flow of skilled migrant workers and the "focus is shifting away from professionals in specific occupations, like doctors, engineers, scientists, architects, bankers, to information technology (IT) professionals embodying, in a way, more generic skills. In other words, it is the generic applicability of information and communications technology (ICT) which has led to large-scale migration of Indians skilled in IT" (Khadria, 2001). Coincidently, "India emerged as a real powerhouse, responding to global demand for information technology (IT) workers and turning out some 70-85,000 software engineers and 45,000 IT graduates every year, with government plan to treble those member by 2002" (Lowell and Findlay, 2001).

The abundance in IT professionals led to the rise of a new kind of migration in India during the post-1990. Large numbers of Indian professionals and skilled personnel, particularly IT workforce have been migrating to the countries such as the US, the UK, Australia, Canada, Germany and other Continental Europe. This is primarily due to the fact that all these countries have in the recent past introduced various temporary employment programmes to admit migrants and meet their specific skill needs, especially those specialized in IT and IT-enabled services.

(a) The United States

Information Technology has played vital role in the US economy since the last four decades. During 1970-1998, its contribution almost doubled from 4.2 percent to 8.2 percent and further accelerated to one third of the real growth between 1995 and 1997. This is due to increasingly high intensity of the IT in the US economy as evident from the statement of Allan Greenspan, the Chairman of the Federal Reserve Bank addressing the industrialists in California, "The United States is currently confronting that can best be describe as another industrial revolution. The rapid acceleration of computer and telecommunications technologies are major reasons for the appreciable increase in our productivity." This increased the demand for IT professionals as well as the workforce required in IT-enabled services in the US economy.

Table 7.5 demonstrates the rising demand for computer scientists in the US economy which almost tripled during the tree temporal periods. In 1970, the computer scientists occupied 12 percent of the total US workforce, which increased to 23.4 percent in 1990 and reached to one third (33.2 percent) of the total workforce demanded in the US in 1997. In terms of native vs. foreign composition of the computer scientists, the foreign born almost increased by 5 times whereas native born rose three times during the same periods. Table 7.5 also reveals that during the same period the demand for the engineers in the US economy declined in same portions of the rise of the computer scientists, from 66 percent in 1970 to 44 percent in 1997. It means that the US economy increased the resiliency to absorb more computer scientists than the engineers.

A review of some studies reveals that in the coming decades the US would be one of the hottest destinations for the IT workforce. There exists significant gap between the native production of the IT workforce and the jobs expected to be created in the coming decades. According to the estimates of the Bureau of

Labour Statistics (US) "between 1994 and 2005 more than a million new computer scientists and engineers, system analyst, and computer programmer will be required in the US- an average of 95,000 per year".

Table 7.5: Scientists and Engineers in Total Labour Force in the US by Occupations and Nativity, 1997, 1990, 1970 *(in percent)*

	1970			1990			1997		
	Native Born	Foreign Born	Total	Native Born	Foreign Born	Total	Native Born	Foreign Born	Total
Computer Scientists	12.5	7.6	12.1	23.6	21.3	23.4	33	34.5	33.2
Life Science	6.1	6.7	6.1	5	5.2	5	5.8	6.3	5.9
Physical Science	8.8	14.8	9.3	7.5	9.4	7.7	6.5	4.7	6.3
Social Science	6.3	6.8	6.4	11.5	7.2	11.3	10.8	9	10.5
Engineering	66.3	64.2	66.1	52	56.9	52.6	43.9	46.5	44.1
Total	100	100	100	100	100	100	100	100	100

Source: Espenshade, 1999.

Of the three occupations, the largest job growth is accounted for by systems analysts, which are projected to increase from 483,000 in 1994 to 928,000 in 2005, a 92 percent jump. This compares to a projected increase of 14.5 percent for all occupations. The number of computer engineers and scientists is expected to grow by 90 percent, from 345,000 to 655,000 over the same period, while the number of computer programmer positions is expected to grow at a much slower 12 percent from 537,000 in 1994 to 601,000 in 2005. However, while only 65,000 new computer programmer jobs are projected to be created during this period, 163,000 new programmers will be required to replace those existing in the occupation (DoL, US, 2005).

H-1B – Workhorse of IT Industry – and Indian Migration

The most notable case is that of the United States which introduced in 1990 the H-1B visa programme to admit migrants to perform services in 'specialty occupations', [13] based on professional education, skills, and/or equivalent experience. Under the H-1B visa programmme, specialty workers are permitted to be employed for as long as three years initially with extensions not exceeding three years. Changes enacted in 1990 made it possible for immigrant workers on the H-1B to have the dual intent of working on a temporary basis and adjusting their status to permanent residency. However, this conversion process is subject to numerous immigration clauses, requirements, and numerical caps, and the transition is not essentially smooth, efficient, or guaranteed which has led Lowell to state that "the H-1B programme is neither fully a temporary programme, nor is it a permanent visa programme" (Banerjee, 2006).

Although, specialty occupations include an array of professions/occupations such as education, medicine & health, managers & officials, architect, engineers, surveying, mathematics & physical sciences, life sciences, social sciences, etc., since 1996, the computer-related occupations particularly computer systems analysts, programmers, etc. grabbed the lion share of all the visas approved under the H-1B progrmame. Evidence shows that the IT industry in the US has increasingly used H-1B visa to meet their labour requirements which Ayers and Syfert (2002) dub it as "Workhorse of IT industry". The IT industry in the United States since the 1990 led a major effort to increase the H-1B visa cap on the ground that: [14]

- the IT industry needs more skilled foreign workers to help meet the skill shortages;
- an inability to find workers has limited growth in the IT industry and other parts of the economy that need IT workers;
- the IT industry needs an international work force to meet the needs of the international markets;
- the IT industry needs to be able to attract the best and brightest workers from around the world; and
- the alternative to bring foreign workers to the United States is to move work overseas.

In 1995, only about a quarter of the temporary skilled workers were in IT-related fields, within two years, by 1997, this rose to the half. Evidence shows that this was largely due to the increased use by the IT industry and in August 1997, the H-1B visa reached to its full cap for the first time. The H-1B workers also account for a substantial share of employment and employment grows in the IT field. Lowell (2000) reports that H-1B Workers make up about 10 percent of the IT labor force, and Lowell and Christian (2000) estimate that new H-1B visa workers contributed about one-fourth of annual growth in the core IT workforce between 1995 and 2000. In 1999 alone, H-1B workers made up 54 to 60 percent of IT labor force growth (Lowell 2000). According to the OECD *Observer* (2002) between 1990 and 2000, an estimated 900,000 skilled workers entered the USA on the H-1B visa. Workers on the H1-B comprise about one-sixth of the total IT workforce in the USA. Table 7.6 shows that the H-1B workers are concentrated in information technology fields. Almost 58.1 percent of H-1B visas in 2000 were issued to workers in computer-related occupations which rose from 25.6 in

1995. The specialty workers of the other disciples such as architecture, engineering, education, medicine, life sciences, social sciences, etc. got eclipsed. As a result of dot.com bubble we notice that although the share of IT in H-1B declined to 38.3 percent in 2002, it revived again to 45.3 percent in 2005.

Table 7.6: Leading Occupations of H-1B Arriving Workers (percent)

Occupation	2005	2003	2002	2001	2000
Computer-related	45.3	38.5	38.3	58	58.1
Architecture, Engineering & Surveying	11.1	12.4	12.8	12.2	12.3
Administrative Specialization	9.5	10.7	10.8	7.2	7.2
Education	11.1	11.1	10.5	5.3	4.9
Medicine & health	6.2	7.2	6.6	3.4	3.9
Managers	3.1	4.8	5.4	3.8	3.0
Life Sciences	3.3	3.8	3.5	2.0	2.0
Social Sciences	--	2.6	2.8	1.9	2.0
Others	10.4	8.9	9.3	6.2	6.6

Source: INS (US)

Large number of Indian professionals, need not to specify IT professionals, have availed H-1B visa route to seek employment in the United States during the past decade. In 1999, 60 percent of all H-1B visas went to those hired for the IT field and about 75 percent of all IT workers on H-1B visas were from India (U.S. GAO, 2000). As Lowell (2000) neatly put it: "The information technology industry is the leading growth sector of the US economy and it is drawing on the readily available source of trained information technology engineers from India. Indian H-1B grew steadily from 1989 clearly becoming the largest category in 1994, then doubling in size by 1996, and quintupling by 1989 –a remarkable pattern of growth and accounting for nearly half of all visas issued in 1999 (47 percent). No other country during the lifetime of the programme has so completely dominated the H-1B visa".

Table 7.7 shows the rising proportions of Indians in the total approvals of the H-1B visa applications during 1989-2005. The share of Indian migrants under this category continuously kept on rising from a minuscule 4.4 percent in 1990 to almost half of the total H-1B approvals by the end of 2001. Major jump we notice occurred after 1997; this was perhaps due to the looming Y2K problems which the US industries were severely expecting to encounter. We also notice a sudden dip in Indian share in 2002, from 48.8 percent in 2001 to 32.9 percent in 2002;

this was the dot com bubble effect which had been universally experienced in the world in general and the US in particular. This convinced the US Congress to bring back the H-1B cap to its original level of 65,000 from 1,95,000.

Table 7.7: Proportion of Indians in Total Approvals of H-1B Visa Applications in USA: 1989-2005

Year	Share of Indian in Total Approvals	Year	Share of Indian in Total Approvals
1989	4.4	1998	44.0
1990	4.6	1999	47.2
1991	6.9	2000	48.4
1992	10.7	2001	48.8
1993	18.0	2002	32.9
1994	22.9	2003	36.4
1995	26.3	2004	43.0
1996	32.0	2005	49
1997	39.3	--	--

Source: Office Immigration Statistics, US (various annual reports)

If we club the total H-1B visa allotted to the IT sector as well as issued to the Indian IT professionals, we notice an astounding parallel between the two. The proportions of the Indian workers entered the US through H-1B visa route have closely followed to the proportions of the total approvals granted to the computer-related occupations (IT) under the H-1B visa programme during 1995-2005.

The two lines show that both are moving almost in the same proportions with exceptions in 1995 and 2005 when the share of Indians exceeded the total H-1B visa approved to the computer-related occupations.

This is also evident from the Table 7.8 that the majority of Indians holding H1-B visas are computer related professionals. This clearly demonstrates the prominence of the Indian IT professionals in managing and operating the major IT activities in the world as the US is the global leader in this field.

The dominance of India under this category of visa admission becomes all the more pronounced when we look at the shares of other sending countries in the H1-B admissions (Table 7.9). The second largest sending nation, China currently accounts for less than 10 percent of the total H1-B visa admissions as compared to nearly 45 percent of India's share. Nearly half of the total admissions are, in fact, accounted by more than twenty major labour sending countries.

Table 7.8: India's Share in H1-B Visas According to Occupations: 2000-2002
(in percent)

Category	2000 India's share in Total	2001 India's share in Total	2002 India's share in Total
Computer-related	68.18	71.39	63.21
Fashion models	0.33	0.11	0.67
Managers and officials	12.28	13.92	11.42
Miscellaneous, professional, technical and managerial	11.49	15.98	13.97
Administrative specialization	13.88	17.14	12.74
Architecture, engineering and surveying	26.01	27.87	22.94
Art	3.36	4.64	3.90
Education	8.36	9.45	9.26
Entertainment and recreation	17.37	6.74	8.89
Law and jurisprudence	4.77	5.02	5.01
Life sciences	11.06	10.52	10.52
Mathematics and physical sciences	10.03	12.21	12.73
Medical and health	18.00	20.34	19.58
Museum, library and archival sciences	8.06	5.65	3.49
Religion and theology	4.41	6.02	5.93
Social sciences	16.37	13.46	13.30
Writing	6.18	7.73	5.23
Unknown	18.88	23.66	13.22
Total	48.42	48.78	32.89

Source: Statistical Yearbook of the Immigration and Naturalization Service (various years).

This axiomatically proves that during 1990-2005, the H-1B visa route has not only speeded up the process of the migration from India to the US in IT industry but also fostered as well as opened avenues to other professions. In fact, H-1B visa route provided a window to the young Indian professionals to establish their calibre and high standard services all across the world.

(b) The UK Work Permit System and Indian Migration

Another country which felt severe shortages of IT workforce, including IT professionals, was the United Kingdom. Here too, the Indian IT professionals such as computer programmers, engineers, analysts, and low end IT-related professions, maintained their super-dominance among more than ten major contending countries.

Table 7.9: Country–wise H1-B Visa Approvals: 2004-2005

Country	2004	Percent	2005	Percent
India	123567	43.0	118520	44.4
China	26258	9.1	24561	9.2
Canada	13412	4.7	11780	4.4
Philippines	11300	3.9	9965	3.7
Korea	8159	2.8	8072	3.0
UK	8198	2.9	6796	2.5
Japan	6234	2.2	5727	2.1
Taiwan	4860	1.7	4285	1.6
Pakistan	4174	1.5	4120	1.5
Mexico	3884	1.4	3605	1.3
Germany	3981	1.4	3449	1.3
Colombia	3892	1.4	3444	1.3
France	3735	1.3	3116	1.2
Venezeulea	3027	1.1	2930	1.1
Russia	3421	1.2	2694	1.0
Turkey	2796	1.0	2665	1.0
Brazil	2688	0.9	2506	0.9
Argentina	2384	0.8	2299	0.9
Israel	2173	0.8	1804	0.7
Malaysia	1975	0.7	1791	0.7
Other Countries	47147	16.4	43002	16.1
Total	287418	--	267131	--

Source: Office Immigration Statistics, US Department of Homeland Security.

Table 7.10 records the total IT Work Permits issued and the proportion of the IT work permits availed by the Indians during 2000-2004. It shows two distinct trends: one, the bulk of the IT work permit is issued to the Indian IT professionals, around 68 percent throughout the period under consideration; and second, it has kept on increasing over the four years and grabbed more than 76 percent of the IT work permit issued in 2004. The significance of the Indian IT professionals in the UK is further confirmed that under the new *Tier1,* the UK authorities have given special preference to the Indian.

Table 7.10: Total IT Work Permits Issued and Proportion of IT Work Permits Availed by Indian in the United Kingdom: 2000-2004

	2000	2001	2002	2003	2004
Total IT Work Permits Issued	18,242	26,819	25,819	25,635	18,689
Proportion of Indians Availed IT Work Permits (percent)	62.8	63.8	67.7	71.4	76.6

Note: Year 2004 carries only 9 months data.
Source: Work Permit, Home Office (UK).

Another channel of the entry of IT professionals in the UK is the sponsorship undertaken by the UK based MNCs. Amicus The Union (UK) has assessed the intensity of the Indian IT professionals in the UK by studying the work permits issued to the sponsoring companies having their headquarters in United Kingdom. Table 7.11 reveals that out of ten, six companies having Indian backgrounds were issued approximately 81 percent of the total work permits issued during 2000-2004. Among the remaining four sponsoring companies, two each from the US and the UK shared 10 percent each.

Table 7.11: Top Ten IT Work Permit Sponsors in UK: 2000-2004

Sponsor	HQ	2000	2001	2002	2003	2004	Total
TCS	India	1587	1968	2012	2135	2918	10620
Wipro UK Branch	India	279	873	781	1177	1852	4962
Mahindra-British Telecom	Indian	573	496	863	958	789	3679
Mastek (UK)	India	427	574	585	483	911	2980
Infosays technologies	India	315	371	483	456	989	2614
Xansa	UK	0	522	458	247	664	1891
Accenture	US	151	256	345	398	604	1754
Satyam Computer Services	India	--	--	460	551	542	1553
Cognizant Technology Solution	US	--	--	460	551	542	1553
Redbrigade Corporation	UK	431	306	--	--	--	737
Totals		3763	5366	6447	6956	9811	32343

HQ: Head Quarter
Source: Work Permits and the IT Industry, Institute for Employment Studies, August 2005.

This evidently demonstrates two valuable trends: one, the Indian IT firms and industries are increasingly developing their base in the UK; and second, the rise in Indian IT companies in the UK would naturally prefer Indian IT workforce which indicates a permanent influx of Indian IT professionals in the United Kingdom in the coming decades.

(c) German Green Card and Indian Migration

Another country which attracted the Indian IT professionals and IT-related workforce most is Germany, one of the major countries in Continental Europe. In fact, for the last decade, the German labour market is facing acute shortages of IT workforce. By an estimate, approximately 60,000 jobs are created in IT and IT-enabled sectors and 25 percent remain vacant due to low output from the educational institutions. On the request of employers the government in 2000 introduced special Green Card programme to attract 20,000 IT professionals in the coming three years. Under this, the authorities showed clear preference to the Indian IT professionals.

It is estimated that at present more than 60 percent of those who have been admitted under Green Card Scheme are Indians.

There is also increased in-take of Indian IT specialists in East Asian countries like Japan and Malaysia under specialised temporary employment schemes. For instance, nearly 10 percent of the total IT engineers admitted to Japan during 2003 were Indians. (Table 7.12)

Table 7.12: IT Engineers Admitted to Japan by Nationality, 2003

Origin Country	Number	Percent
China	11,079	53.2
Korea	3,019	14.5
India	2,001	9.6
Philippines	789	3.8
US	568	2.7

Source: Kamibayashi, 2006.

It is obvious from the above delineations that during the 'second' phase of skilled migration (post-1990 period) from India to the developed world are powered by some of the following developments which clearly demonstrate the strong role of globalization and Information Technology on Indian migration system:

- The revolution in Information Technology since mid-20th century and emergence of knowledge-based economy required specific kind of labour force, hence, mobilized greater volume of cross border emigration of skilled and knowledge workers. In this context, the developed countries received greater flow of knowledge workers from the developing countries and India emerged as frontline State.

- The emergence of population imbalances in developed countries, and ageing as well as shortages of skilled worker force increasingly further necessitated the outflow of skilled and talent workers from abundant regions to the deficit regions.
- Discussion on the movement of the services of natural person under Mode 4 at WTO forum further confirmed the significance of the rising needs of the labourers coming mostly from the third world countries. This provided an opportunity to the big labour origin countries such as India, Brazil, Philippines, etc. to raise the issue of discriminatory labour treatment to mobility vis-à-vis the capital and finance.

Section III [15]

Implications of Mobility of the High Skilled Workers for India

It is quite evident that a number of countries are increasingly resorting to temporary immigration schemes and programs to admit high skilled migrant workers in order to meet the emerging skill shortages. Majority of these programs have been targeted to admit IT and health care professionals. Our analysis of the trends in international labour flows from India clearly shows that India is emerging as one of the important source country for high skilled migrants in sectors such as IT and health care. This has been enabled primarily by two factors: One, India possesses an abundant and high quality pool of talent workers which has been made possible by a large, expansive and established network of academic infrastructure. At the end of March 2005, there were 343 institutes of higher education in India and 16,000 colleges with a total enrollment of nearly 9.3 million. As far as technical education is concerned, there are currently 1522 approved engineering colleges with an in-take capacity of nearly 5, 50,986 students. There are nearly 1000 institutions providing Post-Graduate Computer Application courses with an intake of 56,805. Also, with English being a widely accepted medium of instruction in the Indian education system, a large proportion of the graduate pool is proficient in English. Second, an unabated flow of Indian professionals during the last half a century to the traditionally immigrant receiving countries like US, UK and Canada have ensured that Indians are now one of the most highly educated ethnic groups with much higher median income than any other community in these countries. This has had a major 'demonstration effect' in stimulating further outflows of high skilled labor from India.

As noted earlier, one of the prominent features of high skilled labor mobility from India during the pre -1990s was that such flows were primarily 'permanent' in nature and there was hardly any return associated with them. However, there has been a reversal of these trends since the 1990s mainly due to the fact that the new schemes and programs related to the entry of the high skilled migrants aims to admit them purely on a temporary basis. Such a trend has had vivid and varied implications for the Indian economy.

There has been a significant increase in remittances to India, especially from countries which deploy high skilled migrants. In fact, India now is the largest recipient of workers' remittances, which amounted to nearly US$ 30 billion dollars in 2007-08. It is striking to note that more than 60 percent of these remittances originate from the developed countries (Sasikumar and Hussain, 2007).

The rapid expansion of IT and IT enabled industry in India during the last decade is also encouraging a large number of Indians to return and set up business ventures, especially in key IT centres in India like Bangalore. Such a trend of to and fro movement of professionals and ideas require that the migration of high skilled labor be possibly situated within the framework of 'brain gain'. However a severe dearth of empirical information both in relation to the number of high skilled migrants returning to India and on their business ventures prevents us from probing into the 'brain gains' in detail. However we highlight some aspects of these gains. It is estimated that at least 35000 IT professionals have returned to either work in IT companies or set up business enterprises in Bangalore alone. In their comprehensive survey of India's software industry, Commander, et al. (2004) confirmed the evidence of return of skilled migrants with 30-40 percent high level employees had relevant work experience in developed countries. A Newsweek article in 2004 pointed out the multiplication of native software patents in India as some 35,000 expatriate Indian information-technology workers reportedly have returned home from the United States since September 11, 2001. In our limited discussions with some of those who have taken up new assignments in IT companies, it was revealed that their career prospects in India seem to be even much better than in the developed countries from where they returned. It is also noted that during the recent decades, the emergence of 'brain circulation' has added significance to the growth of multi-chain production system all over the world.

For instance, the flow of Indian IT expertise to the Silicon Valley and their return has developed what many term as silicon 'hump' to various parts of the country such as Bangalore, Hyderabad, Gurgaon etc. Our discussions with those who have established new IT and IT enabled enterprises indicated that the professional stint which they had in developed countries was a key stimulant for them to establish business enterprises. Many of them informed that they are currently taking up outsourcing work from the major IT companies in which they worked earlier. We strongly feel that there is an urgent need to undertake in-depth investigation on the nature and extent of return migration of high skilled migrants, by looking at the major industries and professions and also the national and international factors facilitating such flows.

Another significant implication of international migration of the IT professionals has been felt on the emerging pattern of education in India. It has been noticed that over the period, students are increasingly opting for IT and its related professional courses. This is evident from the fact that India had only 7,000 IT specialists in the mid-1980s, but the multinationals recognized their skills and began moving some Indian IT specialists from India to their operations in other countries and India became the major source of migrant IT workers. The demo effect of these IT migrant workers stimulated the virtuous cycle and pushed the IT specialists to 700,000 and making India a leading provider of low-cost, high-quality IT specialists and services (Martin, 2004).

Considering that India has had a long history of international migration, a number of steps have been taken since independence to recognize and honour the significant contribution of overseas Indian community in India's social and economic progress. However, in most cases these steps were taken in an *ad hoc* manner without considering the modalities to sustain them in a long-term perspective. This was surprising in view of the fact that the sustained international migration from India and its consequences provided massive potential for addressing different developmental concerns. The establishment of a separate Ministry, Ministry of Overseas Indian Affairs, in May 2004, to deal with all matters pertaining to overseas Indians, comprising Persons of Indian Origin (PIO), Non-Resident Indians (NRIs) and Overseas Citizens of India (OCI), was a historic step to acknowledge the fact that the overseas Indian community constitutes a significant economic, social and cultural force and needs mainstream attention. The basic mission of this Ministry is to "promote, nurture and sustain a mutually beneficial relationship between India and its overseas community" (MOIA, 2007). Several schemes and programs have been introduced in recent years to tap the full potential of Indian diaspora. 'Day of Indian Emigrants' is being celebrated ever year in recognition and appreciation of the constructive, economic and philanthropic role played by the Indian Diaspora. In response to a long and persistent demand for "dual citizenship" particularly from the Diaspora in North America and developed countries and keeping in view the Government's deep commitment towards fulfilling the aspirations and expectations of Overseas Indians, the government of India has decided to grant Overseas Citizenship of India (OCI) to all Persons of Indian Origin where local laws permit "dual citizenship" in some form or the other except Pakistan and Bangladesh.

The Indian government has also recognized the rising potential of these return migrants in terms of their cross continental business and commercial linkages and, thus, has facilitated the establishment of the technology parks at various places in the country. The development of outsourcing business activities in India is the results of the enterprising abilities of educated youth which attracted the attentions of the developed countries.

Conclusion

The above detailed discussion clearly establishes that the nexus between globalization, technology (IT) and international migration has been deep rooted and it is spinning off ever since it has aligned to each other. The role of technology on international migration has become more emphatic and decisive since the mid-20th century as the world witnessed the ICT revolution. The cheap and widely accessed network of the IT has not only helped the far flung people to connect themselves but also enabled them to pass on the relevant information as well as link up with within no time. The real phase of globalization, in fact, took place since internet, a major instrument of IT, entered common man's life. This enabled the global creative forces to access information and move to those regions and countries where such skills and talents were either scare or welcomed. This is more likely to increase in the future as a result of expansion of global labour markets, shortages of highly educated workers in the information-age economies, the ageing workforce in developed economies, just-in-time demand from industries eager to get on the front of technology curves and so on are the prime factors signaling the propulsion of skilled mobility in the coming decades. India with its vast reserve of high quality low cost human capital resources, particularly the IT workforce, would remain one of the major or perhaps the most preferred county in the world.

The implication of the movement of skilled professionals from the abundant countries like India has resulted into both kinds of debates, positive as well as negative. The positive side of the debate underscores that the movement of skilled personnel from India to the developed countries has transferred the much needed capital, new ideas, technology know-how and more over working as a 'bridge community' in fostering trade and commerce between India and the developed world. The fruits of 'brain circulation' and 'brain gain' of the IT workers has already been reaped by some of the origin countries, for instance, the rise of silicon 'humps' at Bangalore in India, Honshu Park in Taiwan, Technology Park in Hong Kong and China are the results of the brain circulation and brain gains of the IT professionals who emigrated to Silicon Valley as early as the 1950s, 1960s and 1970s. This has also facilitated the emergence of transnational networks between India, China, Hong Kong and the US. So far the negative side of skilled emigration is concerned many consider that the drain of the skilled personnel particularly in the field of IT, care workers, and specialty occupations like engineering, bio-technology, etc would surely produce cascading effects on our own systems, but this notion is not so popular now as it was during the 1960s and 1970s. Today, the receiving countries are themselves conscious to return back the origin countries the fruits of their talent pools (brain gain). They have resorted to a number of policies in this direction such as ethical recruitment, encouraging return migration with lucrative packages (social benefits), allowing dual citizenship, facilitating brain circulation, etc. They have also resorted actively to temporary entry schemes emphasizing 'return' as one of the important criteria, hence, leading to a circulatory movement of fresh workers as their terms expire. These measures will surely encourage the current as well as potential migrants to utilize their hard-earned income for productive purposes. Hence, these micro level efforts would definitely assist macro level developmental activities in the labour the sending countries. In order to maximize the net effects of the emigration of the skilled workers, at policy front the country like India needs to emphasize on a couple of things: (i) promoting mostly the returnable migrants through mapping out the emerging skills in developed countries and matching it with our potential supply; (ii) harnessing the skills of the return migrants by maintaining proper records, and their skill specificities; (iii) utilizing of the diaspora platform for developmental purposes; and (iv) reorganizing our emigration policies in view of the transforming immigration policies of the developed countries.

Despite all the above elucidations, however, it is also a fact that there exists a diacotomy that globalization of the world economy has not led to the freer flow of the workers. The flows of the trade and investment aspects of globalization have moved faster towards the minimal amount of control whereas international migration has provoked greater blocking and tightening of the overseas contract workers. This is perhaps inherent in the speculate attendant implications as well as the very nature and direction of the flows of labour forces which is taking place from low cost developing economies to the high cost developed economies. Nonetheless, our conjectures clearly mandates that the enduring intimate nexus between globalization and IT can usufruct unprecedented scale of socio-economic and political benefits to both the labour sending countries as well as the labour receiving countries in the twenty first century than hypothesize merely the one aspect that is speeding up the process of cross border mobility.

Notes

1. Author is a Researcher in Institute for Defense Studies and Analyses, New Delhi. Views in this paper are his own.
2. In this paper technology and IT has interchangeably been used.
3. Although, of late, Prof. Patrick Manning (2005) believes, technology is one of the significant components which helped the early hominids to expand and grow out of African Continent. The term *Homo Habilis,* a genus of *Homo*, is identified as "able man" or "man the toolmaker" is perhaps associated with the art of developing effective weaponry which led to the beginning of this expansion. Similarly, *Homo Erectus,* who emerged in East Africa nearly million years ago, in fact, started the out-migration journey, what Prof. Manning terms "colonization", spread out from East Africa to as far as Jordan and the Caucasus, Java, and South China, 1.5 million years ago. The new technology, the Acheulian of "hand axe" which developed 1.5 million ago, helped the *Homo Erectus* in this venture perhaps the most. The Acheulian hand axes were so effective that remained the tool of choice for over a million years, until just over 100,000 years ago. He further records that the *Homo Erectus* gradually experimented with a second major new technology, the 'control of fire' by 200,000 years ago. If we believe Gordon Child's observations that 'since 10,000 B.C man developed aggressive attitude against the nature', we find that technology has played key role in establishing and expanding man kind from one end to other end of this *tera firma*. For more details on this aspect see Jared Diamond (1997), *Guns, Germs, and Steel:The Fates of Human Societies,* New York.
4. Pitchappan, R.M, "Indians Might be the First Human Migrants", *Times of India,* April 4 2008.
5. For detail see L. Potts (1990), *The World of Labour Market: A History of Migration',* Zed Books, London.
6. For details see Peter Salker (1994), *The Work of Strangers: A Survey of International Labour Migration,* International Labour Office, Geneva.
7. Main reason of the shortfall of the IT specialists was the policy pursued during the cold war period which discouraged students from studying engineering and scientists on one hand, (Martin,2003)and the sudden abnormal rise in demand for IT workers at a sudden due to ICT boom and its attendant problems such as Y2K, Millennium Bug, etc, on the other.
8. For detail see Lila Gulati (1993), *In the Absence of their Men: The Impact of Male Migration on Women,* Sage Publication, New Delhi.
9. The idea is developed on two works: Patrick Manning (2005), *Migration in World History,* Routledge, New York; Jared Diamond (1997), *Guns, Germs, and Steel: The Fates of Human Societies,* New York.
10. The data published by MIOA hardly covers all spectrums of labour force migrating outside the country. Till date, it records only those who require ECR while workers migrating under ECNR category and to 154 countries where ECNR is not applied are not recorded.
11. The Indian High Commissioner to Singapore, Mr. S. Jaishankar, raised a pertinent query during the IIMPact function, organized by Singapore chapter of the Indian Institute of

Management Alumni Association, branded as IIMPact. (Source: 'Indian Professionals Hailed as Alternative Global Voices', *The Hindu*, April 22, 2008).

12. It is noted that during this period Australia entered as a regular country to attract and invite Indians on permanent basis coupled with temporary visa schemes, particularly since 1996. This was not so regular during the earlier phases. In fact, during initial phases, Australian immigration policy was Euro-centric and hardly interested in attracting Asian immigrants.
13. The INS 214 defines a specialty occupation is one which requires a "theoretical and practical application of a body of highly specialized knowledge". The basic requirements are a bachelor's degree or higher in the specialty and a full state licence if such a licence is required.
14. See Information Technology Association of America, Winter (1999) available at http://www.itaa.or/gov/pubs/ppr.htm
15. This section is based on previous combined work of Sasikumar S.K and Zakir Hussain, 2007.

References

Abella, M (2003), "*Balancing Migration Policies in a Globalizing World*", www.scidev.net

Alarcon, Rafael (2000), "*Migrants of the Information Age: Indian and Mexican Engineers and Regional Development in Silicon Valley*", Working Paper 16, CCIS, University of California.

Ayers, Kristen Ness and Scott D. Syfert (2002), "US Visa Options and Strategies for the Information Technology Industry", *International Quarterly 14 (3), pp.535-65.*

Banerjee, Payal (2006), "Indian Information Technology Workers in the United States: The H-1B Visa, Flexible Production, and the Radicalization of Labour", *Critical Sociology*, Vol. 32, Issue 2-3, Sage Publication, pp-425-445.

Berman, E. Bound, J. and Machin, S. (1998), 'Implication of skilled-Based Technological Change: International Evidence', *The Quarterly Journal of Economics,* November.

Braddock D. (1999), "Occupational Employment Projections to 2008", *Monthly Labour Review, November*, pp 51-77.

Brinkley, Alan (1997), *The Unfinished Nation: A Concise History of the American People*, 2nd edition, New York: Alfred A. Knopf.

Castells, Manual (1996), *The Rise of the Network Society,* Malden Society, M.A., Blackwell

Castells, Manual (ed.) (1985), *High technology, Space, and Society*, Beverly Hills, Sage.

Chakravartty, Paula (2006), "Symbolic Analysts or Indentured Servants? Indian High-Tech Migrnats in America's Information Economy", *Knowledge, Technology, & Policy,* Vol. 19(3), pp.27.43.

Chammartin, Gloria Moreno-Fontes (2006), "Female Migrant Workers in the Era of Globalisation", in UNFPA-IOM (2006), *Female Migrants: Bridging the Gaps Throughout the Life Cycle,* Selected Papers of the UNFPA-IOM Expert Group meeting, New York, 2-3 May.

Clarke, Colin and Ceri Peach (1990) "*South Asians Overseas Migration and Ethnicity*", Cambridge University Press, Cambridge.

Commander, Simon, R. Chanda, M. Kangasmieni and A. L. Winters (2004), "*Who Gains from Skilled Migration? Evidence from the Software Industry*", Manuscript, Center for Economic Performance, London School of Economics.

Davis, Kingsley (1951), "*Population of India and Pakistan*", Princeton University Press, Princeton, New Jersey

Dinnerstein, Leonard and David M. Reimers (1999), *"Ethnic Americans: A History of Immigration"*, 4th edition, Columbia University Press, New York

Espenshade, Thomas J. (1999), *"High-End Immigrants and the Shortage of Skilled Labor"*, Population Research and Policy Review 20-135-141, Kluwer Academic Publisher, Netherlands.

Hugo, Graeme (2003), *"Migration and Development: A Perspective from Asia"*, Paper prepared for the International Organization for Migration, 2003.

Hugh Tinker *(1977)*, "The *Banyan Tree*: Overseas Emigrants from India, Pakistan, and Bangladesh" *Oxford University Press, Oxford*

ILO (2004), "Towards a fair Deal for Migrant Workers in the Global Economy", Report IV, International Labour Office, Geneva

Jain, P.C (1993), "Five Pattern of Indian Emigration", in Jagat K. Motwani, Mahin Gosine and Jyoti Barot-Motwani (eds.) *Global Indian Diaspora: Yesterday, Today and Tomorrow*, Global Organization of People of Indian origin, New York, USA.

Kamibayashi, Cheko (2006), "Current Migration of IT Engineers to Japan: Beyond Immigration Control and Cultural Barriers", in Christiane Kuptsch and Pang Eng Fong (eds.) *Competing for Global Talent*, ILO & ILLS, Geneva.

Khadria, Binod (1999), *The Migration of Knowledge Workers: Second–Generation Effects of India's Brain Drain*, Sage Publications, New Delhi.

Khadria, Binod (2001), 'Shifting Paradigms of Globalization: The Twenty-first Century Transition Towards Generics in Skilled Migration from India', *International Migration*, Volume 39 (5), Special Issue 1.

Kolb, Holger (2005), "The German Green Card", *Focus Migration*, Policy Brief No.3, November, Migration Research Group, Germany.

Lowell, B. Lindsay (2000), "The Demand and New Legislation for Skilled Temporary Workers (H-1Bs) in the United States", People *and Place* Vol. 8 (3).

Lowell, B. Lindsay (2001), *"Policy Responses of the International Mobility of Skilled Labour"*, IMP 45, International Labour Office, Geneva.

Lowell, B. Lindsay and Allan Findlay (2001), *"Migration of Highly Skilled Persons from Developing Countries: Impact and Policy Responses"*, Synthesis Report, International Labour Office, Geneva.

Lowell, B. Lindsay (2007), *"Trends in International Migration Flows and Stocks, 1975-2005"*, *W*orking Paper No. 58, OECD, Paris.

Madhavan, M.C. (1985), "Indian Emigrants: Numbers, Characteristics, and Economic Impact", *Population and Development Review*, Vol 11(3) September, pp.457-481.

Manning, Patrick (2005), *"Migration in World History"*, The Themes in World History, Routledge, New York.

Martin, Susan and B. Lindsay Lowell (2004), *"Competing for Skills: US Immigration policy Since 1990"*, IRPP Conference on North American Integration, April 1-2, 2004.

Ministry of Overseas Indian Affairs (2007), *"Annual Report 2006-2007"*, MOIA, Government of India, New Delhi.

Minocha, Urmil. (1987), "South Asian Immigrants: Trends and Impacts on the Sending and Receiving Societies" in James Fawcett and Benjamin Cariño, (eds.), *Pacific Bridge: The New Immigration from Asia and the Pacific Island*, Center for Migration Studies, New York.

Motwani, Jagat K. (1993), "The Global Indian Diaspora: An Overview", in Jagat K. Motwani, Mahin Gosine and Jyoti Barot-Motwani (eds.) *Global Indian Diaspora: Yesterday, Today and Tomorrow,* Global Organization of People of Indian origin, New York, USA.

Nayyar, Deepak (1994), "*Migration, Remittances and Capital Outflows: The Indian Experience"*, Oxford University Press, New Delhi.

OECD (2002), "*International Mobility of the Highly Skilled"*, (Policy Brief), *OECD Observer,* July, online www.oecd.org/publication/Pol.breif).

OECD (2006), "*International Migration Outlook",* Organization for Economic Co-operation and Development, Paris.

Philip Martin, (2003a) "*Highly Skilled Labour Migration: Sharing the Benefits",* IILS, Geneva.

Philip Martin, (2004), "*Migration and Development: Towards Sustainable Solutions",* DP/153/2004, IILS, Geneva.

Philip Martin, . (2003b) "*Managing Labour Migration: Temporary Worker Programmes for the 21st Century",* ILO-IILS, Geneva.

Philip Martin, Manolo Abella and C. Kuptsch (2006), "*Managing Labour Migration in the Twenty- First Century",* Yale University Press, London.

Sandhu, K. S (1969), "*Indians in Malaya: Some Aspects of their Immigration and Settlements (786-1957)",* Cambridge University Press, Cambridge.

Sasikumar, S.K. (2001), *International Labour Migration from Independent India,* NLI Research Studies Series No.:022/2001, V.V.Giri National Labour Institute, NOIDA.

Sasikumar, S.K. (2003), "Indian Labour Overseas: Trends and Prospects", *"Labour and Development"* December, Vol. 9(2).

Sasikumar, S.K. and Zakir Hussain (2007), "*Migration, Remittances and Development: Lesson from India"*, NLI Research Studies Series No. 083/2007, V.V. Giri National Labour Institute, NOIDA.

Saxenian, AnnaLee (1999), "*Silicon valley's New Immigrant Entrepreneurs",* Public Policy Institute of California, San Fransisco.

Siddiqui, Tasneem (2005), "Migration and Millennium Development Goals: DFID-RMMRU Partnership", in UNFPA, *International Migration and the Millennium Development Goal: Selected Papers of the UNFPA Expert Group Meeting,* UNFPA, Morocco.

Simmons, Alan (2005), "Globalization and Migration", in UNFPA, *International Migration and the Millennium Development Goal: Selected Papers of the UNFPA Expert Group Meeting,* UNFPA, Morocco.

Solimano, Andres (2002), "*Globalizing Talent and Human Capital: Implications for Developing Countries"*, Paper prepared for the 4th Annual World Bank Conference on Development Economics, Oslo, 2002.

Srivastava, Ravi and S.K.Sasikumar (2003), *"An Overview of Migration in India, its Impacts and Key Issues"*, RRMRU-DFID, Dhaka.

Stalker, Peter (1994), "*The Work of Strangers: A Survey of International Labour Migration",* International Labour Office, Geneva.

Stanworth, Celia and Janet Druker (2004), "Temporary Agency Labour in the UK", in John Burgess and Julia Connell (ed.), *International Perspectives on temporary Agency Work,* Routledge, New York.

UNFPA (2005), "*International Migration and the Millennium Development Goal: Selected Papers of the UNFPA Expert Group Meeting",* UNFPA, Morocco.

UNFPA (2006), "*A Passage to Hope: Women and International Migration*", State of World Population, United Nations Population Fund, New York.

United Nations (2006), "*High- Level Dialogue on Migration and Development*", Agenda Item 55 (b), United Nations.

US Department of Labour (2005), "Bureau of Labour Statistics", *Monthly Labour Review*, November.

US GAO (2000), "*H-1B Foreign Workers: Better Controls Needed to Help Employers and Protect Workers*", Washington, DC.

Wells, Spencer (2002), "*The Journey of Man: A Genetic Odyssey*", Princeton University press, Princeton.

8

Split-Unit Approach for Demand and Supply Management of IT Resources

Krishan Kumar

Introduction

Demand and supply are two major forces determining prices and hence the level of production and employment. Concepts of demand and supply are related to the economic activities of the consumer and the producer. Demand denotes these elements viz. volume, price and time. Be it noted that price and time are important to any discussion about demand. It is meaningless to say that demand for a particular model of a car is so much, unless its price range and the requirement per month are quoted. For information technology services the demand means volume of services and IT (Information Technology) goods required, which the users are willing to acquire at a given price and in a given period of time. Supply of IT unlike the common concept of supply used in economic terms, is the availability of entire skill to the field at a given period of time. Supply price tend to rise in accordance with growing demand and at that moment IT professionals generally tend to switch around for want of higher perks. Supply of hardware components can be understood as in the same manner as supply of any other common physical good. Both demand and supply of IT are subject to the concepts of 'increase', 'decrease', 'extension' and 'contraction' phenomena due to price and non-price factors.

The present paper initially focuses on assessing the demand for IT resources and the main areas of concern for the information technology sector. The next section then highlights an approach that has attempted to find a viable way for creation of a stable demand-supply framework for IT resources through splitting of business organizations into demand and supply units.

Demand and Supply in IT Industry

The concepts of demand and supply in IT industry are of immense importance. Dividing IT into demand and supply organizations helps the concerned companies realize the full potential of their IT investments. In zeal to improve its ability to meet business needs, many companies have brought teams of IT developers into the business units they serve. Although moving IT into the business helps to align development efforts with business goals, it also has the unfortunate consequence of fragmenting IT developers across business and functional units, so coordinating and prioritizing projects becomes harder. Consider the case of a company that has two different business units using separate teams and different vendors and technologies, may simultaneously be creating similar applications. Business units at their level may be satisfied with the short-term results; however, the company as a whole may be suffering due to high development cost, poorly managed performance and difficulties in developing cross-group functionality. Companies that have focused on efficiency through centralization have struggled with agility and speed in developing applications. Business units can become frustrated by long delays in the deployment of needed capabilities and IT may be viewed as an unresponsive bureaucracy, a black hole for business requests.

The Principal Concern

In most of our writings, lectures and discussions on evolution and development of information technology, the United States is always talked about, having already contributed so much to the IT World. According to much of what one hears in media reports and policy debates, the US never seems to have enough high-tech workers. Warnings of shortage of engineers, scientists, and now information technology specialists have appeared regularly since the 1950s. On most occasions, such professionals have been sourced from other countries and India has always been a big contributor.

With the advent of globalization the trend has spread over the developing world. Emergence of countries like India and China to the fore of the Info-Tech has brought in new issues. Countries like these are now concentrating on more and more self- reliance rather than depending on the overseas support. While China has emerged as a strong global production base for the hardware, the emergence of India as a software master and producer of human IT talent has drawn everyone's attention. Countries like India are now concerned more about the developmental steps into other related areas. Such concerns are understandable; technological strength is seen as a crucial ingredient of economic prosperity.

Now the country itself feels being short of trained and skilled IT talent in many areas. The major problem has been the movement of best of the talent to other countries. Students from the premier institutes are lured by the overseas hirers. Other factors add to the continual pressure to maximize the country's supply of technical talent. Large labour pools help employers contain their payroll costs and improve the odds for a steady supply of occasional brilliant performers, the technical wizards whose contributions have been shown to be crucial. Training all of these specialists keeps academics busy; large academic establishments also contribute to research and development capabilities, fueling the emergent technologies that keep this system going. For many, questioning this state of affairs is tantamount to arguing with success.

However, there is a problem: in all these years, there is no evidence that any serious shortages of technical professionals – engineers in the past, information technology specialists now – have ever occurred. To be sure, there have been ups and downs in supply and demand as normal business cycles of boom and recession affected general levels of employment and as demographic trends influenced supplies of newly trained workers. In the early 1970s and again in the early 1990s, there were serious problems of weak demand for technical people. But on the national level, data on trends that reflect supply and demand for technical professionals have been mainly notable for their stability.

This applies with special force to measures of pay, the best indicators of actual shortages of workers. If shortages exist, compensation should rise sharply as employers bid for scarce talent, but when adjustments are made for inflation, technical pay has remained remarkably consistent since the 1950s, even with allowances for such factors as bonuses and stock options.

Demand for IT Workers

In conjunction with co-sponsoring convocation on the supply of IT workforce, the US Department of Commerce [1] prepared a report to explore the possibility of a shortage of IT workers in the United States. To assess the analysis of IT workers' demand and supply and to evaluate the basis for its conclusion that there is shortage of IT workers, the department compared the data presented in the report with the data available from other sources such as Bureau of Labor Statistics (BLS), the National Science Foundation (NSF), and the Information Technology Association of America (ITAA).

In its report that has covered a period of eleven years from 1994 to 2005, BLS has indicated that the greatest job growth has been in the field of System Analysts that has grown at a huge 92 percent over past eleven years. Very close to it has been the growth of Computer Engineers and Computer Scientists that has grown at 90 percent, while the growth in the field of Computer Programmers that has been low at 12 percent. BLS [2] also projected 1.3 million new jobs for 2007 as a result of growth and net replacements. This demand for IT workforce is demonstrated in Table 8.1.

Table 8.1: Growth of IT Workers Demand

Occupation	No. of Workers in Thousands		Percentage Change
	1994	**2005**	
IT Occupants	1365	2184	60
System Analysts	483	928	92
Computer Engineers and Scientists	345	655	90
Computer Programmers	537	601	12
All Other Occupations	125649	142524	13
Total IT Occupation	**127014**	**144708**	**14**

Source: Bureau of Labor Statistics [3]

Supply of IT Workers

The supply of IT workers in the United States has also been indicated by the US Department of Commerce. They identify the supply of IT workers as the number of students graduating with bachelor's degree in computer and information sciences. Their report [4] presents data from the Department of Education showing that 24553 students earned bachelor's degree in Information Science and Computers, which was 40 percent lower than the 1986 figure. Since there are a number of professional who acquire the required skills through various other means – special courses, certificate courses, hands-on experience etc., there can be no universally acceptable way for making such predictions, however, it has been thought that the employers almost always seek college graduates. The supply of IT workforce hence has been lower than the demand. In this connection the Commerce Department report has cited four pieces of evidence that an inadequate demand of IT workers has been emerging. They are:

- Rising salaries of IT workers.
- Reports of unfilled vacancies of IT workers.
- Offshore sourcing and recruiting.
- The fact that the estimated supply of IT workers is less than the estimated demand.

The Department of Commerce has stated that the upward movement in salaries is an evidence of short supply of IT resources. The report of a survey conducted by the Maryland Department of Labor, Licensing and Regulation [5] can be cited in this reference. The report that was published in August 2006, elucidates that while total employment in Information Technology sector has decreased by 1.42 percent, the payroll have registered a growth of 7.74 percent. This is further explained through Table 8.2.

Table 8.2: Employment and Wages in Information Technology Cluster Industries, 2002 and 2005

Industry Description	Employment			Payroll		
	2002	2005	%age change	2002	2005	%age change
Computer and Electronic Product Manufacturing	11424	9208	-19.3978	948600612	899817696	-5.1426
IT Wholesale Distribution	12629	11811	-6.4772	1035679369	1058220232	2.1764
Information Technology Services	15156	13773	-9.1251	806833813	759669675	-5.8456
Computer System Services	50952	54064	6.1077	3874638356	4459003137	15.0818
IT Equipment Repair and Maintenance	695	708	1.8705	31896667	39405987	23.5426
IT as a Whole	90856	89564	-1.4220	6697648817	7216116727	7.7410

Source: Maryland Department of Labor, Licensing and Regulation, August 2006

Despite that employment in the IT sector has decreased over the indicated period, the rising salaries denote the short supply of IT resources. In cases where the salaries have decreased, the reduction has been lower when compared with reduction in total employment. These wage increases, however, may not be conclusive evidence of a long term limited supply of IT workers, but may be an indication of a current tightening of labor market condition for IT workers. According to BLS, increases in wages have been less substantial when viewed over a longer period of time.

As this paper shows, this general picture of high-tech employment trends seems to apply to information technology today just as it has to other kinds of technical professions in the past: spot shortages may exist, and strong demand can be seen for some kinds of people, but on the whole there is no compelling evidence to suggest a national shortage of IT workers, either now or in the near future. To assess the nation's need for technical talent, it helps to start by understanding where we already are. The number of engineers, scientists, and other technical professionals in the United States has been growing rapidly for 60 years. At present, information technology includes the fastest growing occupations in the country. To take a US example, even the "Baby Bust" of the

1960s and 1970s, when annual numbers of live births dropped 25 percent, did not stop growth in the technology sector's share of all jobs, although the rate of growth did slow down. Now the numbers of college-age youth are rising again, which will fuel further increases in the share of the labor force that is composed of information technology workers and other high tech specialists. In Indian context, NASSCOM had estimated the demand for the software developers in the region of 112000 for the year 2005. [6] As per data available around 20000 software developers moved to the US in the year 2004 and the number is estimated to have increased since then. This implies a total demand of around 920000 developers. By comparison the total supply in 2005 has been about 70000 as shown in Table 8.3. This figure includes around 15000 graduates in Computer Science, Electronics and Telecommunications, another 10000 Engineering Graduates with degrees unrelated to software development, around 10000 MCA graduates and 10000 graduates with B.Sc. (Computer Science) and B.Sc. (Maths). In addition, there are about 20000 software developers with diplomas in Computer Science or Electronics, with certificates in specialized software development areas from private institutions such as Aptech, NIIT etc. and further around 10000 software developers with PG degrees such as M.Sc., M.Tech., and MBA. [7] If the current hiring practices continue, this implies an excess demand of 22000 in 2005.

Table 8.3: Number of Software Developers Potentially Available for Employment in Software Development Firms in India

Qualifications	Degree	No. of Software Developers
Graduate	Doctorate (Ph.D.)	95
Graduate	Master of Technology (M.Tech.)	2,130
Graduate	Master of Science (M.Sc.)	2,800
Graduate	Bachelor/Master of Computer Applications (BCA/MCA)	7,700
Graduate	Post Graduate Diploma	5,000
Under-graduate	Bachelor of Technology (B.Tech.)	16,160
Under-graduate	Bachelor of Science (B.Sc.)	3,200
Technical	Diploma	16,700
Technical	ITI/Certificate Holders	14,000
	TOTAL	**67,785**

Source: The Software Industry in India: A Strategic Review. NASSCOM, 2004-05

India has a rapidly growing technical workforce and that growth is going to accelerate. Thus the question is not whether the supply of high tech people is growing; rather, the issue is whether existing growth is sufficient to meet national interests. Experts in specialized trades within IT are still being felt short while there is glut in certain other fields.

To cite an example consider the case of Internet Security (IS), which is one of the major segments of IT market and also one of the fastest growing one as well. With more and more industries, departments, financial institutions and organizations turning to IT, the security of their invaluable data has becomes a major issue. The global demand for IS services that was approximately US$8 billion in 2001 has grown to US$24 billion by the end of 2006. The Indian market is also growing at a compound growth rate of approximately 27 to 30 percent per annum. This growth has fuelled for various specialized security professionals – firewall analysts, incident handlers, cyber law experts, security trainers, cryptographers, cyber insurance analysts etc. Statistics by different placement consultants and job portals reveal a growing demand for internet security professionals in the coming years. According to the recent Nasscom-JobsAhead.com study [8], conducted by the research firm IDC, the current demand for IS professionals is 18000 and 60000, in India and worldwide respectively. This is expected to grow to over 77000 in India and 188000 worldwide by 2008. Since every field of IT has its own significance and importance that the industry cannot do without, it comes to the core of their interests. To concentrate more on the regular business, a steady demand-supply set up is needed.

Industry says those interests are threatened by inadequate supplies of IT specialists. Creation of regular demand and an adequate supply of IT resources is matter of concern. There have been attempts to derive a meaningful approach to achieve the objective of stabilizing the demand and supply of IT. In the next few paragraphs of the present paper, one such approach has been highlighted that advocates for splitting the demand and supply units of an IT organization as it would help each unit work in a objective specific manner.

Splitting Demand from Supply

The demand for IT workers has fluctuated in variegated form worldwide. The fluctuated demand, as in any general approach, affects the agility and efficiency of application development and also the overall growth. There have been occasions when the supply of IT has been inadequate and also at times the demand has been below the supply level. It may be due to the reason that in case of most business organizations both the demand and supply of IT are handled by the same business units. Therefore, in order to achieve both agility and efficiency in application development, some leading-edge organizations are disaggregating the problem by splitting supply from demand. Creation of supply-units such as centers of excellence or shared application-development groups and demand units to take up the services; act as tech-savvy intermediaries between the business units and the IT to coordinate requests across the two.

While splitting demand from supply, some companies focus on improving the supply organizations rather than attempting to get the demand right. This is untraditional and distinguishes this model from the conventional set up. In most cases, business customers and IT are quick to see the potential benefits of a demand organization, so making the necessary changes is usually easy. Defining priorities and areas of responsibility can take more time, but these often evolve as the demand group begins to work with IT and its internal customers.

To succeed in their mission, the business organizations are suggested to adopt following practices:

- Aligning of the demand organizations with business units to act as the true voice of customers in clarifying what they mean, which at times might be different from what is being said. Demand managers should have a deep knowledge of their customer's processes and a solid understanding of application development. The ideal background for such a manger includes experience as an application developer and a desire to pursue a management career track. The development of these mangers should involve ongoing training in general management skills, such as project management and developing a business case.
- Supply organizations, in contrast, may be structured around broader competencies and business processes such as sales, operations, back-office functions or around technologies like online transactions and data warehousing rather than business units. Since supply organizations work with their demand counterparts that coordinate similar requests among business units, they can make broad decisions in the best interests of the entire organizations rather than a single business unit.
- Letting the demand organizations own business processes can be of a great help as business processes are increasingly shaped by IT solutions. Hence business should assign the responsibility for work-flow designs to a group that understands the underlying technology as well as the key business pain-points. The demand organizations have expertise in defining business needs and requirements, so it is well suited to work with business leaders to use IT

realistically and effectively to enable processes, avoiding the costly fixes that become necessary in case of process failure.

- Most companies have too many applications doing similar things, so that they periodically have to prune their application portfolio – a costly and time consuming process. Demand organizations can help prevent this situation from developing in the first place by minimizing the number of IT projects and applications at the time of funding. Managers should avoid silo-effects by holding regular meetings among demand organizations to review the project pipeline and identify the opportunities. In doing so, they help the organization spread investments responsibly. Without such fiduciary behaviour, profitable business lines may spend too much on IT but fail to share solutions with other parts of the organization.
- Demand organizations should be empowered to manage suppliers also. By doing this, business units will become free from handling the complex task of managing a broad range of IT suppliers that could be both internal as well as external. Typically, people with the right skills are already in IT supply organizations, where they manage internal teams and outsourced resources. The best option usually is to bring such people into demand organizations, although this approach is not always straightforward, where such people will require new training in the needs of the business and skills in managing a broad range of relationships.
- In addition to providing for these success factors, the organization must refine the metrics for success and some key processes of project management. Metrics for the demand organizations should centre on their effectiveness and customer satisfaction. For supply organizations, metrics continue to address cost efficiency and quality. The organization also must change its processes, especially those for project funding and software development.

Traditionally, the business units frequently finance their own application development projects. While they are often satisfied with their return on investment, the company as a whole may be spending more than what it should have. Introduction of demand-management organization helps the business and its demand group work together to define a capable strategy to show how the application portfolio must change to achieve business goals. The entire demand organization should then coordinate with the business units on this strategy, which includes not only individual projects but also longer term decisions on architectures and application portfolios.

The demand-supply model also requires changes in the software-development process. Typically a project involves five phases viz. envisions, design, build, test, and deploy. In traditional approach, the business customer envisions and then IT designs, builds, tests and deploys. Here it is endangered that unfortunately, what has been delivered may not be what the business envisioned. In the demand supply arrangement, the demand organization drives this envisioning phase regardless of whether the original ideas came from the business, the demand organization or the supply organization. The demand group brings the business and the supply organization into one conversation to explore ideas, which the demand organization ultimately translates into a business requirement document. The supply organization then manages the next phase – technical design, building, and testing. The demand organization will be involved during those phases and draw in the business as needed. For developers, this model provides the best proxy for the business customer, whenever they have the ideas about the requirements of the customer. The demand organization also provides expertise during deployment phase at the time of carrying out acceptance testing.

Patience and persistence are key words for any company creating a demand organization. To ensure that IT demand gets off perfectly, the companies should clearly define the role and scope of the demand group and make sure that the details are clearly understood by everyone involved i.e. the business units, the demand groups and the supply groups.

Despite taking every care there could be chances of occurrence of certain common flaws. There could be mistakes that include letting the demand organization focus too closely on a single business unit or on the supply organization's processes. Either pitfall can distract the demand group from its higher mission of enterprise-side efficiency. Transparent communication is of key metrics, of course, critical not just between demand and supply but also with senior managers who will be tracking the new organization's success. As the organizations implement and experiment with the demand supply approach, extension of the model can be expected. Customer companies with complex products could introduce a customer advocate organization that would represent the needs of the customers to the supply organization. What is more significant about the approach is that it can be extended to areas beyond IT. A sound demand management could help many organizations improve their total productivity in an effective manner.

Conclusion

On the basis of above discussion it can be ascertained that fluctuating gaps in demand and supply trends of IT resources, of which the IT specialists are the core is a concern of the industry. The shortage of IT specialists that hold the key to producing of future inventions in the field of information technology and hence of economic growth and development, is a key issue. There has been shortage in certain fields while at times there has been overflow of resources in the other. Creation of a regular demand and a steady supply of IT resources is the major concern and is in the interests of the industry. In order to achieve this objective, one way is to split demand from supply. By doing this the business organizations with separate demand and supply units will be better-off than before since the introduction of demand-management organization will help the business and its demand group work together to define a capable strategy to show how the application portfolio must change to achieve business goals.

Notes

1. Annual Report on 'Technology Transfer: Approach and Plans,Activities and Achievements', US Department of Commerce, 2006.
2. Bureau of Labor Statistics: www.bls.gov
3. US General Accounting Office Report dated March 20, 2006.
4. Annual Report on 'Technology Transfer: Approach and Plans, Activities and Achievements', US Department of Commerce, 2006.
5. Career and Workforce Information, Maryland Department of Labor, Licensing and Regulation, August 2006.
6. Study on Domestic Services (IT-ITES) Market Opportunity, National Association of Software and Science Components (NASSCOM), September 2006.
7. Supply and Demand of Software Developers in India, Working Paper, H John III School of Public Policy Management, Carnegie Melton University, Pittsburg PA, 2006.
8. Demand for IT Experts on the Rise, NASSCOM-JobsAhead.com, available at: www.nasscom.in/Nasscom/templates/NormalPage.aspx?id=1740, 2006.

References

Alpar, P. and Kim, M. A (1999), "Comparison of Approaches to the Measurement of IT Values", in proceedings of the Twenty-Second Hawaii International Conference on System Science.

Arora, Ashish, Arunachalam, V.S., Asundi, Jai, and Fernandes, Ronald (2006) "The Indian Software Industry" Heinz School Working Paper, Carnegie Mellon University, Pittsburgh.

Basic Background Report (BR-3) for the National Task Force on IT and SD submitted to the Prime Minister of India: 18th March 1999. http://it-taskforce.nic.in/it-taskforce/

Brooke, G.(1999), "Information Technology and Productivity: An Economic Analysis of the Effects of Product Differentiation," Published Ph.D. Thesis, University of Minnesota.

Career and Workforce Information, Maryland Department of Labor, Licensing and Regulation, August 2006.

David, Mark and Eric, Monnoyer (2004), "Next Generation CIOs," McKinsey on IT, Number 2, Spring, pp 2-8.

David, Mark and Rau, Diogo P (2006), "Realizing Full Potential of IT Investments," McKinsey on IT, Web exclusive, Fall 2006.

NASSCOM (2006) "Study on Domestic Services (IT-ITES) Market Opportunity" National Association of Software and Science Components (Nasscom), New Delhi.

Phukan, Punita Jasrotia (2007), "Demand-Supply Gap Widens for IS Professionals," http://www.expressitpeople.com, New Delhi.

US Department of Commerce (2006), "Annual Report on Technology Transfer: Approach and Plans, Activities and Achievements".

9

Business Process Outsourcing (BPO) Industry in Sri Lanka and India

M.F.M. Nusry

Business process outsourcing (BPO) is the contracting of a specific business task, such as payroll, to a third-party service provider. Usually, BPO is implemented as a cost-saving measure for tasks that a company requires but does not depend upon to maintain its position in the marketplace. BPO is often divided into two categories: back office outsourcing, which includes internal business functions such as billing or purchasing, and front office outsourcing, which includes customer-related services such as marketing or tech support. Before proceeding further, it is interesting to have a look at the Sri Lankan economy.

Economy of Sri Lanka: An Introduction

Sri Lanka has the highest literacy rate in the South Asian region and has a large pool of educated young persons with skills in IT, computer programming, accountancy, medicine and engineering. At the higher end of the skill spectrum, qualified professionals offer levels of supervisory skill that are among the best in the world. The country's leaders believe that outsourcing, software development, systems integration and a vast range of Information Technology Enabled Services (ITES) related services are the passport to the future. The Sri Lankan Government has actively encouraged the spread of computer literacy by providing duty concessions on hardware imports. This environment is further enhanced by modern data transmission systems and complementary hardware. Low cost offshore data entry operations have also thrived in Sri Lanka since the early 1980s. Sri Lanka already serves as the headquarters of one of the largest software development houses in South East Asia.

Sri Lanka has faced natural as well as man-made disasters in the recent past. But it has been receptive in overcoming these obstacles. The political instability and the insurgency in the north are creating obstacles in the economic growth of the country. The country has a Gross Domestic Product (GDP) growth of 5 percent during 2004-2006.

The economy is primarily based on Agriculture, agriculture sustains about one-third of the working population and directly contributes to around 20 percent of the GDP. The main cash crops are tea, rubber and coconuts, which provide for over 75 percent of export earnings. Forestry and fishing are also important.

The main industrial sectors are the mining and manufacturing. Iron ore, limestone, clay and uranium ore are the minerals that are present in commercially-exploitable quantities and the important manufacturing industries include cement and textiles, both of which are valuable export earners.

The service sector is dominated by banking and insurance both of which are performing well. Tourism is a major sector which the government is trying to boost but it is stunned time and time by the civil war.

The governments followed the usual prescription of market-oriented policies of privatisation and deregulation, at the same time seeking to build up potential export earning industries. The perseverance shown by the Sri Lankan government on these long term plans has started to show up now. Even with a country waging a war against terrorism for more than a decade the country's economy seems to have become quite resilient and on the track. The government is now hoping to consolidate its progress by further deregulation, fiscal reform and privatisation.

There are seven universities in Sri Lanka producing graduates in the fields of electronics, engineering and computer studies. In addition, there are four institutes providing training in electronics. Sri Lanka has extensive deposits of high quality minerals such as kaolin/feldspar, silica sand, quartz, etc., which could be used as a base material for electronic products.

Characteristics of Sri Lankan BPO Industry

According to a baseline analysis of the BPO sector carried out by Information and Communication Technology Agency (ICTA) of Sri Lanka partnered with Learning Initiatives on Reforms for Network Economies (LIRNEasia) in 2006 [1] the first BPO unit an offshore US operation, was set up in Sri Lanka in 1983, but the BPO sector only started emerging from the year 2000. Through the BOI (Board of Investment) initiatives the sector gained momentum in 2004 with the establishment of 5 companies. Growth continued in 2005 with 8 additional entrants and in 2006 another 6 entrants with more on the pipeline for commercial operations. Most of the BPO companies are established under the Board of Investment of Sri Lanka. This is due to the attractive package these companies enjoy through Board of Investment projects.

The main export markets for Sri Lanka based BPOs are the United States and the United Kingdom. These BPOs are primarily attracted to Sri Lanka due to low labour and operating costs. Average turnover of employees is around 6 percent and around 6000 are employed in this sector and employment figures are growing year on year.

Of a population of 19.2 million, the total workforce is approximately 6, 7000,000. The population growth rate averages 1.2 percent per annum, ensuring a steady pool of manpower resources for both existing and new industries. The Government provides free education from Kindergarten to University. Sri Lanka is also ranked amongst the East Asia tiger economies and first in South Asia on the Human Development Index.

The local software industry believes that Sri Lanka can achieve USUS$ 1.0 billion in total ICT related export services by 2012 – with the right mix of industry coordination, support from academia and establishment of a policy environment conducive to ICT growth. The ICT Industry Cluster, which has now been formally incorporated as the Sri Lanka ICT Association (SLICTA) is committed to transforming Sri Lanka into a sustainable, world-class center for high value, rapidly evolving IT services.

According to a survey carried out by LIRNEasia the formation of Sri Lankan BPO sector began in year 2000 with the regular emergence of BPO units. There is a spur of BPO activities in the recent years. With the ending of Sri Lanka Telecom's exclusivity on international connectivity in August 2002 and the subsequent issuance of External Gateway Operator Licenses from March 2003 onwards this has fuelled further growth. In addition to the telecommunications liberalizations, tax incentives introduced in the early 1990s to pull in IT related investments, may have also had spillover effects on the BPO sector.

Sri Lanka's BPO sector comprise of offshore BPOs (52 percent), onshore BPOs (38 percent) and those that classified themselves as both onshore and offshore operations (10 percent). The largest number of offshore investments is from the US with four BPO units in Sri Lanka set up by principals in the US. The marketplace is also evenly divided between captive (48 percent) and non-captive (48 percent) operators. Two out of the four companies that are offshored from the US are non-captive operations and a majority of the companies that are headquartered in Sri Lanka (63 percent) are also non-captive units.

Investments in Sri Lanka

Investment flows into Sri Lanka's BPO sector is relatively regular since year-2000. The inflow is in all three forms: foreign investments, local investments and joint ventures. Foreign investments in joint ventures ranged from 30 percent to 95 percent. Most BPO investments (86 percent) are under the Board of Investment of Sri Lanka.

The total investment value of the surveyed companies established during 2000 to 2006 is about USD 13.2 million. The value of individual BPO investments ranged from USD 15,000 to USD 3.5 million. The survey results also indicate that investments by offshore BPOs (USD 4.3 million) are lower than onshore investments (USD 7.8 million). This could be due to less international recognition of Sri Lanka as an off shoring destination and also due to lower capacities to cater to large-scale offshore ventures.

Service Offerings

Accounting services is the most popular product offering in Sri Lanka with 43 percent of the companies listing it as a main service. Call centre services and medical insurance processing are the next most popular with 19 percent and 14 percent of the companies offering these services. Most BPOs offer more than one type of work and in general, are geared to undertake at least two different types of services. Non-captive operators appear to provide a greater variation than captives, with the number of services offered averaging to 3.5, compared to the average of 1.4 services offered by captive BPOs. In addition, more onshore than offshore provide accounting services, indicating that local companies are tapping into the opportunities presented by the large pool of young, qualified accountants in Sri Lanka.

Export Markets

The industry is predominantly export oriented with 71% of the companies exporting 100% of their services. The leading consumer of the Sri Lankan BPO services is the US, which is catered to by over half the BPOs. The UK is targeted by 63% of the BPO companies and 31% are exporting to Australia. Other export markets are Sweden, New Zealand, Hong Kong, Singapore and Japan.

Operating Costs

The total operating costs for a BPO unit averages to around LKR (Sri Lankan Rupees) 4.9 million per month (approximately USD 48,000). Due to the high labour intensive nature of BPO services, labour costs account for the largest share of overheads averaging to about 44 percent of total operating costs. On average, labour costs at an offshore BPO unit is about 51 percent of total costs as compared to 35 percent at an onshore BPO.

Training and development, on average, account for about 8 percent of the total operating expenditure of a BPO. The survey findings also indicate that offshore BPOs spend more on training and labour than onshore units. This could be due to the fact that both average employment numbers as well as average salaries are higher at offshore BPOs than onshore BPOs. However, the price of labour is not considered a cause for concern by most BPO establishments and only 14 percent of the companies cited labour costs as a potential industry constraint.

After human resources, the second highest expense faced by a BPO is for telecommunications, with costs averaging to around 22 percent of total monthly costs. The electricity expense is a much lower 8 percent of operating costs on average.

In terms of office space, the World Bank Report [2] found that rents in Colombo on average, are lower than in other major Asian cities. The average monthly cost per square meter is estimated at around USD 11.60 in Colombo compared to USD 15.45 to USD 26.15 in New Delhi while the price of office space in Shanghai, is nearly three times the cost in Colombo.

Sector Management and Workers

Sri Lanka's BPO sector only recently acquired critical mass to gain recognition as an export oriented service distinct from the island's IT and software export industries. The total BPO workforce is currently estimated to be slightly over 6,000. However, the sector is expanding, offering new opportunities for young Sri Lankans.

Half the BPO employees have GCE (General Certificate of Education) Advanced Level(AL) as their highest qualification, 15 percent are graduates, 9 percent are diploma holders and 7 percent have an Ordinary Level(OL) qualification. Despite the trend towards a large intake of Advanced Level holders BPO companies require 'fresh graduates' as an industry requirement.

Salaries and Incentives

The average monthly salary of a BPO worker is around USD 300, inclusive of bonuses and commissions. However, remuneration packages ranged from USD 120 to USD 1,000 per month. Salaries for services like investment research are about USD 1,000 per month while payments for accounting services are around USD 175. Offshore operations displayed marginally higher average monthly salaries than onshore BPOs.

Non-financial incentives are given in the form of transport and food during non-traditional work hours and also medical insurance coverage. Most of the BPOs provide transportation for the employees. Some other benefits that the BPO companies provide are club membership, professional subscriptions, and overseas trips. Cars are geared towards management in some BPO companies. Labour cost does not seem to be an obstacle for BPO companies for growth of their operations in Sri Lanka as of yet.

Employee Turnover and Recruitment

The average employee turnover for the industry is 6 percent per year. But for some companies this is near zero but for some this is higher. Average turnover at captive BPOs is lower than at non-captive BPOs. The no-captive turnover may be due to the nature of the operations rather than due to attrition.

BPOs compete for talent in the labour pool through most of the regular recruitment channels. The most popular method of hiring employees is conventional advertisements in newspapers. Although Internet usage in Sri Lanka is increasing, only few of the companies use this medium to recruit new employees. Informal channels, such as 'head-hunting' and the 'grapevine' are more popular than Internet usage. References and 'word of mouth' are also used by many of the companies.

Skills in Demand

Sri Lanka is noted for its large and qualified pool of accountants and primary skills on accounting qualified personnel seems to be not in a shortfall but it may be possible that supply limitations of qualified personnel for BPOs may arise due to societal attitudes, where non-traditional working hours may be perceived unfavorable. And there might be low public awareness about the nature of BPO workplace, salaries, career opportunities and other factors may also limit the number of potential candidates applying for BPO jobs. For the growth of the sector it is necessary to address these constraints which may negatively impact the long term competitiveness and growth of this sector.

Major Obstacles for Sri Lanka in Realising Results

According to the survey done by ICTA in 2006 the following has been identified as the major obstacles to growth for leading BPO companies in Sri Lanka.

Domestic Stability: The overwhelming concern of the BPO sector is with regard to domestic civil and political stability. The most frequently cited threat to successful operations and growth is domestic instability. This could be due to the fact that BPOs mostly service foreign clients and the industry could be impacted by external perceptions of domestic instability. BPOs may also be concerned that domestic instability could disrupt their ability to deliver uninterrupted services. However, the emphasis on the subject could also be a result of current country situations when domestic security concerns re-emerging strongly and wide international media coverage was focused on the peace process and ceasefire violations.

Domestic Infrastructure: Transport, telecommunications and electricity are the three facilities important for the BPO sector. Mass transport services are necessary for labour mobility since many BPOs cater to outside world and work hours are most of the time based on the times of the client time zones. To give real time services to different time zones availability of labour is necessary at unconventional hours. For operational efficiency BPOs require cost-effective electricity and telecommunication services and it is essential to be competitive in the market place.

Transportation: BPO companies consider mass transport services as the most inefficient of all the services and see it as the greatest infrastructure related bottleneck to BPO activities. So Sri Lanka needs to concentrate in this area on a long term basis. Most BPO service providers work in shifts and larger portion provide transport for employees. But transport facilities are one-way in general during non-traditional work hours or when employees are required to work late after normal work hours.

Telecommunication: Telecommunication services are an essential part of BPO operations since for quick delivery of services overseas these are of primary requirement. This primarily includes broad band, leased line facilities and also video conferencing and basic telephone services. On average BPOs spend about 20 percent of their operational costs on telecommunication services, therefore, it is critical telecommunication services function smoothly and reliably. All of the BPO companies use common amenities such as telephones, emails, internet and some have invested in VoIP (Voice over Internet Protocol) and other types of facilities. A majority of BPO service providers almost 71 percent are satisfied with the reliability of basic telephone services in Sri Lanka rating it as 'very reliable' or 'reliable'.

For broadband access BPOs utilize dedicated and shared leased lines, Asymmetric Digital Subscriber Lines (ADSL) facilities, Very Small Aperture Terminal (VSAT) and other facilities with bandwidths ranging from 256 Kbps to 4 Mbps. A majority of the companies use dedicated leased lines for their business with bandwidths ranging from 45 Kbps to 2 Mbps (Table 9.1).

The availability, bandwidth, reliability and costs associated with leased line facilities impact the ability of BPOs to deliver their services in a timely and efficient manner. An independent study on regional domestic leased line costs on the other hand, suggests that costs in Sri Lanka are considerably higher compared to other countries of the region like India and Pakistan. Sri Lanka's Domestic

Private Leased Circuit (DPLC) prices for 2 Mbit/s is around USD 166 for a 2 km distance. Comparative DPLC costs in India are USD 21 and in Pakistan the prices are USD 4.8.

Table 9.1: Types of Broadband Facilities Used by BPOs

Type of connection	Bandwidth (Kbps)	Percentage of companies
ADSL	256 - 2048	24
Shared leased line	256 -4096	19
Dedicated leased line	45 - 2560	67
VSAT	72 - 512	10
Other	Not mentioned	5

The high costs and low penetration of leased line services could adversely impact expansion of BPO activity. Given that leased line penetration is low outside of the greater Colombo area, BPO businesses are restricted in their ability to expand service centers outside Colombo and thus leverage potentially lower labor costs. Only two of the surveyed companies reported branches outside the greater Colombo area. One of the companies reported branches in Kandy while the others reported branches in Kandy, Galle, Matara, Nuwara Eliya, Ratnapura, Chilaw, Badulla and Kaduruwela.

Significant changes have taken place in the local telecom sector in Sri Lanka since the 1990s. Fixed access (both wire-line and Wireless Local Loop (WLL)) telephone connections have risen from 121,388 in 1991 to 958,832 in June 2004 – a growth of over 700 percent. The number of WLL connections by the two operators; Suntel and Lanka Bell stood at 121,606 as of June 2004. However, the most significant trend in the local telecom sector has been the phenomenal growth in cellular connections, which increased from a mere 1800 in 1991 to a staggering 1,801,392 by June 2004. The overall tele-density, which stood at around 3-4 percent a few years back, had increased to 14.2 percent by June 2004.

The opening of the international market has seen two more companies - MTN Networks (MTN Networks refers to Dialog Company of Sri Lanka's parent company) and VSNL of India gaining access to the international under-sea cable SEA-ME-WE-3. In addition, Sri Lanka Telecom is also in the process of commissioning a link with BSNL to India. The SEA-ME-WE-4 cable would provide a third international landing point for the country. All of the above would enable Sri Lankan operators to provide true dual diverse redundancy connectivity to their operational sites.

Electricity: Electricity costs as compared to telecommunications and labour costs represent a smaller share of operating expenditure and most BPOs do not consider electricity an obstacle to daily operations and future growth. There are no complaints of frequent power outages as most BPO companies have invested in generators as well. This is due to the critical nature of the BPO operations they need backup power in case electricity fails.

However, there is discontent with electricity prices. Sri Lanka's electricity tariffs are at around USD 0.08 per kilowatt-hour is noted as among the highest in the world for business operations due to cross subsidization of domestic consumers. Businesses in Sri Lanka pay more for electricity than companies located in many East Asian countries and companies in the US, Indonesia, India and the US charge electricity rates below USD 0.06 per kilowatt-hour while electricity tariffs are below USD 0.07 in Thailand.

Capacity Constraints: Sri Lanka requires further investments in education, particularly in English language skills. Irrespective of a company's services, English language is a requirement across the board. Therefore, improving English language skills is a blanket requirement for BPO sector growth. Measures to improve English education and other types of skills should primarily target the youth, as the BPO sector is an emerging opportunity for this demographic.

While the irregular work hours and company-funded transport provided by BPO establishments may help to somewhat alleviate the effects of the poor mass transportation services, the need for further improvements is quite clear.

Telecommunications liberalisations, especially the opening of the international gateway to competition in 2002, may have stimulated BPO sector growth. However, the higher costs associated with leased lines, their relative unreliability and the low penetration of broadband facilities outside of Colombo must be addressed for BPO sector expansion. In lieu of the sector's emerging potential, the government should assess the feasibility of setting up a dedicated BPO zone with enhanced transport facilities and telecommunications infrastructure.

Regulatory and Procedural Constraints: Complex regulatory procedures not only lead to delays in realising investments but also add to operating costs. It is necessary to streamline regulations that govern the industry to facilitate the process of setting up and conducting BPO operations.

Sri Lanka as a BPO Destination: Appreciating Rupee and Costs in India

India's outsourcing companies have thrived by winning work from companies in the US and Europe that sought to tap the country's low costs and large employee pool by handing over jobs ranging from answering customers' calls to risk management and financial analysis. But costs are surging in the prime cities in India, which has earned a reputation as the world's back office, as property values and rentals rise and wages increase at an annual pace of more than 15 percent amid a shortage of skilled employees.

India's currency is also appreciating against major currencies and this in turn is affecting the bottom line of BPO companies. Indian outsourcing firms are feeling the pinch from the appreciating rupee, which dents dollar-billed earnings, forcing them to cut costs by expanding to less expensive locations or otherwise. With the depreciation of the Sri Lankan currency the advantage if any for India-based companies hoping for a foothold in Sri Lanka would be much. Because the dollar billed earning will more than compensate BPO's operating costs in Sri Lanka. Nevertheless there are interesting opportunities for any Indian players. If one had an operation in Chennai and Bangalore, a new centre in Colombo made more sense than one in Gurgaon.

With the demand-supply gap widening, newer cities will play a critical role in re-engineered globalisation models. The destinations need to provide greater level of cost effectiveness and operational efficiency for the BPO companies. However, Sri Lanka's strategy is not to compete on price but on quality and value addition. Some of the internationally known names that have already set up operations in Sri Lanka include WPA, HSBC, Astron, Hellocorp, Office Tiger, WNS Global Services, etc. Sri Lanka provides an option for these BPO companies with low cost labour, lower taxation, tax holidays, cheap floor price and low attrition rate. With the country having a high literacy rate, the attractive investment offered by the Government for foreigners and the date line advantage are key factors BPOs can consider in having operations in Sri Lanka. Since Sri Lankan employees also have a high loyalty towards the company and this too is an added advantage.

Another important factor is the Sri Lankan attrition rate is around 18 percent while the Indian rate is over 30 percent. The world trend is between 30 to 50 percent. In comparison to some of the other countries, the labour for the BPO industry too is low in Sri Lanka while the appreciation of the US dollar and the British Pound make the industry attractive.

Sri Lanka's vicinity to India with many connecting flights with frequent transport facilities is another very important factor for a BPO company based on India in having their second or backup operations in Sri Lanka. There are around 120 flights from Sri Lanka to various points in India. This level of connectivity with stream lined visa processing could bring about more collaboration and travel between the two countries.

The Indian government has announced new tax on fringe benefit tax on Employee Stock Option Plans (ESOP), 12 percent on rental space and 11 percent Minimum Alternative Tax (MAT). This in turn forces the BPO sector to lose jobs to countries like China, Sri Lanka, the Philippines and South Africa where the governments are handing out tax breaks. The Indian companies are facing the risk of losing the first-mover advantage. The policy will make the sector less productive compared to other countries. Sri Lanka is offering a 15 year tax holiday, while Philippines and China offer a 10 year tax holiday each.

Sri Lanka is strategically located along the east and west sea lanes. The country also offers preferential access and duty concessions to the Indian market as a result of the Indo-Lanka Free Trade Agreement. The conclusion of the agreement was an important milestone in the trade liberalisation process between India and Sri Lanka. The objectives of the agreement are to promote bilateral trade through the development of economic relations, fair and open competition and the removal of trade and investment barriers between the two countries. Sri Lanka is also signatory to the South Asia Preferential Tariff Agreements (SAPTA). Under the agreement, tariff concessions for 226 items are available with an envisaged increase of up to 2000 items in the near future. SAARC member states have also resolved that SAPTA should eventually progress towards the South Asia Free Trade Agreement (SAFTA). These developments coupled with further negotiations with the two countries are helping both countries to realize mutual benefits in the new emerging sector and other spheres.

New Initiatives

The steps to set up an IT park in Katunayake close to the International Air port is a major step towards attracting BPO companies. The increased tourism arrival and stream lining of visa and other formalities will help in executives to travel with more ease between Sri Lanka and India, with the already existing 120 flights per week for various points in India this will further enable travel and accessibility for personnel to travel here and forth. Flight availability and visa on arrival are two major problems that existed which have now been sorted out. The plan is to have more flights to and from India for the Techno Park. Further Infrastructure concerns are being addressed through negotiations with Telecom to have intermediate costs brought down. BOI is negotiating with Telecom regulatory commission (TRC) to have rates revised or to give some type of incentive so that the cost of broadband telecom solutions for BPO/IT companies is reduced in Sri Lanka. The Techno Park will provide all the facilities for a BPO/IT company to set up operations as fast as possible people can rent the space and start operations immediately. It is strategically placed near to the air port so travelling is eased.

New telecommunication initiatives are underway to explore the possibility of establishing a national fibre optic telecom backbone. The TRC initiative has got the support of the World Bank which has provided a grant for the consultancy report. The objective of the preliminary study is to assess the possibility of harnessing the existing infrastructure of the Ceylon Electricity Board (CEB) as well as the Ceylon Government Railway (CGR). The current telecom network revolves around microwaves quality of which is impacted due to several factors including bad weather. According to analysts a fibre optic backbone would considerably enhance the rapid growth of the Information and Communication Technology usage in Sri Lanka.

The Government has recently launched an initiative to use ICT to develop Sri Lanka's economy, alleviate poverty, improve the quality of life and the opportunities for all Sri Lankan people, which is widely known as *e*-Sri Lanka.

Government Relations and Industry Bodies

Sri Lanka's BPO sector has strong links with the state, with almost the entire trade coming under the purview of the Board of Investment of Sri Lanka (BOI). Most of the BPOs are registered under the BOI and are recipients of various BOI incentives. The BOI's strategy to provide BPO-specific incentives for the first time clearly indicates the importance that the government is placing on the sector. Most BPO companies were able to make a relatively smooth entry into commercial operations.

Payment of taxes is not considered a hindrance to growth and operations by the companies. This is due to the fact that over half of the BPO companies enjoy tax holidays under the prevailing BOI Incentive schemes. Tax administration is considered as an obstacle by some companies. Customs, trade, environmental regulations and business licensing and operating permits are in general not considered problematic.

Many companies enjoy various government incentives in the form of Tax holidays, duty free imports, status of BOI, Value Added Tax (VAT) exemptions and so on. BPO companies also expressed the opinion that national and international marketing and promotional activities to position Sri Lanka as an outsourcing destination, should be augmented to support individual and industry level efforts.

Conclusion

Following accelerated telecommunication liberalisation in 2002-03, Sri Lanka's BPO sector is expanding both in terms of increased activities of existing players, as well as new entrants. This growth of BPO activity has so far been fuelled through low labour and low operating costs. Government incentives too, seem to have been successful in encouraging BPO investments. However, Sri Lanka needs to address capacity constraints in the three areas of English language skills, mass transport and telecommunications infrastructure (specifically improvement of broadband facilities) for any significant expansion of BPO activity to occur.

Building Capacity in the form of increasing the number of IT-BPO professionals available in a timely manner is very critical for the industry growth. The capacity issues need to be effectively handled so as to not to hamper growth potential of the industry. The aim should be to maximize the opportunities that lie ahead for the industry and the approach Sri Lanka needs to follow to exploit the potential will be through focusing on a few niche areas. Overall the industry needs to attract people by improving willingness of the student population to enter into IT and BPO jobs, expand capacity of the education system and take measures to keep education relevant and business aligned.

Some of the short term initiatives that could be done are to launch a nationwide awareness campaign for IT/BPO careers, organizing career fairs and creating awareness through media about it. Another initiative is to make advanced ICT education affordable and accelerate ICT teaching capacity developments and encourage investment in education.

On the longer run its necessary to have IT literacy for all and English for all because these two are key factors that will be necessary for the IT/BPO sectors to move forward and create new opportunities or to find new frontiers. This coupled with an eco system of innovation will help in achieving greater heights for the country as a whole as well as individuals in the country.

Cooperation for Growth

Sri Lanka is working on a larger plan to seek partnership with India in attracting investments into the IT and Business Process Outsourcing (BPO) sectors gradually. Already both countries have signed a Free Trade Agreement (FTA) to promote trade and investment. This partnership is bearing fruit at a moderate level now. The Islands BPO and IT sectors are attracting foreign Investors. Sri Lanka's proximity to India helps investors to put up an alternate business hub to their Indian operations in case of any exigency.

The Sri Lankan government with the funding of World Bank and support of ICT (Information and Communication Technology) are conducting Capacity Building Programme (CBP). Its total cost will be US$5.3 million. Information technology enabled services today represent a US$100 billion market that is growing at 30 percent per annum. In India, IT exports amount to close to US US$20 billion and almost half of total export revenues. The development of ITES and offshoring is the priority of developing countries like Sri Lanka as it will help create more jobs and export opportunities and simultaneously will boost the initiative for education, technology transfer and an improvement in the quality of locally provided services.

Since a large number of multinational firms explore the opportunity to diversify their service sources to prevent dependence on a single provider, more countries are competing to gain a bigger share of the expanding market. Other popular location for offshoring in Asia includes China and Philippines. Geographical proximity certainly contributes to the choice of offshoring destinations, particularly when the functions being outsourced increase in complexity.

Marketing and Promoting the BPO sector

Successfully leveraging Sri Lanka competitive advantages of lower labour and operating costs requires strong promotional and branding exercises that would benefit from greater government participation. There are concerns about maintaining Sri Lanka's image abroad, so more needs to be done to promote Sri Lanka as a viable outsourcing destination. Therefore, national promotional campaigns could attract further offshore investments and also encourage non-captive operators by attracting outsourced workflows. However, other domestic factors need to be addressed in parallel, for promotional campaigns to show adequate results. Although large, captive, offshore BPOs tend to be more cost effective, generate higher employment and may also provide greater employment stability, promotional activities to draw such investments must be accompanied by measures to address capacity constraints to successful supply operational needs of BPOs.

BPO activities are directly affected by factors such as civil and political stability and law and order. These factors are of particular importance for non-captive operators as external perceptions of domestic instability could adversely impact inflow of outsourced work. Further, offshore investments could be hampered by perceptions of domestic instability and security concerns. National decision makers need to pay greater attention and facilitation to nurture and realize the potential of the BPO sector at this important stage.

Notes

1. ICTA partnered with LIRNEasia (September 2006) "A Baseline Sector Analysis of the Business Process Outsourcing Industry of Sri Lanka", LIRNEasia, Sri Lanka.
2. Ismail Radawan, Gihani Fernando (August 2005), "Sri Lanaka, Offshoring Professional Services, A Development Opportunity", Staff Paper ,World Bank.

References

Definition of BPO (http://en.wikipedia.org/wiki/BPO), World Wide Web

Divakar Goswami (March, 2007), "Grasping the Internet Measuring the Beast", WDR Expert Forum, Singapore.

HCL Technologies BPO Services Ltd (2003), A White Paper on "BPO India", HCL Technologies, Noida, India.

National Association of Software and Services Companies (2005) "Extending India's Leadership of the Global IT and BPO Industries", The Nasscom-McKinsey Study (2005), New Delhi.

NeoIT(October 2006), "Offshore Insights - Market Report Series", Title: "*Global City Competitiveness*", Volume 4, Issue 8, NeoIT, India.

Part II: Diffusion (Use) of IT into other Sectors of the Economy

10

Role of Information and Communication Technologies (ICTs) in Enhancing Cooperation in South Asia [1]

Somesh Kumar Mathur

There are common set of problems faced by ICT sectors in the South Asian countries. These problems include limited access to ICT technologies and low tele-density especially in rural areas, inadequate quality and skills of graduates, rising salaries not commensurate with increase in productivity, weak infrastructure which result in frequent power shortages, low level of personal computer (PC) use and internet penetration, low level of domestic technology development, limited bandwidth, inadequate availability of venture capital for the small scale units (taxed in India), high attrition rates in BPO companies, e-security, improper content design of e- governance programs, and limited domestic market for knowledge based technology and products. The South Asian governments have an important role to address them in times to come and they need to pool their resources to overcome them. This paper suggests ways and means to overcome the problems faced in the ICT sector by the South Asian Countries and suggests some future policy directions.

In particular, the South Asian Cooperation in ICT sector could cover the following areas: (a) Education and Training through building institutions of research and higher learning (enhancing national innovation system for long term growth of IT sector) (b) Joint development of Hardware and Software in local languages and open source platforms like LINUX (for e-governance programs and increasing ICT usage) (c) Exchange of experiences in Electronic Governance programs for poor and the under-privileged. It could be organized through NAM (Non-Aligned Movement) Information Clearing as proposed in RIS Policy Brief, 2006, No. 27, August. It would indeed be useful if a common R& D fund is created for the South Asian countries and utilized by identified centres of excellence in different countries so that research findings and developments can be freely shared on ICT. Government and Industry of the various countries can perhaps fund this research, along with contributions from research organizations.

Kumar and Agarwala (2006) discuss a number of examples of ICT for pro-poor use. These are community owned computers with visual or graphic interface, use of public fixed line and mobile telephone bureaus as sources of rural employment as well as for increasing the virtual tele-density, use of mobile phones by fisher folks, use of internet for e-governance such as computerization of land records, among many other applications. Internet governance can help create an information society in which all South Asian nations have an equitable stake to share in the benefits from the IT revolution.

Cooperation among developing countries can be in the form of designing effective models of regulation and e-governance for growth and development in the developing countries. Also, to reduce poaching of client data base by employees in the BPO (Business Process Outsourcing) sector a uniform regulation need to be designed by the governments. This will also reduce attrition rates among employer and employee in the high growth BPO sector. In addition, legally regularizing movement of IT professionals among SAARC(South Asian Association for Regional Cooperation) countries can be effective tool in promoting efficiency in BPO sectors particularly in Bangladesh and Sri Lanka. Above all, big size companies in India can think of outsourcing some of the jobs to LDC of the South Asian region. Also, cooperation in new areas like bioinformatics, disaster management, human genome analysis through customized IT software, e-services like personalized consultations for traditional and non-traditional medicines on the net and information of custom-designed garments, among others can promote growth and development in the South Asian region. The e-services and custom-manufactured products offer several advantages for South Asian countries, including better wages for knowledge workers, a higher productivity of the workforce and knowledge gained by working on world class projects for leading firms.

We discuss below various specific programs for developing cooperation in ICTs for furthering the goals of South Asian cooperation in regional connectivity.

E-Governance Programmes

The E- Governance role could be:

- Technical role, in terms of automation of tedious tasks earlier done by humans.
- Facilitating role, leading to participatory and all encompassing decision-making and implementation processes.
- Innovative role, involving new services and mechanisms to deliver these services

Computerization of the Indian Railways' Passenger Reservation System one form of e-governance may be branded a success (Heeks, 1996). Not only did it significantly increase the efficiency of the reservation process, but it also reduced corruption (though did not eliminate it), increased rail staff morale and improved the quality of customer service. Beyond these reform components, it also gave Indian Railways (and India more widely) a more modern image and it helped to build information age capabilities within the country. Various works of other government ministries and departments is made accessible to the general public in India through their own websites (Bhattacharya, 2004). These include, allocation of Permanent Account Number (PAN) for income tax payers, processing of passport application, conduct of public examination, custom clearance, Regional Transport Offices, schemes under implementation by the NGOs, vigilant information, VSAT (Very Small Aperture Terminal)-based money orders under the Department of Post, Supreme Court, land records, Parliament questions, debates and deliberations, among others (see www.mit.gov.in for more details of computerization in the Central and State governments).

There are more than 800 e-governance programs running all over India. These projects are at different stages of their implementation (see for details: www.egovdatabase.gov.in) Also, the other South Asian countries have realized the importance of e-governance programs and they are also in the process of implementing such programs in their respective country. More important is the continuous availability and access of quality ICT infrastructure in rural areas including internet for all for the success of e-governance programs. Such infrastructure can be made available through community service centres or kiosks (government owned or public private partnerships), village panchayat offices, schools and colleges premises, temple premises or any place of worship and post offices. The services available at the kiosk (in interactive version) may include online application for caste certificates, domicile certificates, income certificates, driving license, employment registration, pension forms, registration of birth and death certificates, advice on agriculture, veterinary, health and administrative services, details of work available in village and district, examination details for students, information on new agriculture practices and rates of seeds, government tenders, telemedicine, important telephones and addresses of hospitals, schools and colleges and other government functionaries, grievance redressals, educational scholarships, land records, among others. These services should be interactive with the availability of both wire and wireless technologies (fixed line services, mobile services and Television). These community service centres should be ideally handled by entrepreneurs from the local area. Entrepreneurs should also be encouraged to start their own tele-centres. Appropriate training needs to be given to the entrepreneurs by government functionaries from the district so that they can handle the internet kiosk or the community service centre efficiently. The quality of infrastructure should be monitored by the appropriate government authorities for its constant and continuous services. The software should be in local language and easy to understand. Open source software environment should be used for being less prone to software virus attack. The role of service engineers is also likely to be greater in the coming years for handling faults in computers and application software. The use of gram phones as stated above will be useful for increasing ICT usage. High speed communication through broadband policies/ graphic interface/touch screen will go a long way in improving social cohesion, economic vitality and in improving information networks in the society. Creating ICT awareness among common citizens will help in building an efficient interactive model of e-governance.

Sustainable Model of E-Governance

The Department of Information technology in India recently embarked on a programme under its National E-Governance Plan to establish 100,000 tele-centres which are called community service centres. The connectivity will be provided by State Wide Area Network (SWAN) and the content will be provided by various public sector agencies as well as private players. The structure is three tiered one, with the village level entrepreneur at the bottom, a service agency centre managing a cluster of community service centres (for one or more districts), and the state designated agency in charge of providing the requisite policy content and support to the service centre agency. This three-tiered structured model can be adopted by other countries in South Asia. Currently there are around 12,000 to 13,000 tele-centres (community service centres) spread across India. Of these 45 percent to 50 percent are government initiatives or public private partnerships. The remaining ones are private initiatives like the e-Choupal programme [2] run by private commodities trading company ITC.

For sustaining e- governance programs running all over Asia it is important to address issues of accountability (block development officer or panchayats) and revenue generation of the e-governance programs (through Universal Obligation Funds or self sustaining models or combination of both) and content development of services to be provided through net and wireless technologies.

Electronic Governance models will have to be designed to provide governance information which is of "value" for end-users, instead of providing information that can be readily supplied by the governments. Governance information will have more "value" for anyone, when information is:

- Customized: Is useful for anyone's information needs.
- Timely: Updated information is readily available when required.
- Trustworthy: Is usable, and individuals/ institutions can be held accountable for any wrong information provided.

Electronic Governance models should try to reach those sections of the society who have been bypassed and remained out of governance sphere due to high transaction costs of accessing governance information. And yet the need for governance information for these sections of society may be the highest.

Due to high transaction cost of gaining governance information, several sections of the community, for instance, small and landless farmers, urban poor, tribal and backward communities, minorities, unemployed rural youths, often remain outside of governance sphere, and remain unaware of governance information which could be useful for them- for instance, information about government schemes and subsidies, credit and loans availability, employment opportunities, new bills and notifications, etc. When electronic governance models are specifically designed to serve information needs of these marginalized communities, they bring more number of people into governance sphere and thereby increase the "public value" of information being provided by the governments.

IT Training

One area in which South Asian countries need special attention is IT education. [3] The IT sector is labour-intensive and the lack of trained professionals could have deeper impact on the outsourcing business, e-governance & e-commerce programs in these countries. The biggest problem is the lack of quality IT teachers, course content not in tuned with the requirements of outsourcing business and lack of skilled IT professionals and service engineers. To counteract this, creation of convergence institute may be initiated under the aegis of SAARC, which will take creative young people who have just finished school and college, and train them using a highly hands-on approach. They will get a strong grounding in computer programming, web designing, service engineering, animation and other computer and management skills that are highly in demand. Recognizing that information technology is not just computers, the basic training will include the electronic and print media as well. Students will be encouraged to start their own radio, TV stations, community service stations, producing programmes that could be aired through terrestrial and satellite broadcasting channels, as well as through the Internet to cater to needs and requirements of general public. Practical Training in ICT education for developing extensive pool of trained manpower at all levels is needed to meet local and export requirements. Block Development Officers should be trained well so that they become efficient in constantly apprising the public of government policies and programs available for use using ICT technology.

In this regard, the Indian Government set up a Task Force on Human Resource Development for the IT sector, whose main objective was to prepare a long-term strategy to increase the number of trained IT professionals. In addition, initiatives by the All India Council of Technical Education (AICTE) include industry-institute interactions through collaboration with industry associations, and revision of curriculum to ensure quality and industry relevance. The initiatives taken by the University Grants Commission (UGC) include establishing digital repository of research and training material, information, communication and computer education, teaching innovations and career-oriented education programmes. Indian Institutes of Information Technology (IIITs) have also been set up by the Central and State governments together with the private sector to provide qualified IT professionals to the industry and more such institutes and central universities are likely to be set up in the coming time (Manmohan Singh, 2007).

South Asia's relatively unsafe e-security environment is not only costing the BPO/ITES Business Process Outsourcing/IT Enabled Services industry but is also becoming a national security risk with accessibility of broadband ICTs by terrorist groups. The national legislations need to crucially define cyber harassment, phishing, cyber terrorism and cyber stalking to take care of cyber crimes. Provide incentives for network owners, cyber cafes and operators to implement security measures and compensate them for adopting such practices. IT/BPO companies need to invest in upgrading security measures for sustaining competitiveness. Necessary cyber training of government officials should be mandated under the aegis of the SAARC. Qualitative progress cannot be made without enacting comprehensive data protection legislation. National legislations though should not intrude the privacy of individuals through harsh regulations.

Export Success Model

The fact of the matter is that in case of India most of export of services from India is consumed abroad in the US and the UK (and diversifying) and this has an impact on technical efficiency of the firms (Mathur, 2007). This phenomena is somewhat different from what Porter (1990) had hypothesized about the reasons why firms move up the ladder in terms of its competitiveness. Porter (1990) had argued that local demand conditions, higher intellectual property levels and R&D intensity levels and domestic rivalry promote innovation and competitiveness of firms. This is not evidently clear from the Indian IT success story. In a general sense, the software export success model has proven useful as a way of understanding the experiences of developing and transitional economies. It offers a template against which to analyze national strengths and weaknesses. It also offers some more general guidance for countries seeking to increase their software exports

Cluster Development

Universities and R& D labs (both government and private) in conjunction with Government and venture capital funds can provide seed money for start up firms who would be ingrained in a cluster where sustained regional development, network building, skill transfers and innovation can take place with efficient knowledge exchange.

As a start, focus can be on 5 main areas that lay foundation for a vibrant IT economy

- **E-infrastructure:** Success will depend on the links between the main city hubs. The focus is to strengthening such links through the development of Trust Marks and Public Key Infrastructure, which could support the greater use of online transactions and facilitate e-commerce.
- **E-Markets:** The focus would be on strengthening business relationships between the IT industries of different cities through the promotion of internet based facilitators, such as e-business inter-operability standards. For example, the Pan Asian E-Commerce Alliance is an industry alliance among private companies from Korea, China, Taiwan, Singapore and Japan. They are collaborating in areas such as secure cross border transaction services, Mutual Recognition of Public Key Infrastructure, and the creation of pan-Asian portal focusing on trader directories.
- **E-Capital:** This focuses on free investment flows to give entrepreneurs greater access to funding support from across South countries. It may be feasible to create incubation centres funded by venture capitals for innovative start ups through the South countries. Thus, for example, Temasek Capital's In Cubate Programme in China and India provides start ups with seed funding, facilities, management services, branding and network support.
- **E-Talent:** The focus here is on facilitating of cross border IT talent movement, such as mutual recognition of IT qualifications, in order to facilitate free flow of talent and develop human resources of South countries. One example is cross certification of IT skills.
- **E-Ideas:** Leveraging on South's large talent pool and cultural diversity, the focus can be on developing South Asian Countries as a pool for new ideas. An example would be where industry associations form networks to stimulate progress in areas such as ASP/IDC, wireless applications and e-learning

India can learn from the Chinese experiences from their software and hardware industry. Since the middle of the 1990s, China's software industry has achieved advancements as demonstrated in its world highest increase in number of papers on software in the international journals and IT software spending. At the same time, rapid development in China's software industry induced dramatic increase in software outsourcing from Japan to China, which in turn stimulated further development of China's software industry. Lessons to be learnt are, focus on developing application software and exporting hardware. At another level, the strength of Chinese policy towards research and innovation has allowed Chinese companies to compete with the US and other countries on Linux, translation and even security software. India has not really provided such strong research support till recently.

Further, co-operation initiatives need to focus on a series of thematic areas. These include Embedded Systems, e-Applications, e-Government, Research Infrastructures & GRID Systems & Services for Business together with 4G [4] and Communication Network Technologies.

Use of Open Source Software

One major advantage of developing a full spectrum of open source applications is that the flexibility in them can be fully utilized by the user communities. The user communities need not get bound to proprietary software. As source code is available, it is easy to customize the software and modify it according to the needs of the user. The Open components, Open APIs [5] and Open Data standards can be used to develop Open source applications as in bioinformatics.

Since this needs human resources and access to net and computers, developing nations which have the capability of capacity building like India and other nations which have a well developed R &D system to undertake research should explore the possibility of using open source to meet their further needs. Building standards for open source interoperability especially for e-governance programs will be one area for technologists to work on in future.

Many of the South Asian governments does not have a formal policy on Free and open source software for fear of antagonizing the industry as the source code gets transferred to users. However, the governments tend to indirectly favor of Free and Open Source Software (FOSS) by creating websites using Plone (Plone is an open source management system).

South Asian Governments needs to come up with direct policies of investments in open source software and also using FOSS environment in local language which have definite advantage for users of ICT. The latter in turn can improve ICT penetration rates.

Building Norms for Corporate Social Responsibility

Intel the world's largest micro chipmaker has tied up with state governments in India and Infrastructure Leasing and Financial Services, India in a programme to spread computer literacy especially among the youths in the countryside. The company has recently unveiled a portable personal computer designed for school children. Intel will also provide technology support, educational content and wireless connectivity to 100,000 rural community centres. Intel is hoping that children in the villages who experience first hand experience on working with computers will be better acclimatized to take up a job or go into business in future. The same project can be implemented in other South Asian countries.

ITC [6] in India has spearheaded its e-choupals to benefit farmers while getting raw materials for its edible oil business at cheap rates through doing business directly with the farmers. The list of farmers doing effective business is a potential client for other corporate businesses. The list should be shared at common platform for benefit of farmers and corporate businesses so that intermediation in society reduces.

National Employment Database

There seems to be lack of skilled people in IT Industry in India. According to a report on talent pool by NASSCOM-KPMG, 2004 India will have a requirement of 1.12 million qualified workers for the IT software and service areas by 2009 and the estimates of supply indicate the availability of only about 885,000 by then. Also, only 25 percent of the engineering graduates are employable. Identification and selection of talent in this scenario becomes critical. Therefore assessments are important. National registry of entrants need to be made at the level of national association of software companies in the respective countries and this information should be made available to government for planning and implementation of their IT programs. University Grants Commission, university academia along with the national associations need to develop IT course curricula which helps students orient themselves better for jobs in the IT Industry.

Provision for Stable and Quality ICT Infrastructure

Reliable electric power, efficient and inexpensive telecommunications and access to venture capital are essential infrastructures for the IT sector. Although telecommunications infrastructure has vastly improved in South Asia in the last decade or so, there are still regulatory issues relating to the pricing and content of wireless technologies (including mobile, internet telephony, broadband policies) and issue of broadening spectrum and enhancing commercial interest vs national security (defense forces concerns), among others. The latter can be probably be solved through use of encryption technologies which helps in securing information flows.

The situation regarding electric power continues to be abysmal in most of the South Asian countries. In fact, the large Indian IT firms, like other large enterprises, had to invest in their own captive power generation facilities. To the extent that the unit cost of power from small scale captive plants is much higher than what it would have been had their supply come from an efficient large-scale utility, the failure of India's public power system adds an avoidable cost and dampens the competitiveness of its IT sector.

In the Budget 2007-08, Indian Finance Minister had announced taxes on the IT sector- Fringe benefit taxes will be imposed on employees stock option (these may induce employees to demand higher wages), tax on dividend distribution and venture capital and reduced exemptions on operating in the software technology parks. Taxes on venture capital may affect competitiveness of small companies as they try to catch up with their richer counterparts.

A project that could be of interest to the West, South and East Asian region is the setting up of east west information highway in order to have broadband connectivity from Iran/UAE to Vietnam via Pakistan, India, Bangladesh, Myanmar, Thailand, Malaysia and Cambodia. Apart from serving the region, this has further advantage of acting as a standby route from Iran/UAE to Vietnam to the SEA-ME-WE [7] 3 submarine cable and also to forthcoming SEA-ME-WE-4 cable.

Tackling Digital Divide through Setting up of Regional Information Centres

To reduce digital divide, ICT projects: (a) should be developed in local language prioritizing the local needs and content (b) be a model of low cost solution so that poor people can replicate this model or can own or share the system(c) be owned and participated by the community in general (d) be sustainable in long term (e) be able to adopt and utilize innovative ICT means that might be low cost and alternative solutions to access problems (f) be sensible to gender issues such as infant mortality rate, family size, family income, etc has direct relationship with women empowerment (g) be supportive to local and public access points as in rural areas where the divide is the widest ICT solutions need to reach the people through public access points.

IPRs (Intellectual Property Rights), ICTs and South Asian Countries

To facilitate and strengthen the Patent Cooperation Treaty process of the WIPO (World Intellectual Property Organisation) India has launched the e-filing system of Patent and Trademark Applications through its Department of Industrial Policy & Promotion (DIPP). These initiatives would definitely strengthen the Intellectual Property Rights in India (IPRs in India). India is also at an advanced stage of joining the Madrid Protocol on Trade Marks. The Bill for making the necessary changes in the Trade Mark Law of India will be introduced in the forthcoming session of the Parliament. With the launch of e-filing facilities, applicants would now be able to file their Patent and Trademark Applications right from their homes and at their convenience through Internet. The requisite fees or payments could be made through the payment gateway of authorized bankers and this would save time, money and efforts as well as the hassles involved in physical visiting and filing of applications at the offices. Presently, there are about a dozen countries which have e-filing facilities. These countries include U.S., Japan, South Korea, China and the European Patent Office, etc. The modules for managing e-filing facility and on-line processing would be developed by the National Informatics Centre (NIC). Similarly, the payment gateway is managed till now by State Bank of India (SBI).

If the South Asian countries focus their policies more on developing software products and their applications in future then it will require strengthening of IPR regimes to invite foreign direct investments further – a change from the past policies. The trend shows that domestic market growth is now equivalent to growth in export services in India at least in 2007.

Digital Rights Management

Digital Rights Management (DRM) is a term used for technologies that define and enforce parameters of access to digital media or software. DRM is now built in many copyrights law. DRM is used to enforce the copyright of manufacture or the copyright holder. However, DRM may be extra statutory. DRM may allow for copyright holders to restrict access to digital media or software under terms which would be currently permissible under copyright law. Further more, DRM will have significant impact on innovation. An example is the invention of simputer 2 [8] which was build on reverse engineering.

With the introduction of DRM and the possibility of its circumvention low cost locally relevant and contextually appropriate computer software and hardware may never become available to the public at large. If adequate policy response is not given to technology enforced international property restrictions, the internet may loose its egalitarian character. Also DRM is likely to restrict usage of open source software.

Software Patents

Software embedded in hardware may get patented in future. This would mean virtually opening all software to patenting. The South Asian countries need to decide quickly and see to it that atleast the machine language (0 & 1) does not get patents.

Building Synergies for Dealing with E- Waste

The technological advancement has led to very short lives of PCs and laptops (may be five to six years) in terms of speed and other features. This is leading to accumulation of large waste of computer hardware and software. Many developed nations are also dumping these outdated obsolete IT products in developing countries. The developing countries need to guard themselves against such export- import policies and also develop appropriate technology for recycling the e- waste.

Conclusion

Information and Communication Technologies (ICTs) has a ubiquitous presence everywhere and the governments and society needs to keep looking for its better and beneficial use and application in the society. The paper has outlined on how South Asian cooperation can be enhanced by building up on the strengths of ICTs and by pooling resources for dealing with downside impact of ICTs. Further, ICT can promote cooperation and development among developing countries by promoting regional connectivity through setting up of new telecommunication networks, second, by harmonizing technical and regulatory standards in ICT policies (WIPO intervention may be useful), third, by strengthening the capacity for human resource development in the region by pooling resources for appropriate IT training, by facilitating cooperation in ICT applications by creating more awareness of the benefits of ICT, among others.

The low ICT penetration rates in South Asian countries can improve if the telecommunication charges in South Asia falls further. It requires deeper reforms in telecommunication sector (improving competition among producers) by keeping national and regional objectives in mind including decisiveness on regulations pertaining to communication sector. Further, the Governments can influence the IT Industry and ICT penetration by embarking on economic policies aimed at affecting not only supply side but as well demand side factors. Supply side factors consist of laying down telecommunications networks, stable power, access of venture capital funds to small and medium sized companies, human resource development, PC penetration and developing computer keyboards in local languages, internet for all (say where power and net can be accessed at the same time) and improving bandwidth and export- import policies pertaining to communication services. Demand side policies may be fiscal and monetary incentives for easy financing of computer hardware and software by individuals.

Despite Moore's law [9] predicting doubling of computing power and speed of computer and Gilder's law [10] predicting doubling of communications power every six months, we do not see major and discernable upsurge of computer and internet usage especially among the rural communities. This may be because of lack of basic knowledge of ICTs [11] and cost of acquiring the computer. Therefore, South Asian governments under the aegis of the SAARC need to focus their energies and pool resources for creating ICT awareness and benefits of using the same among the common citizens, develop human resources (with proper and focused IT training) and also produce low cost ICT technologies for use by everyone. This may be one of the ways to tackle the digital divide. Commensurate with it is the continuous need of vitality in economic growth rates in South Asian region which are correlated with tele- densities. Also, growth of internet in South Asia will be boosted if not only ICT infrastructure but also rural and urban infrastructure improves. For example, the impact of improved ICT access on farm earnings through increased knowledge of market prices will be of no use if there are no roads to carry crops to markets, or no markets because of unreformed agriculture sector. Higher ICT awareness will help facilitate cooperation in the development of ICT applications for the South Asian region.

For South Asia to fully capitalize on the opportunity and sustain its lead in the global IT-ITES space, it needs to focus on six key areas: Enhancing the talent pool advantage – focus on skill development to better leverage the world's largest working population. Second, strengthen the urban and rural infrastructure and continue to emphasize on proactive regulatory reform in telecommunication sector and in facilitating greater ease of doing business. Third, drive a philosophy of operational excellence amongst industry players (across the board) to ensure that South Asia based delivery sustains world-leading benchmarks in performance. Fourth, one needs to catalyze domestic market development with particular attention paid to hardware and software product development. This will help the South Asian IT industry to grow dynamically by establishing more close links with industrial and commercial users. This will alsofoster intensive learning in the area of product development for a large and rapidly growing domestic market. Fifth, reach out telecommunication and communication infrastructure including internet to remotest places for dealing with the digital divide problem. Last but not least, improve e-governance and e-commerce models for the benefit of all.

We have focused on ICT despite its low penetration rates in South Asian countries due to the revolution it has brought in business and society by taking e-governance programs to the poor, through empowering individuals with more information, reducing intermediation in business and society, and by re-organizing in the way institutions tend to work and interact. The ICT usage is bound to increase and the future is bright as lot many people are realizing the importance of internet and wireless technologies. The internet (including the World Wide Web) is one of the most important technologies to affect not only communication but also computerization. However, downside of ICT technology is the use by terrorist for their own means. Appropriate ICT laws with flexibility need to be framed for tackling cyber crimes.

India has been able to achieve so much in terms of exports of IT and ITES services (earning foreign exchange and changing its image abroad) despite low level of per capita income and low ICT penetration. The other countries can well emulate the example by building national innovation systems and institutions (North D.S, 1990) through long term planning and vision. The South Asian countries have realized that the main growth drivers of ICT industry in South Asia are mobility, convergence (between data, text and voice and between mobile and fixed line internet) and IT management and thereby focusing their respective communication policies in the same direction. Further, concrete actions are needed to improve the use of ICT through e-commerce, e-governance, e-democracy and e-tourism.

Specific actions at the national and regional levels may be undertaken to create such an information society which can mainstream ICT for achieving the Millennium Development Goals. [12] Further, since the actions required at the regional level are cross sectoral in nature, India must take the lead in cooperating with organizations like the International Telecommunication Union & Society, the Asia-Pacific tele-community, the United Nations Development Programme, the United Nations Educational, Scientific and Cultural Organization and others. The suggested regional priorities may include preparing a regulatory, legal and security framework, facilitating and promoting a regional broadband network and preparing and developing special programmes for weaker countries.

The core question in context of South Asia is, whether economic growth and globalization together with the ICTs will reduce poverty. Conceptually, ICTs tend to have impact on transparency, voice and accountability and thereby impact poverty defined in broader sense. The effectiveness of ICT in combating poverty will depend on (not exhaustive): (a) complementarities with other local, regional and multilateral development efforts, (b) responding to local community needs and (c)involving stakeholders in application development. ICTs does not bring sustainable development automatically. It may bring economic growth by enhancing upward mobility by empowering individuals with more information, but not to all. Polarization and social exclusion may become worse without targeted policy intervention and accessibility problems may arise as well. It is upto future research to find whether ICTs are positively correlated with economic opportunities, tackle social problems and reduce societal inequalities, contribute to accessibility of amenities, reduce digital divide between regions; between rural and urban areas; between educated and illiterates; between men and women & between young and old and improve quality of life of individuals so that governments are in better positions to formulate regional policies in their respective countries. This will require some coherent policies among South Asian countries like strengthening of regional information networks, councils for business to business contacts among IT entrepreneurs and core research bodies to look into the deeper questions of relating ICT with national and regional developments. It is up to future policy research and regulations to focus on balancing the rising commercial interests and new technology policy of ICT producers with concerns of government and citizens namely national security and market dominance.

Notes

1. Somesh K Mathur, PhD is presently Assistant Professor of Economics, HSS Department, IIT Kanpur. skmathur@iitk.ac.in. This paper was presented at the seminar organized by the Department of Economics, JMI on ICT technologies while the author was on deputation to the RIS (www.ris.org.in) from the Department of Economics, JMI (http://jmi.nic.in). I wish to thank Dr Prabir De, Dr Nagesh Kumar, Professor M.S. Bhatt, Professor Joseph, among others and above all Professor Ashok Parthasarthy for very helpful discussion on ICT and India. The Complete paper was presented at the Conference on ICT technologies at Curtin Business School, Perth, Australia in 2007.
2. E-Choupal is a system of village internet kiosks which provide information, products and services for improving farm productivity, reducing transaction costs and improving farm gate price realization.
3. There are four major IT components from which various roles (employment) emanate; data bases (data base developer/ administrator), front end tools (application developer, e-commerce/ web developer), net working (network engineer), operating services (operating system administrator). Some roles are common across the segments, such as computer support specialists, project managers, system analysts and system designers. An IT professional competency is evaluated at four levels- a) Technical(programming languages, databases/RDBMS, Application/ Web servers, software engineering tools),(b) Academic (reading, writing, mathematics, communication, analytical thinking, active learning), (c) workplace (teamwork, creative thinking, problem solving, planning and organizing) and (d) personal (interpersonal skills, dependability and willingness to learn). However, these competencies assume different levels of importance depending on the hierarchy at which a person is working and the nature of the job.

4. 4G is the short term for fourth-generation wireless, the stage of broadband mobile communications that will supercede the third generation (3G). While neither standards bodies nor carriers have concretely defined or agreed upon what exactly 4G will be, it is expected that end-to-end IP and high-quality streaming video will be among 4G's distinguishing features. Fourth generation networks are likely to use a combination of wireless technologies like WiMAX and WiFi. (from Wikipedia).
5. An **application programming interface** (**API**) is a set of declarations of the functions (or procedures) that an operating system, library or service provides to support requests made by computer programs.
6. ITC's Agri Business Division, one of India's largest exporters of agricultural commodities, has conceived e-Choupal(an e-governance program for the benefit of farmers in India) as a more efficient supply chain aimed at delivering value to its customers around the world on a sustainable basis. The e-Choupal model has been specifically designed to tackle the challenges posed by the unique features of Indian agriculture, characterised by fragmented farms, weak infrastructure and the involvement of numerous intermediaries, among others.
7. **SEA-ME-WE 3** or South-East Asia - Middle East - Western Europe 3 is an optical submarine telecommunications cable linking those regions and is the longest in the world, completed in late 2000. It is operated by India's Tata Indicom and 92 other investors from the telecom industry. It was commissioned in March 2000 in India. It is 39,000 km in length and uses Wavelength Division Multiplexing (*WDM*) technology to increase capacity and enhance the quality of the signal, especially over long distances (this cable stretches from North Germany to Australia and Japan (definition from Wikepedia).
8. Simputer 2 (advanced version) is a low price pocket size computer intended to be simple to use, uses multilingual software and is linux based (open source) developed by the Indian Institute of Science, Bangalore .The computer has applications for education and literacy. Other applications are in the field of health and telemedicine; micro banking; police work (FIRs, information retrieval), land records, meter reading, e- governance and ticket collection. With falling prices of PCs it may have eroded some of the early advantages of using the Simputer. The latter development re-enforces the need of government interventions on supply side (providing more computers) but also interventions through demand side policies like monetary and fiscal incentives which makes finance easily available to buy cheap computers.
9. **Moore's law** describes an important trend in the history of computer hardware: that the number of transistors that can be inexpensively placed on an integrated circuit is increasing exponentially, doubling approximately every two year The observation was first made by Intel co-founder Gordon E. Moore in a 1965 paper. The trend has continued for more than half a century and is not expected to stop for another decade at least and perhaps much longer. Almost every measure of the capabilities of digital electronic devices is linked to Moore's law: processing speed, memory capacity, even the resolution of digital cameras. All of these are improving at (roughly) exponential rates as well. This has dramatically increased the *usefulness* of digital electronics in nearly every segment of the world economy. Moore's law describes this driving force of technological and social change in the late 20th and early 21st centuries (definition from Wikipedia).
10. An assertion by George Gilder, visionary author of Telecosm, which states that "bandwidth grows at least three times faster than computer power." This means that if computer power

doubles every eighteen months (per Moore's Law), then communications power doubles every six months.

11. The typical internet user worldwide is male with university education, high income earning, urban based and English speaking member of society (UNDP, 1999).
12. The eight Millennium Development Goals (MDGs) an agreement done under the aegis of the UN – which range from halving extreme poverty to halting the spread of HIV/AIDS and providing universal primary education, all by the target date of 2015 – form a blueprint agreed to by all the world's countries and all the world's leading development institutions. They have galvanized unprecedented efforts to meet the needs of the world's poorest.

References

Bhattacharya, M (2004), "Telecom Sector in India: Vision 2020", in *India Vision 2020*, The Report of the Committee on India Vision 2020, Planning Commission, Government of India and Background Papers, Academic Foundation, New Delhi.

Heeks, Richard and Brian Nicholson (2002), "Software Export Success Factors and Strategies in Developing and Transitional Economies", Institute for Development Policies and Management Development Informatics, Working Paper no. 12, University of Manchester, UK.

Heeks, R, (1996), "India's Software Industry: State Policy, Liberalization and Industrial Development", Sage Publications, New Delhi.

Kumar, Nagesh and Ramgopala Agarwala (2006), "NAM More Relevant than Before", The Hindu, September 15, 2006.

Manmohan, Singh (2007), "Independence Day Speech", 15 August from Red Fort, Delhi.

Mathur, S.K(2007), "Indian IT&ICT Industry: a Performance Analysis Using Data Envelopment Analysis & Malmquist Index", Global Economy Journal, May, The Berkeley Electronic Press, US.

NASSCOM KPMG Report (2004), "Choosing a Location for Offshore Operations in India", NASSCOM Website.

North, D.S. (1990), "Institution, Institutional Change, and Economic Performance". Cambridge University, New York.

Porter (1990), "The Competitive Advantage of Nations", The Free Press, New York.

RIS (2006),"Seizing New Opportunities for South -South Cooperation: Some Proposals for the 14 NAM Summit, RIS Policy Brief no 27, August 2006.

UNDP (1999), "Human Development Report", New York.

Wikipedia (http://en.wikipedia.org) for some of the definitions of ICT terms in the text above.

11

E-comfortability and Digital Divide amongst Retail Investors: The Indian Scenario

Prathiba Wasan

Introduction

India is considered to be one of the foremost destination for investments and is well on its way to becoming a dominant player in the world market. It is a time of 9.2 percent growth, booming stock markets and ultra positive sentiment. As per Reserve Bank of India (RBI) the Indian economy grew at 9.2 percent in 2006-2007 as against 9 percent in the previous fiscal. The growth was fuelled by services and manufacturing, which are expected to continue recording double digit growth. The spurt in capital market activities led to a voluminous growth in the transactions for the financial sector services. It may be mentioned that during this period the Sensex touched new heights, signifying a bullish trend in the capital market.

"Psychologically, it keeps India in a different league in terms of investors confidence," Andrew Holland, MD of DSP Merrill Lynch Securities (Hindustan Times, September 2007). In the year 2007, India's stock market capitalization stood at US$ 944 billion (Rs. 38,70,400 crore) just shy of the US$1 trillion mark. This booming growth in the financial service sector or capital markets in the recent years, whether or not has received a positive impact from the Securities and Exchange Board of India's (SEBI) decision to go electronic on January 2001, was the motive behind the conduction of this research.

Since the latter half of the 1990s, the Internet and other innovative information technology (IT) have affected the financial system greatly. Since mid-2000 there has been a correction in public perceptions about Internet-related activities in general, as reflected in the sharp falls in the price of high-tech. stocks and disillusion with the earlier E-commerce euphoria. Use of the Internet for retail purchases lags far behind more traditional channels of distribution, and uptake has been much higher in the United States than in other countries. However, there was less speculative mania surrounding applications of the Internet in the financial industry. Perhaps the area which Internet technology has most transformed is the retail broking market. Costs here have been slashed and customers can now access vastly more information cheaply from the Internet. As a result, e-trading, now accounts for over half of retail stock trades in the United States and a new category of investors - online day traders - has emerged. This has been associated with an unbundling of research, advice and transactions services by brokers. It is a global belief that Internet will fundamentally change retail securities trading.

Witnessing the superior performance of electronic securities market from the overseas, the SEBI approved the report on E-trading brought out by the SEBI committee on Internet based trading and services in January, 2001 (SEBI circular no.SMDRP/POLICY/CIR-06/2000). However, E-trading has not been able to gain popularity with the majority of the retail Indian investors who have indulged into self-exclusion from this trading platform. The Digital Divide between the retail traders is monopolized by the presence of traditional traders who trade via the broker's intervention. This has implications that, the market participants, unlike global markets, have been unable to exploit the value created by E-trading technologies. The reasons for this tilted Digital Divide are attributed to the psychological concept of E-comfortability. Indian investors are not much E-comfortable in using electronic technologies for the purpose of making transactions. E-trading in Indian markets are found, not be in pace with the global trends and therefore, the market participants need to put in efforts to improve the situation, by providing a secure, stable and reliable infrastructural support for trading over the Internet.

Table 11.1: Facts and Figures of E-trading in India

Year	Enabled Members	Registered Clients	Trading Volume (Rs. crore)	Trading Volume (US$ million)	Percent of Total Trading Volume
1999-00	3	-	-	-	-
2000-01	68	1,23,578	7,287.81	1,562.57	-
2001-02	82	2,31,899	8,138.81	1,667.79	1.59
2002-03	80	3,46,420	15,360.76	8,233.84	2.48
2003-04	70	4,63,560	37,945.08	8,745.12	3.45
2004-05	78	8,49,696	81,033.81	18,522.01	7.11
2005-06	142	14,43,291	1,83,428.52	41,118.25	11.68
2006-07	74	22,48,711	2,32,856.90	53,419.80	11.97

Source: NSE Fact Book, 2007

The contribution of Internet trading to total trading volume has been growing at a very small pace. In a span of six years post introduction of internet trading, the contribution has reached merely 11.97 of the total trading volume. In comparison markets like South Korea where internet trading started in 1997, the contribution to the total trading volume was as high as 50 percent in 2001 *(Annual Report, Year 2001)*, the Korea Securities Dealers Association (KSDA). US had an online trading penetration of nearly 40 percent (Yiu, 2000) in a span of six years. In such a case it would be hard to contend that in India all stock transactions may be made in cyberspace in near future.

The number of e-accounts has been increasing year by year since the introduction of internet trading. But as it can be inferred from the graph, the rate of increase in e-accounts has been decreasing from early 2004 onwards till the latest date for which the data is available.

Review of the Literature

In general, E-trading creates value for the market participants. The way stocks were traded in the past was very institutionally driven, but now individuals may also control the prices of the stocks, because E-trading contributes to market efficiency. A capital market is said to be efficient if it fully and correctly reflects all relevant information in determining security prices. Most empirical studies show that E-systems are more efficient than traditional trading venues. The automated market absorbs information much more quickly than the floor (Grunbichler et al., 1994). This makes the market more efficient. Market Efficiency is important because lack of confidence in market efficiency tends to induce investors and fund raisers to focus their attention on exploiting the inefficiencies and walk away from a more positive recognition of the message contained in the market prices (Reilly and Brown, 2000). E-trading empowers the individual investor with more pre-trade and post-trade information and as a result, the information gap between institutional investor and individual investor has narrowed considerably. Conversely, they can operate with minimal information leakage in a manner that trading based on personal contact could not achieve. E-trading system can more readily meet different user's preferences regarding information, some of which were once too complex to put into practice. For example, the basic demand for anonymous trading is now met through electronic systems. Therefore E-trading increases market transparency. The major motivation for individual investors to trade over Internet directly, comes from the fact that one can make quick money by taking advantages of minute to minute price fluctuations in some of the most volatile stocks (Jonathan, 2001). E-trading therefore promotes day trading. It reduces brokerage cost as well as investor's search and transaction costs. Thus, E-trading is more cost-efficient than traditional broker based trading, and will contribute to higher trading activities in the long run. Also, spreads are smaller, or at least no larger, in the automated market than in the floor market (Pirrong, 1996). The high trading activity and narrowing spreads because of E-trading, increases market liquidity.

E-trading is also cost efficient for the brokers. It does away with intermediate manual intervention so that not only are the overhead costs for back office reduced, but the risk of errors in trade reporting and record keeping is also minimized (Bank for International Settlements, 2001).Technological innovations permit the financial intermediaries to reach out into the market place and provide these services at a lower price than building the offering themselves. In the long run, such efficiencies and cost reduction will translate into a sustainable competitive advantage for the Broker.

Despite the value proposition of E-trading, Most researchers and practitioners believe that disintermediation is unlikely to occur and financial intermediation is still essential in the age of Internet (Sato et al. 2001; Beck, 2001). However, it is globally believed that Internet will transform the securities market and one thing appears certain: in general, E-brokerage will continue to grow in popularity.

Impact of E-trading Technologies

Low Barriers to Entry: Internet technology has enhanced the opportunities to harness efficiency gains, better market information, handle higher volumes lessen physical constraints and costs, lower operation costs and switching costs on trading practices and participation. This indicates that the firms find it easier and attractive to enter into E-broking. Also, the increasing confidence of investors in the perceived transparent, convenient and efficient electronic markets has resulted in channeling a greater share of their surplus income in the markets. Realization of this acts as an additional motivator for the brokers to enter into E-broking.

The most important difference between E-brokers and traditional brokers is that E-brokers lack a physical presence (including the branch office and financial consultants) and the associated cost of development such as building costs, overhead and employee wages. The second key difference is the broker's strategic approach. While established companies have attempted to retain an established customer base, electronic brokers must follow a customer acquisition strategy. This means they must differentiate themselves, whether through low costs or value added services.

Investors' Greater Bargaining Power or Consumer Empowerment: 'Internet technology provides buyers with easier access to information about products and suppliers, thus bolstering buyer's bargaining power' (Porter, 2001). Diminishing information asymmetry is one of the major forces that are driving this growing consumer power. To a large extent, information asymmetry results from buyer's inability to purchase and/or to access appropriate information due to high search costs. The emergence of the electronic marketplace can affect both factors' (Varadarajan and Yadav, 2002). Buyers cannot only learn how to evaluate a product, but may also access the evaluation of other buyers who may already have purchased and used the product. As a consequence of the Internet, consumers have access to 'accurate, recent, unbiased information' and they are readily able to talk to many other consumers (Pitt et al., 2002).

Brokerage firms must be aware of the modern investor who has exposure to the Internet and technology. The investor of today has greater challenge of balancing personal and professional priorities and has less time to select brokerage service providers. The Internet provides access to a much wider potential client base than a physical brick and mortar structure, but also makes it much easier for investors to compare different service providers. Investors have both less time to devote to selection of brokerage service provider and much greater control over the selling process. This can make the cost of attracting and retaining customers become more significant as stiff competition forces E-brokers to offer more innovative and better products and services to win business.

Shifting Competitive Landscape: The competitive landscape can shift rapidly. For example, outcomes of competing but interoperable systems are likely to become more common in securities trading (as seen in, for example, telephone networks). This occurs as network interoperability and linkages between systems contribute to the lessening to the physical hurdles to entering markets, for example removing the disadvantage of requiring separate screens for alternative systems. Moreover, while it is never easy to compete successfully with a dominant system, the effects of electronic markets open up more possibilities for tipping to occur, a prospect reinforced by the lower cost base of electronic networks. Overall, electronic linkages might in principle open up markets to competition and enable more "level playing fields". E-brokers have delivered revolutionary products to the market with E-trading. However, with more new participants coming to the market, it is extremely difficult for any player to differentiate itself from others through distinctitive products. Moreover, E-brokers are also facing increasing competition from the existing traditional full-service players, who can offer more comprehensive services for dependent investors. The inevitable maturing of the consumer market for E-trading will thus force E-brokers to shift gears and move onto non-trading investment business or try to appeal to a wider market.

Consumer's Value: The net impact for a trader is that he now gets a bundle of value-added services rendering him the quality of service and performance and returns that he perceives are more valuable than what he used to get with the traditional system. In a modern theory of financial intermediation, intermediaries should be viewed as independent market parties who create financial products and whose worth to their client is transformation of risk, term, scale, location and liquidity (Scholtens and Wenseveen, 2000). These could include easy and low-cost access to research reports, resource links, tools and analysis, charts and news, customizable portfolio views, and checking and banking services. This medium can help the investor do their own research, and get informed about new initial public offerings (IPOs). It could also provide asset management tools. In general, the Internet makes transaction more apparent with its speed and availability. In this environment, the investors' expectation from the E-brokerages are more demanding than ever before. The most common belief is that E-brokerages must provide hassle-free automated transaction execution along with other products at a reduced price. Those products and services should be available 24-hours a day, seven days a week.

Moreover, with the high automation capability of the medium, the services must be customized to match the taste and styles of each investor. These could include easy and low-cost access to research reports, resource links, tools and analysis, charts and news, customizable portfolio views and checking and banking services. The way stocks were traded in the past was very institutionally driven, but now individuals may also control the prices of the stocks. E-trading empowers the individual investor with more pre-trade and post-trade information and as a result, the information gap between institutional investor and individual investor has narrowed considerably. Direct connectivity to the market enables electronic trader to place trades that can take advantage of even 1/8 price movements. Electronic markets have leveled the investing field for retail investors and also supplemented it with a complete advanced tool kit aiding him to take sound decisions and increase the value of his investment and cut down his search cost as well as cost of investment.

Broker's Value: Finally, the Internet is providing brokerages cost saving in overhead, labor, transaction and customer services. It reduces the cost of providing customer support services by providing a cheaper communication medium (cheaper than telephone support) and lowers transaction cost by eliminating store fronts and improving business processes. Opportunities to streamline payment/receivables processes become a reality, thanks to the rapid development of Internet transaction technologies. With the brokerages growing drastically down, competition getting severe one might wonder what value a broker can get for his investment in E-system. Broker's can attain value in volumes. With the increased liquidity of the market and investors confidence broker's can derive economies of scale and electronic system helps them to handle bulk of volumes very conveniently and relatively at a much lower cost than the traditional system. Also they can maintain a good customer relationship with their customers by providing them value-added services. Internet also may be of a lot of help to brokers in their customer acquisition strategy as it is much easier and cheaper on Internet to reach a vast number of potential customers.

Value Creation by E-Trading: Conceptual Framework

E-trading impacts market quality by improving efficiency, liquidity and transparency in the markets and reducing trading costs. Investors can benefit from the reducing brokerages and search costs which reduce their cost of investment in terms of time as well as money. E-trading has also increased the investor's access to the markets. They can now trade from any place of their choice using a computer. Improved liquidity in an efficient and transparent market would ideally attract traders and cause them to trade frequently, creating further volumes and liquidity in trading. Therefore, markets now get a higher share of the investor's investible income. On the other hand, E-trading provides convenience of operations and reduces the cost of operations for the brokerage houses. An increased volume of trading lets them to earn economies of scale and compensates for the reduction in brokerages. Increased volumes, and lower switching and operations cost, allures new entrants in the E-broking industry. As the entrants increase, it will result into greater competition in the market between substitute products and services and will also empower the investor with more choices and greater bargaining potential. This will further make the competition fierce and the survival of the firms will demand them to explore innovative business capabilities and come up with a strategy which is strong enough to combat competition, retain the existing customers and acquire new customers. Therefore, e-technologies create value for the market which results in the creation of value for the investor as well as the broker.

Digital Divide

If Digital Divide refers to the perceived gap between those who have access to the latest information technologies and those who do not, then in case of E-trading digital divide is increasingly being reduced with more and more brokers providing latest infrastructure for their clients to trade on. But if users often give up in frustration and confusion, do not take up at all for lack of need felt or due to their various perceptions about it, then merely having access does not mean that a digital divide has been solved because a divide remains in their capacity for effective use of the Internet. Access to the Internet is not universally desired as some people do not inevitably find the Internet something they wish to use. Results from a number of studies from different countries indicate that substantial numbers of people are not using the Internet because they feel they have no need for it rather than perceived lack of access. In the year 2000, US Pew Internet study, 51 percent of people felt that they were not missing anything by not having access to the Internet and around 37 percent of the study conducted in Texas in the same year said they were not interested in Internet use (Strover, 2000). When a Japanese survey (Kubota et al. 2001) asked about disadvantages and bad experiences due to not using the Internet, over 80 percent of non-users did not actually feel themselves to be disadvantaged in not having access on a number of issues, despite the concerns from government, etc. about social exclusion.

Following telecom liberalization, the Government of India (GoI) allowed private sector organizations to provide Internet service to end users. Though Internet was introduced in 1995, it took about half a decade to reach individuals at an affordable price and henceforth it is the most popular channel of communication. Internet serves three roles to its users namely: i) Internet as an information channel; ii) Internet as a communication channel and iii) Internet as payment and delivery channel. Generally speaking online users in India use Internet vastly as an information and communication channel. Therefore, e-mailing, surfing for news updates, searching for articles or any other information of interest, online chatting and discussions etc. find the most common use amongst Indian online users. More and more people have started using the web for communication and information retrieval. However, when it comes to Internet as a payment and delivery channel, they become highly skeptical of its capabilities and reliability. E-trading was introduced in India in January, 2001.We investigate whether the same story stands true for retail traders in India. For this we have considered the definition of Digital Divide, in the context of securities market, from a different and an important perspective i.e. the divide between retail traders who are actually using the access to the E-trading facility for trading and those who are not.

E-comfortability

Behavioral studies suggest that those who derive less comfort in any of their deed are less likely to perform related behaviors in the future. There is evidence that non- use or restriction of Internet use may also be related to lack of comfort with technology in general and the Internet in particular. Lack of comfort may be through unfamiliarity but also through concerns related to worries about privacy, security, computer viruses, protection of children and quality of Internet information (Schauer, 2002). Although these concerns may not stop people from ever using the Internet they contribute to restricted use, for example, privacy and security apprehensions have been identified as key barriers to the use of e-commerce. Removing the access barrier merely fully exposes the other barriers to use. The fundamental disparity is that the Internet makes different cognitive demands on the user compared to other communication technologies. It is not as intuitive to learn as the television or the telephone, especially since "these media can be adequately operated with cultural techniques that are learned in childhood and used daily. When the user sits in front of a computer nothing is clear on the basis of his everyday experience" (Kubicek, 2000). Thus, for many the Internet may present as an unfamiliar experience behind which is a complex technology that appears to have almost magical powers. Moreover, the cognitive demands needed to search, navigate and interpret written information are novel skills for many users.

E-comfortability is the belief in one's capabilities to organize and execute actions on Internet leading to a positive attitude for the use of Internet required to produce profitable trades, is a potentially important factor in efforts to close the digital divide that separates E-traders from traditional traders. *E-comfortability* is not a measure of skill, it is a psychological concept that influences decisions about what behaviors to undertake, the amount of effort and persistence put forth when faced with obstacles, and finally, the mastery of the behavior. E-comfortability is a strong positive reinforcer for behaviors related to Internet use. Lack of human element in the system, security apprehensions, complexity, knowledge barriers to initial Internet adoption, lack of reliability on the performance of the medium, negative attitude towards change, reluctance to learning by the traders may be construed as E-comfortability deficits. Those with low E-comfortability opt for self-exclusion and should be less likely to adopt and use the Internet for trading than those with high degrees of E-comfortability. It is found that the E-comfortability variables are moderated by age factor. Traders in high age category are found to be less E-comfortable than those in the young or early middle age group.

Research Findings

The paper explores whether E-trading in India is on its way to evolve along the global lines and whether the investors are actually seeking the benefits of technology, by opting to trade over the Internet. It investigates the digital divide existing between the traders and the causes for the divide.

The findings depicted here are based on telephonic interviews with 40 brokers located in Delhi. The sample was randomly selected from the list of brokers registered with SEBI and available on SEBI website (www.sebi.gov.in). The questions posed to the brokers were structured and close ended. This research was taken up as a preliminary investigation to formulate a valid research problem and conduct a more comprehensive study on a wider sample of respondents. Therefore the findings are taken to be mere guiding elements for the researcher in formulating her research program.

We find that the divide exists in the Indian securities market and is highly skewed with a very large majority choosing to be on the ignorant side of the divide. In fact in our interviews with the brokers it was found that the number of clients choosing to trade traditionally without the use of Internet is so very high (90-95 percent) that it would be unjustified to call it a divide. Given the facility, clients voluntarily choose to remain non-digital. They might surf the Web for price updates, news on stock or receive a price alert via Internet, but when it comes to placing orders and executing them, they revert back to their conventional mode of trading, which involves the use of a telephone but not Internet. The results of the study are presented below:

1. Only 35 percent of the interviewed brokers were found to be providing E-trading platform to their clients.
2. According to 78 percent of respondents E-clients for direct and active trading as a percentage of total number of clients was found to be within a meager range of 1 -2 percent. The remaining 23 percent did not find it popular at all.
3. The most common age group of traders was 35-55 years.
4. 69 percent of respondents felt that E-trading clients fall in the age category of 25-35 years.
5. Approximately 9 percent of the respondents felt that traders are users of Internet for the purpose of seeking trade related information.
6. Given the situation of withdrawal of physical presence in the market majority of the respondents felt that the loss of clients would be more than 80 percent.
7. Given the situation of withdrawal of E-trading facilities, the loss of clients predicted by majority of the brokers was marginal, falling in the range of 1-10 percent.
8. About 80 percent of the respondents felt that 90 percent of their e-clients use the online facility for the purpose of making long term investment rather than active trading.
9. About 90-95 percent of active trading clients prefer coming to the broker's physical office and trade for the day.
10. The respondents were sure that the Indian investors are either low or poor on the scale of E-comfortability for the use of Internet as a payment and delivery channel.

11. The E-comfortability deficits (as revealed by the research and in the sequence of descending importance) responsible for the skewed digital divide amongst traders are listed below:

- Security apprehensions regarding Internet.
- Lack of reliability on Internet performance in terms of speed and connectivity.
- The clients already have a lucrative model based on human relationship (Screen-based trading), why bother to change it?
- Client's perception that broker's intervention provides more flexibility and market intelligence than do computer screens for the purpose of investing and trading in securities.
- Traders have fear of change.
- Lack of technical knowledge/awareness in the area of computers and Internet.

Also the study revealed that 90 percent of the interviewed brokers are not actively involved in promoting direct E-trading by their clients and 40 percent of them feel that their promotion may help in clients taking up E-trading. 55 percent of brokers preferred their clients to be trading with their manual involvement.

Analyzing the E-comfortability Deficits for Reasoning the Digital Divide amongst Traders

Security Apprehensions: Security apprehensions have emerged out to be the prime barrier to limiting trading online for Indian investors. Investors are apprehensive of using Internet as a payment and delivery channel. This probably explains the finding, that while, 70 percent of the clients are actively seeking trade related information on Internet, E-traders fall in the range of 1 -5 percent. *"on the Internet any hacker could get into our data*" –this common perception of the traders about Internet has to be converted into the perception of Internet as a system which is transparent, giving the trader control of the entire trading process and with magical powers 'protecting' the user, ensuring their privacy and security, so that they can feel safer trading over the Internet. While concerns about security can reasonably be predicted to be allayed by advances in secure technology, investments needs to be made by the brokers in acquiring secure technologies. If one is waiting for a justifiable noumber of E-clients, before making an investment in security installations, the hope for a better situation is hard to materialize.

Complexity, Technical Knowledge and Learning Attitude: The NASSCOM survey provides detailed demographics of Internet users in India. According to it, 36 percent of the world's population accesses the Internet today and India, with 3.7 million users, accounts for a mere 0.37 percent. Of the 3.7 million users in India 49 percent belonged to the 18-24 age group, followed by 28 percent in the 25-39 year age group. People above the age of 40 constitute only 11 percent of the total Net population (http://www.india-today.com/ctoday/20000901/net.html). The average age of the head (and financial decision taker) of the Indian equity-investor household, as revealed by the SEBI-NCAER study of Indian investors in 2000 is 45 years (www.sebi.gov.in).The older, experienced equity investor is not online today and the fact that older, mature investors are not 'tech-positive' and hence unlikely to move to online trading is a major barrier to the growth of e-broking in India.

Behaviorism and Cognitivism view knowledge as external to the learner and the learning process as the act of internalizing knowledge. The majority of traders are active on Internet for the purpose other than transacting and 90 percent of the e-traders do not use Internet for active trading but for making long term investments, implying that, the traders internalization of the Internet learning has been partial and restricted to the use of Internet as an information channel only. Adoption literature highlight that demographics are related to new media adoption and usage patterns (Dutton, et al. 1987; Atkin and LaRose, 1994). Studies have found that this is true of computer adopters (Dickerson and Gentry 1983; Lin 1998). Trading in securities exists from a time when learning was not impacted through technology. Over the last twenty years, technology has reorganized how we live, how we communicate, and how we learn. The research revealed that the majority traders are in the age group of 45-55 years. Behavioral studies indicate that with increase in age the learning tends to get slow and is avoided if possible Learners as little as forty years ago would complete the required schooling and enter a career that would often last a lifetime. Information development was slow. The life of knowledge was measured in decades. Today, these foundational principles have been altered. Knowledge is growing exponentially. Traders in their late age are likely to find this new learning more messy and complex than the younger ones. Greater the complexity of knowledge more is the reluctance to learn and use. This is probably why the age group for e-traders is found to be 25-30 years.

Fear of Something "Hidden" or Resistance to Change: One of the reasons for the usage of internet trading is the mental acceptability factor. Chaos is a new reality for knowledge workers. Chaos is the breakdown of predictability, evidenced in complicated arrangements that initially defy order. Chaos states that the meaning exists – the Trader's challenge is to recognize the patterns which appear to be hidden. Meaning-making and forming connections between specialized communities are important activities. It highlights a real challenge: "sensitive dependence on initial conditions" profoundly impacts what we learn and how we act based on our learning. Decision making is indicative of this. If the underlying conditions used to make decisions change, the decision itself is no longer as correct as it was at the time it was made. The ability to recognize and adjust to pattern shifts is a key learning task that traders must know how to perform.

Internet is full of information. The ability to draw distinctions between important and unimportant information is vital. The ability to recognize meaning lying behind every piece of information is also critical. For instance, investing decisions taken on the basis of online expert advice or messages posted on the chat boards, could land up into losses, as online expert cannot individualize its advice and messages can have hidden meanings/ intentions. The traders who possess or build up such skill may enter the comfort zone needed to trade independently over Internet else, already having a lucrative model based on human relationship (Screen-based trading) they would resist any change in their trading practices on account of their fear of "something hidden" which, they are not confident of making out. Another major issue has been awareness. Not only is consumer awareness at a very low level, but even brokers have not been forthcoming in accepting Internet trading. In markets like the US, online brokerages are advertised very heavily. Online trading in India has so far not seen similar levels of aggressive advertising, with the exception of ICICI Direct and Home Trade. Unware customer is definitely less likely to use internet for making monetary transactions.

Also there is no clear incentive offered to the trader to induce him to trade over internet. There is no difference between the conventional brokerage charged and e-brokerage as revealed by majority of the respondents. The brokerages in India are already very low and ranges from 30 paisa to 70 paisa. Therefore the scope for expansion of the market through reduction in brokerage is not a possibility. A reason for the popularity of online trading in the US was that a lot of people took to day trading. In India, speculative trading is already rampant and it is estimated that delivery is taken for only 25-28 percent of the total traded volume. Hence, the surge in volumes due to day trading may not be of the same magnitude.

Lack of Reliability on the Medium Performance: However, even technology bottlenecks are responsible for online trading having not taken off the ground. The usual suspect—poor bandwidth—figures prominently as one of the main reasons why online broking hasn't taken off. Though players like Reliance and Bharti have announced broadband services, till the last-mile problem is solved, online broking will suffer from the bandwidth problem. The three main technology obstacles which have prevented Internet broking from taking off are lack of Internet penetration, bandwidth infrastructure and poor quality of Internet

Service Provider (ISP) infrastructure. Though the Internet use has increased by 1000 percent from 2000 to 20007 the penetration is as low as 5.30 percent (http://www.internetworldstats.com/stats3.htm). Most investors are not sure of executing a trade online. This particular E-comfortability deficit demands special address. Research reveals that about 90-95 percent of active traders' trade from physical office of their brokers. The very advantage of direct connectivity for day trader lies in exploiting minute to minute price fluctuations and cutting down on possible losses by moving out as quickly as possible. This clearly visible advantage has led to the conversion of a number of part time traders in U.S.A, into full-time day traders. However, in emerging markets like India, the required reliable medium support for active trading is clearly lacking. The Internet speed is highly erratic and there is daily power failure for hours hindering the connectivity. So when a trader is typing a buy order for 100 shares priced at Rs.100 and Internet gets disconnected on account of power failure or the speed slows down, he couldn't be sure that on the restoration of connectivity or at the end of order execution, the price would have been Rs. 100 and not Rs. 101. There is no failsafe method for closing a trade if a customer's connection cuts out midway For a small investor this minor fluctuation in price because of medium delays may not value much, but for a volume trader this loss is appreciable. This probably explains why investors on the digitally positive side of the divide restrict the use of this electronic platform for the purpose of making long term investments and not for day trading. Besides, Internet penetration in India is still very low. The system reliability will have to be developed in terms of connectivity and speed. The service providers and Government will have to address the concerns related to Internet performance and availability so as to build up E-comfortability amongst the potential users of Internet for active trading.

Internet-based trading typically takes place through a bank account with an online bank. Here, the number of banks with a strong online presence is very few-again, dominated by new private banks and foreign banks. Both have lesser reach owing to a smaller network in the country. The relative inability of large public sector banks to offer facilities for Internet banking is a barrier in this regard.

Human Element: One common fact on which even online brokers agree on is that a pure play online broking firm cannot succeed on its own. A physical presence is crucial for any brokerage firm in India. Home Trade, for instance, provides a trading platform that offers the flexibility to transact either online or through the more conventional telephone channel. Both modes have a world-class service offering built around them, and are backed by a state-of-the-art technology backbone and centralised call centre. "Online investors prefer multi-channel delivery, even if they intend to execute most trades over the Web." Multi-channel delivery provides both flexibility and security. Clients can opt to call to execute trades when they are unable to access the Web.

Therefore, one of the most important obstacles in the popularity of E-trading is the lack of human relationships. Country like India, where emotional and human ties are considered to be more reliable and comfort giving than a machine, no doubts the findings that 90-95 percent traders are more comfortable with the traditional trading method which involves the manual intervention of the broker and the use of Internet is restricted to the role of a support element to the entire trading process by the traders who are active on Internet. The comfort of voice and personal acquaintance increases their confidence needed to put their stakes in trading. Brokers will have to make efforts to create human touch for the E-traders who trade via their websites. The websites should be made highly interactive and informational. The look of the website should be appealing and the content simple and straight. The user guidance should be readily available online. Online chat boards, discussion forum, expert advice etc. should be included on an elaborate scale. The websites should be designed to suit the specific needs of the user. Physical offices should act as tool to support the use of Internet.

Brokers should use their voice and personal relations to build up and promote E-comfortability amongst their clients. They should not forget that only when they close this digital divide, would they be able to reap the actual benefits (lying in volume trading over the medium) from their investment in Internet technology.

Conclusion

E-trading in India is not evolving along the global lines. Though Internet is popular amongst seekers of trade information and finds minor use with long-term investors, it is least popular amongst day traders. The electronic markets of India have taken a deviation from the rest of the world. There is a huge digital divide and the restricted or no use of Internet for the purpose of transactions is self-imposed by the traders. This self-exclusion is resulting from poor scores on E-comfortability. If the E-comfortability deficits are overcome, one can hope for a better prospect for E-trading in India and this would require the active involvement of all the market participants including the investors. All hope is not lost as players like Reliance and Bharti have started laying optic fibre across the length and breadth of the country and in a matter of years, the last mile connectivity issue will be solved. The solution lies in balancing a technology-centric approach to transactions with the human touch factor which will be a key factor in making the transition from the traditional method of broking to e-broking. Brand building, assurances of security, developing multiple delivery channels with anytime telephonic grievance redressal options are some directions which may be of use for the immediate future. Online trading firms can also market themselves aggressively to students who are entering the professional arena, ensuring that their entry into equity happens online. The bricks-and-clicks networks being developed by players such as ICICI Bank and HDFC Bank should help.

The results arrived at, in this paper, are based on a small sample survey and is, naturally, only indicative at this stage. Rather than providing generalisable conclusions, it offers both insights into an electronic retail securities market as well as a vantage point from which further, broader investigations can begin. Therefore, there is a need to take up further research on a more elaborate scale and on a wider sample to challenge or validate the findings of this paper. Findings of this research can be taken as guiding elements for further research in this area.

References

Atkin, D.J., Jeffres L.W. and Neuendorf K.A. (1998), "Understanding Internet Adoption as Telecommunications Behavior", *Journal of Broadcasting & Electronic Media* (Fall).

Atkin, D. and LaRose R. (1994), "An Analysis of the Information Services Adoption Literature" in J. Hanson (Ed) *Advances in Telematics*..New York, Ablex. 2:91-110.

Jonathan, R.A. (2001), "The New Electronic Trader -Understanding what it takes to make the Jump to E-trading", McGraw Hill, USA.

Bank For International Settlements (2001), "The Implication of E-trading In Financial Market", http//www.bis.org/pub/cgfs16.pdf.

Beck, H. (2001), "Banking is Essential, Banks are not. The Future of Financial Intermediation in the Age of the Internet", Netnomics, vol 3, 2-22.

Boyd, J. and Prescott E. (1986), "Financial Intermediation Coalitions", Journal of Economic Theory, vol 38.

Chitnus A. (1995), "Internet in India. View from the Ground" http://ece.iisc.ernet.in/monster/atul/indianet/index.html.

Dickerson, M. D. and Gentry J.W. (1983), "Characteristics of Adopters and Non-adopters of Home Computers", *Journal of Consumer Research,* 10: 225-235.

DiMaggio, Paul, and Hargittai. (2001), "From the 'Digital divide' to 'Digital inequality': Studying Internet use as Penetration Increases". Princeton: Center for Arts and Cultural Policy Studies, Woodrow Wilson School, Princeton University.

Dutton, W., Rogers and Jun U.H. (1987), "Diffusion and Social Impacts of Personal Computers", *Communication Research,* 14 : 219-250.

Gonzalez C. (2004); *The Role of Blended Learning in the World of Technology.* http://www.unt.edu/benchmarks/archives/2004/september04/eis.htm.

Grunbichler A., Longstaff F.A. and Schwartz E.S. (1994), "Electronic Screen Trading and the Transmission of Information: An Empirical Examination", Journal of Financial Intermediation 3 (1): pp. 166-87.

Kotler, P., Jain D.C. and Masincee S. (2002) "Marketing Moves: a New Approach to Profits, Growth and Renewal", Boston. MA: Harvard Business school Press.

Kubicek, H. (2000), "The Digital Gap: A Challenge for Local Communities", http://www.mediakomm.net/documents/kongress/eslingen/kubicek_en.pdf

Kubota, F., Hashimoto, Y., Mikami, S., and Yoshii, H. (2001), "*Internet Usage Trends in Japan. Survey Report 2000*", First published in Tokyo, Japan, August 2001by the Communications Research Laboratory. http://media.asaka.toyo.ac.jp/wip/survey2000e/report2000e.pdf

Lin, C. A. (1998), "Exploring the Personal Computer Adoption Dynamics", *Journal of Broadcasting &Electronic Media,* 41 (1): pp. 95-112.

Madhavan, A (2001), "In Search of Liquidity in the Internet Era", http://www.frbatlanta.org/news/confrence/efinance00/madavan_final.pdf

Merton, R. and Bodie Z. (1993); "Deposit Insurance Reform, a Functional Approach", in A Meltzer and C Plosser, (eds), Carnegie-Rochester *Conference Series on Public Policy*, Vol 38.

Pirrong, S.C. (1996), "Market Liquidity and Depth on Computerized and Open-Outcry Trading Systems: A Comparison of DTB and LIFFE Bund Contracts." Journal of Futures Markets, 16 (3): pp. 519-43.

Pew Research Center (2000), "*Tracking Online Life: How Women Use the Internet to Cultivate Relationships with Family and Friends",* http://www.pewInternet.org/reports/toc.asp?Report=11

Pitt, L.F., Berthon P.R., Watson R.T. and Zinkhan G.M. (2002); "The Internet and the Birth of Real Consumer Power", Business Horizons, July-August.

Porter, M.E. (1980), "Competitive Strategy", Free Press, New York.

Porter, M.E. (2001), "Strategy and the Internet", Harvard Business Review, pp.63-78.

Reilly, F.K. And Brown K.C. (2000), "Investment Analysis and Portfolio Management", Sixth Edition, the Dryden Press, Chicago.

Sato, S., Hawkins J. and Berensten A. (2001), "E-Finance: Recent Developments and Policy Implications", Forthcoming in *tracking a transformation: E-Commerce and the Terms of Competition in Industries*, Broking Institution.

Scholtens, B. and Wenseveen Van D. (2000), "A Critique on the Theory of Financial Intermediation", Journal of Banking and Finance, Vol 24, pp. 1243-51.

Schauer, T. (2002), "*Internet Refusers: A Risk to the Digital Economy?"* http://www.terra-2000.org/Documents/Prague/Papers/Internet percent20Refusers.pdf.

Shah, A. and Thomas S. (1996); "How Competition and Automation Have Changed the Bombay Stock Exchange", Technical Report, Centre for Monitoring the Indian Economy, Bombay.

Strover, S. (2000), "*Aspects of Internet Use in Texas"*: University of Texas, Austin TX Report no (512) 471-5826; www.utexas.edu/research/tipi/Reports/PUC.pdf

The Korea Securities Dealers Association (KSDA) *(2001).*, "*Annual Report (Year2001)"*, South Korea.

Walid, M. (1998), "*Opening Digital Markets: Battle Plans and Business Strategies for Internet commerce", McGraw Hill, New York.*

www.bseindia.com

www.internetworldstats.com

www.nasscom.in

www.nseindia.com

www.sebi.gov.in

Website of Korea Stock Exchange (*http://www.kse.or.kr/*).

Website of Korea Network Information Centre (http://stat.nic.or.kr/).

Website of KOSCOM (*http://www.koscom.co.kr/*).

Varadarajan, R. and Yadav, M. (2002), "Marketing Strategy and the Internet: an Organizing Framework", Journal of the Academy of Marketing Science, Vol. 30, No.4, pp. 296-312.

Yiu E.(2001) "Online Broking Tally Sets Fast Growth Pace", South China Morning Post, 2 January.

12

IT Solutions for Energy Market

Vrushal T. Ghoble

Be it the freezing Russia, the deserts of the Middle East or the deep waters of the Gulf of Mexico, these are only the geographical challenges faced by the world oil and gas industry. But other challenges namely technology is equally serious and threatening. Global geo-political forces are creating a highly volatile and rapidly fluctuating crude oil and gas market. Some of the technologies that are available to us today are not being widely used. From the many beneficial technologies that we can identify, those which, if applied, can have an immediate and dominant impact on improving recovery from our existing fields.

Technological progress globally has created enormous benefits for the world's energy consumers. It has made possible, the commercially viable resource base to expand and grow, despite the nearing end of the world's oil production. In the light of repeated energy disruptions leading to the emerging environmental issues, the global energy industry and the techno – scientific community are looking forward to the much globalised Information Technology (IT) sector to find solutions to the ongoing energy security concerns. Reducing the complexity of IT and expediting the benefits of IT transformation is a huge challenge for many companies today. It is essential to first identify those IT applications that are most appropriate for rationalization and then to rationalize those applications on to more appropriate i.e. cost - effective hardware and software environments.

IT Convergence

In a globalised era, the Communications and IT industries are rapidly converging with the energy market. Now there is a very real possibility to deliver business oriented data solutions in the emerging market by integrating Information Technology. The alignment of IT and energy industry started in early 1960s and gradually has become critical to the energy industry's survival. Integration of IT systems is essential in order to support remote locations, different applications and to store the massive amount of data that the industry generates. IT solutions to the oil and gas industry could tremendously enhance the hidden potentials of the emerging market. IT solutions of seismic mapping, deeper onshore and offshore drilling and production techniques, reservoir simulators and other advances have increased the accessibility of oil and natural gas. Not only that, technology has diversified supply sources by allowing more energy to be accessed in more locations around the world. The Saudi Arabia's move to store oil in China can be one of the examples. [1]

Ice resistant and ice breaking innovations, new state of the art platforms and sub sea-drilling technologies have the capacity to open rich reserves. As this energy consumption rises worldwide, a huge wall of carbon and sulphur contents will contribute towards reduction in human and environmental wellbeing.

Shareholders are pressuring companies for a return on their investments that is commensurate with other long-term investment strategies. The present paper is an attempt to study some of the technologies like 'Broadband and Satellite technology', IT solutions providing 'high – performance technical computing (HPTC)', 'storage systems', 'data warehousing solutions', 'Supervisory, Control and Data Acquisition (SCADA) solutions', 'Intelligent oilfield solutions', 'Digital oilfield solutions', 'Distributed Control Systems', 'Drilling Information Systems', etc., thus understanding their benefits for the Indian energy industry. These technologies are useful for the oil and gas companies involved in the upstream and downstream activities. While advanced technology, such as 3D and 4D seismic and broadband technologies can provide vast amounts of information that can help companies to be more successful in exploration and production, there is often too much complex information to assimilate and understand in the time needed to make quick and accurate decisions. Hence, new age technologies are very essential in preventing this damage.

Broadband Satellite Technology: Unleashing the Benefits for the Oil and Gas Industry

Major hydrocarbon companies worldwide have evidenced the cost - saving and revenue - enhancing benefits being realized by major oil and gas companies worldwide with the use of advanced Broadband Satellite Networks. For remote businesses, where severe weather and facilities are a challenge or non existent, Broadband Internet access communications services so that they can conduct operations and easily communicate. With the broadband, we can get connected to both upstream and downstream bandwidths so that you can communicate anywhere you want via 2 way satellite Internet. Broadband system provides the following features for the Oil and Gas Industry: [2]

- Broadband Internet access over Satellite
- Voice, Video and Data over Satellite
- Field – to - office, field – to - field and office – to - field communications

- Flexible plans – only pay for bandwidth when used
- High speed, multi megabit connections
- Quick deploy communications for new site development
- Transfer data from isolated sites and remote locations back to home office
- Extend corporate networks and be completely interoperable with landline networks

The customer today at gas pumps expect to swipe a credit card, have their credit payment validated, and the pump is activated in a matter of seconds, purchase a drink or snack and return to the fueled - up vehicle, ready to drive. And before the customer leaves the site, inventory information from the pumps and the snack bar has already updated appropriate data bases in regional and central offices, ensuring efficient, in time, supply chain management from the field to the pump. Broadband Satellite networks are making this scenario a reality.

What are the Profits? "Surveys estimate that the addition of pay–at–the-pump capability increases gas station sales by 2 percent, while availability of automated machines at retail gas stations increases sales by 4 percent. With just these two capabilities, the monthly net sales uplift can amount to several thousands of dollars, representing up to six times return on the investment made in the technology. The potential of broadband technology to unlock new revenue streams at the point of purchase at retail gas stations presents a significant way to uplift sales and increase profits. The result is increased sales of non-petroleum products, which can often be more profitable than sales at the pump itself". [3]

Russia's Satellite Market

Russia holds an estimated of 25 percent of the world's oil and gas reserves, yet has an inadequate network infrastructure. Hence, Russia presents an ideal environment for the industry to leapover the old technologies and reap the benefits of broadband satellite technology. In the spring of 2004, Yukos, a Russian oil company completed a pilot project that validated its custom-developed software for pump station applications. The project was successfully accomplished with the assistance of IPNet, a Moscow based Internet Service Provider, using Hughes Network Systems technology. Later this year, SETTELCOM, a value-added DIRECWAY service provider, which offers its services under the AltegroSky brand name, is planning to run a test for Yukos for various applications such as oil geological surveys, oil drilling and production, oil and gas processing, SCADA(Supervisory Control and Data Acquisition) and retail gas station applications. The project will comprise of 20 locations (approx.). In 2005, the number of sites is planned to increase to 400, with several production and processing sites to be added as well. Lukoil is also actively implementing broadband satellite solutions via Russat.

Supervisory Control and Data Acquisition (SCADA)

The initial use of satellite technology in the global hydrocarbon industry was for Supervisory, Control and Data Acquisition (SCADA) applications at remote exploration sites and for monitoring pipelines. Every oil well is virtually a factory in itself and needs a 24 hour monitoring of its functioning. SCADA is used to monitor critical infrastructure systems and provide early warning of potential disaster situations. [4] One of the most important aspects of SCADA has been its ability to evolve with the changing face of technology that is now referred to as Information Technology. SCADA is mainly used to monitor and control remote pipelines.

High-performance Technical Computing (HPTC), Storage and Data Systems

HP offers a range of solutions to help address the challenges that oil and gas companies today face. These solutions focus primarily on achieving the following goals: [5]

- Replenishing oil and gas reserves faster;
- Increasing production yield;
- Improving the efficiency and cost structure of downstream operations;
- Enhancing IT agility and return on investment;
- Helping to increase customer loyalty for gas - station Chains.

HP's IT solutions for the global oil and gas industry is designed for large companies that want to leverage IT technologies for the purpose of finds, production activities and distribution of hydrocarbon products. These solutions have the following features - high performance technical computing (HPTC), storage systems, and utility data centers. These systems provide an alert, responsive and cost-effective infrastructure to manage evolving energy business needs.

Solutions for Oil and Gas Industry

Oil and gas companies are now focusing and spending on more complex and difficult reservoirs that require models to be built on a finer scale to achieve accurate forecasting. For oil and gas companies focusing on exploration, acquisition and the development of oil field, HP provides solutions to create a secure IT infrastructure that integrates information and provide a single view of the field. HP solutions for the upstream solutions where smart and quick decisions can be made based on information about oil field assets. HP upstream solutions can help the hydrocarbon companies shorten their time to production, increase efficiency and lower costs in their oil fields. In addition, these solutions handle the acquisition, capture and archiving of data from oil fields.

HP solutions for downstream activities are for the oil and gas companies focusing on fuel retail automation and integration. (For example, these solutions help fuel - stations to meet new regulatory compliance demands, address mergers and acquisitions, reduce cost, operations, etc.)

What are the profits? HP's strong partnerships, global presence and support services differentiate HP as a leader in manufacturing today. HP solutions for these upstream and downstream activities do provide some key benefits:

- Faster time to oil through reduced processing times and improved collaboration.
- Business agility through a highly available, scalable, and adaptive infrastructure.
- Increased return on IT investment through increased productivity and integrated information systems.
- Lower total cost of ownership and reduced overall IT spending.

24x7 Refinery Operations, Optimum Sales and Operations and Optimizing Upstream Activities

Refining and transport operations present a lot of challenges including the supply disruptions. Companies can use these technologies to optimize supply and distribution networks and manage risks. Oracle also provides a 24 x 7 refinery operations and avoid the refinery turnaround times with the refinery maintenance operations using mobility solutions for refinery workers. Oracle provides Comprehensive Functionality for Optimizing Upstream Operations. [6] First, Oracle offers a complete data management solution for exploration and production, using business intelligence, data warehousing and the Public Petroleum Data Model [7] (PPDM) industry model. Oracle offers the only solution that integrates 3D visualization with structured and unstructured operational data. Second, Oracle offers the fastest time to value and lowest cost of any integrated oilfield solution, using the enterprise upstream solution from Oracle and P2 Energy Solutions. Third, Oracle offers integrated Customer Relationship Management [8] (CRM), order management, and field service applications for oil service and drilling companies. Increasing global demand for oil and gas acts as a premium on successful upstream operations. Exploration & Production (E & P) companies must act very efficiently and manage the huge amount of data generated by modern search techniques into producing oil wells. This is where Oracle plays an important role, it offers a complete data management system for exploration. Oil and gas companies are continuously challenged of their efficiently manage operations i.e. balancing the costs against the revenue from increased production. Oracle for Oil and Gas enables these companies to optimize oilfield operations by integrating production data with operational and financial data to give a unified view of operations.

Oracle Provides Comprehensive Functionality for Increasing Downstream Revenue Growth: [9]: First, Oracle has an integrated suite of downstream applications for refining and marketing. Second, Oracle's JD Edwards fuels management solution integrates petroleum supply chain systems and manages bulk product storage and transportation. Third, Oracle Process Manufacturing integrates least cost formulations and superior cost accounting, allowing multiple simultaneous costing types. Fourth, Oracle provides process-specific advanced supply chain planning, strategic network optimization (SNO), and production scheduling to optimize sales and operations (S & OP) planning decisions. Oil and gas companies looking to improve profits in commercial and retail operations, then can depend on Oracle for help. Today, more than 160 upstream companies run applications from Oracle. [10] Nearly 150 downstream companies run applications from Oracle. [11] More than 750 oil and gas companies run on Oracle technology. [12]

Intelligent Oil Field

The latest Information Technology innovation has made the concept of 'Intelligent Oil field' possible. With the help of this technology, huge amounts of sensor data can be stored and searched using advanced visualization technology. Complex data patterns can be detected automatically, such as for sanding or carbonate, hence a right person can be alerted and a response can be initiated before a production problem occurs. Visualization, modeling and analysis have made it easier for the companies to understand the wealth of complex information leading to improved reservoir management. The technology groundwork has been already laid down for the intelligent oil field to become a reality. [13]

Early warnings and alerts, along with suggestive solutions can be sent to the control centers through the detection of complex data from sensors. Hydrocarbon companies can access the data generated from the pertaining to assets, such as real - time sensors and trend histories, which can then make them more informed to take appropriate decisions. Wells and fields can be managed remotely, helping to save time and investments, extend the leverage of skilled resources and increase recoverables. [14] Information Technology and Communications can improve production and recovery rates with fewer people on fields. The company involved in upstream activities, need technology to enable evaluation and prospective action on demand. Intelligent oil field for E & P from IBM helps remote, collaborative management of reservoirs, facilities and equipment with accurate information.

A number of urgent business challenges support the idea of 'Intelligent Oilfield':

- Pressure to improve recovery, short drilling time and increased lifting costs.
- Aging workforce.
- Declining reserves in the areas with reduced expertise.
- A need to speed up decisions, while improving the quality.
- Massive amounts of data from new sensor networks being implemented.

What is the Profit? Intelligent Oil Fields for Exploration and Production from IBM brings: [16]

- Evaluate and execute drilling programs for better reservoir management.
- Maximize profit from producing assets.
- Minimize lifting costs through remote monitoring and optimized operations by better allocating your limited resources to the most productive assets.
- Maximize production and return on investment over the life of your assets.
- Reduce costs by improving supply chain processes to spot bottlenecks.

According to a Cambridge Energy Research Associates (CERA) study, [17] "the benefits of the intelligent oilfield can include lower operational costs, earlier and increased production and lower capital investment, increased recovery of oil and gas and lower abandonment costs". The CERA study also notes that "field operator productivity can increase between 100 and 400 percent, operating costs can decline by 10 to 20 percent and average production rates can increase by 1 to 3 percent". [18] Depending on the oil and gas field size, savings can be generated in the hundreds of millions of dollars. This could result in value creation in the billions of dollars each year.

Digital Oil Field (DOF)

'Digital Oil Field' – the term used to describe the high - tech software solutions for facilitating oil and gas exploration and production, which plays a leading role in the industry's effort to target current challenges, the industry faces additional challenges rising from increased health, safety and environmental regulations. Achieving regulatory compliance requires companies to frequently monitor activities, upgrade processes through fresh technology and most importantly to bring information that accurately demonstrates compliance with industry's and the governmental regulations at all levels. Companies can also remotely and collaboratively manage wells and fields, this helps to save time and money, while also increasing recovery and reducing risks.

Despite continuing high prices for their production output, the oil and gas companies continue to seek more cost-effective ways to deal with E & P in new and remote areas. Even today, many of the E & P companies are yet to integrate with the technology, thus unavailing the real - time information. To compete profitably in an increasingly complex and demanding industry, oil and gas companies must have access to accurate, timely information when they need it. Thus, necessitating the 'Digital Oil Field of the Future' (DOFF).

Digital Oil Fields of the Future (DOFF) in Upstream

It has the following advantages:

- Reduce finding and development costs.
- Increase production.
- Reduce lifting costs.

Up to 15 wells can be monitored at one time, 24/7, from anywhere in the world. [19] As the well is drilled, the data are fed back into the earth or reservoir model to update it, further refining the way the well is drilled. Tata Consultancy Services (TCS) provides the DOF solution for the upstream activities and help the oil & gas companies integrate their E & P data for remote decision making processes.

With the advanced 3D and 4D Seismic technology, geoscientists can now inspect and see the inside reality sitting in their offices. Schlumberger's InterACT monitoring and data delivery system is highly efficient in this case, providing Web-based data delivery system. Individual users-from technical team members to senior management, have access via a user name and password to the company's data in real time. Having 3D/4D visualization capabilities also permits more thorough analysis by creating the ability to slice data in arbitrary directions, perform analyses and then do a volumetric calculation instantaneously. "The vision rooms have large screens with bright projectors to provide immersive images that can be viewed in stereographic projection. Head - tracking devices connect the team to the virtual world, making it possible to walk into 3D datasets and models. A handheld, cordless 3D mouse operates the system and supports natural hand interaction with the data using gestures like pointing, grabbing and drawing. The resulting 3D/4D visualization facilitates geologic interpretation and analyses. By analyzing the reservoir in a 3D manner, our staff can be completely immersed in a project, thus providing a better visual perspective of what the structure looks like". [20]

Distributed Control Systems (DCS)

A greater contributor to build efficiency is Distributed Control Systems (DCS) which has been a significant development of information technology solutions. DCS were introduced in the late 1980s, enabling centralized process of monitoring and control. DCS systems replaced integrated circuit board controllers. Inputs from field instruments and outputs to valves and pumps were shifted to high – tech electronic signals. They were generally run in short distances to cabinets. Each DCS cabinet was connected to a main control computer. Process instruments, output to pumps and valves and controller settings driven from a computer located in a central control room. This system also enabled monitoring and control from multiple locations, such as local control rooms, engineering offices or even remote locations. [21] This system is used to monitor and control refineries and petrochemical plants. [22]

Today, most processes in the oil and gas infrastructure are completely dependent on the computerized control systems, from normal operations to complicated ones. Thus where computers are used, there is always a risk of an unauthorised party can gaining access to the system and can manipulate information or disrupt operations. Even though, human lives are not directly at risks, many business applications can have serious economic consequences. Due to the fact that society is highly dependent on the oil and gas industry, carefully planned attacks against critical infrastructures could impact sectors of the economy.

Drilling Information Systems (DIS)

DIS is a tool-kit to develop onsite information systems for drilling projects. The toolkit contains tools to define, generate and monitor data. The DIS features are unique and offer a professional data and information management for scientific drilling projects on land and on seas. The Extended Drilling Information Systems (XDIS) is an upgradation of the previous DIS. The original DIS was designed for the use at the drill location only, the extended version has a special Web interface. The interface allows remote access to the DIS. The XDIS allows researchers to download existing data and reference profiles and to upload new data and documents.

Petron Industries Inc. is a major manufacturer and supplier of drilling instrumentation and control systems to the worldwide oil and gas industry using an integrated systems approach for the design and manufacture of data acquisition, processing, storage and information. Petron has quickly gained a reputation for providing reliable, cost - effective products that increase both efficiency and safety during critical drilling operations. Petron specializes in digital and analog rig instrumentation for newly manufactured drilling rigs. Whether it is a simple pit monitoring system or a cabin drilling, Petron has engineered in instrumentation to meet the hydrocarbon industry's needs.

Petron, the company manufactures a variety of oil industry monitoring equipment ranging from specializing in the drilling rig instrumentation. Petron is one of the few independent suppliers in its field which designs, manufactures, sells and rents instrumentation. The newest of Petron innovations is our recently available broadband satellite service, which allows remote drilling locations the accessibility of a local branch office. This accessibility is paramount in realizing the full benefits of today's information technology.

Petron's comprehensive line of drilling instrumentation and control systems provide the reliable, up–to–the-minute information you need to make critical drilling decisions when millions of dollars are on the line. Petron's systems feature easy to read graphic displays, programmable alarms and the ability to monitor drilling progress from a remote site.

IT Infrastructure for Oil and Gas Industry

Upstream oil and gas companies are ever-challenged with finding new reserves and production as much as possible, minimizing costs in the process. Additionally, the search for new reserves requires the strong global collaboration that enables companies to think globally and act locally. Oil and gas companies are well past thinking of technology as networks of hardware and software. Today, they are focusing on how well these assets work together to elevate productivity, eradicate poverty, minimize costs and contribute to the development of a state. The global nature of the industry means that businesses can no longer afford to work in vacuum but need an information technology (IT) infrastructure that will enhance and support maximum performance.

Strategies for Profitable Growth (SAP) along with its partners [23] chose to focus on the area of digital oil field to tackle industry challenges that span the entire upstream asset lifecycle – from exploration, appraisal, development and production.

JV Technological Solutions with SAP

With the SAP Integrated E & P application, co - developed by Accenture and SAP, the companies can reduce time to market, accelerate the prospect evaluation process and coordinate well planning, rig selection and drilling across the enterprise. LogicaCMG provides SAP customers with comprehensive, real-time allocation in upstream gas production processes. SAP and IBM have had a close relationship for the past three decades, maintaining partnership links to benefit their mutual customers. Both companies are among the industry leaders sufficing to the oil and gas industry needs. By combining our international capability and local market knowledge within the industry, IBM Business Consulting Services and SAP can deliver solutions and services rapidly, reliably and cost – effectively.

IBM offers a broad range of innovative technology solutions that can help upstream oil and gas companies anticipate problems such as equipment and production impairment or failure before they happen, which can help reduce the costs associated with repairs

Implications for India

India has emerged as a net energy importer and with a growing population and enhanced economic growth, its consumption pattern has grown steadily. Self-sufficiency in hydrocarbons cannot be reached. Hence, energy imports have become a priority interest for India. Its imports for the year 2004–05 accounted for 95.86 mmt [25], while International Energy Agency (IEA) projects that, India's oil imports will exceed 90 percent by 2020. [26] The transport sector is alone responsible for 70 percent of the rise in oil demand. [27] In addition to oil, the natural gas consumption has risen more rapidly, from 600 m cu ft/yr in 1995 to 960 m cu ft in 2003. [28]

India imports around 70 percent of its oil and gas. The International Energy Agency (IEA) predicts that, its dependence on hydrocarbons will further grow to 91.6 percent by the year 2010. By 2010, India is expected to be the world's fourth largest consumer of oil, absorbing 3.2 mb/d. This means that there will be increased competition amongst the stakeholders, primarily India, not only for the existing resources but also to unleash new discoveries. Hence, search for new energy technologies to get to these resources have become an imperative for the energy industry.

India's rapid economic progress and increasing energy consumption has brought to the forefront, the very question of sustainable development for all. Adequate and affordable energy supply for all is the solution to the economic development of a state. Energy stands as a nucleus of an improved socio-economic well-being of all. It is vital for relieving poverty, improving and raising the living standards. Hence, emerging energy technologies can only fulfill these goals. India can leverage and strengthen its national energy research and deployment efforts by: [29]

- Bringing the abundance of existing technologies closer to commercial deployment through technology learning;

- Identifying the challenges in bringing viable medium term technologies forward (i.e Enhanced Oil Recovery (EOR), Carbon Capture and Storage, Hydrogen and Fuel Cells); and
- Offering a successful, cost sharing structure for expensive exploratory research (e.g. ITER [30] – International Thermonuclear Experimental Reactor).

Investments Needed

A current rate of investment in this sector is not adequate to meet the increasing difference, which is expected to expand due to the increasing energy demand. Without a sizable increase in the capital investment in the emerging energy technologies, oil and gas production will remain stagnant with prices reaching sky - high.

India also ranks amongst one of the countries, providing best IT professionals. 'IT Brain drain' [31] as it can be called has proved to seriously affect the developing countries, but have bore good fruits for the IT industry worldwide. Indian IT professionals are willing to work on remuneration (Cost to Company - CTC) where, the workforce abroad is not ready to work with. Indian IT force is involved with the energy industry in terms of data processing and management, as geologists, drilling experts and software and hardware professionals. Hence, India's investment in terms of its IT intellect is quiet crucial.

India should participate in the international energy technology ventures along with the western countries. Adequate instruments should be adopted for joint ventures in Research and Development (R & D) and the exchange of expertise and technical know–how should be encouraged.

Infrastructure Development

There is an increasing need for energy technology infrastructure development. Proper planning and successful execution of projects are needed. Delays (official / unofficial) in initiating new projects are a major problem, while the present projects under consideration are not in progressive mood. The timely response to these projects plays an important role in generating support for emerging technologies.

Another challenge is to integrate the present and the new energy technology infrastructure. The current network of production, transportation and distribution facilities needs to be changed or developed. There is a need to rebuild the upstream and downstream technology infrastructure with the integration of new technologies.

Today's upstream oil and gas companies struggle with data management systems. This struggle is going to grow acute as companies establish operations centers in more remote locations in search of resources, thereby increasing their demands for quality information, timely decisions and a quick response from these locations to the chief centers and vice - versa.

To decrease the cycle time from an adverse occurrence in the field to a decision and its proper execution, companies need accurate, access to all data and information related to wells and the associated equipment. Once gathered, the data should be transferred over vast distances, to enable the technology and people to assist in evaluation and decision making. With more raw data constantly arriving, more data-mining time is required. This, in turn, demands a high-bandwidth communications infrastructure and more robust data management. By combining analytics and business intelligence tools with this high-bandwidth infrastructure that helps to turn raw data into useful information, upstream oil and gas companies can increase business benefits from productivity and efficiency gains.

Integrating/Upgrading Technology

In India, much of the current energy supply and use, based, as it is, on limited resources. Alongside the energy chain, right from the resource extraction to the distribution of energy to its consumers, pollutants are emitted and spread to far off places, leaving severe health hazards and the risk of climate change. Integration of the hydrocarbon industry with these emerging energy technologies can significantly prove to be profitable and reduce the above mentioned hazards. Nuclear energy will bring in serious issues such as radioactive wastes and massive investments required. Hence it has become quiet an imperative for India to search the available technological options.

Indian Space Research Organisation (ISRO) was established in1969 for the research and development of technologies related to space and their application to India's development. ISRO can play an important role in providing these technologies for the energy industry. In the process, India can bargain in the international energy market for its technology. Because of the availability of improved technology and an increasing awareness of the effects of energy and the technology related to it, a country like India can progress at a faster pace, minimizing the costs and risks associated with it.

Analysis

There may be no other industry today that demands a more diverse set of skilled manpower and scientific and technological capabilities than the oil and gas upstream & downstream industry. Competition for hydrocarbons has driven companies to explore and produce in harsh, remote and hostile locations, where even the simplest of logistical tasks can be difficult and costly. And, as the environment grows more diverse and severe, the challenges become more complex, while the skilled manpower is aging and growing scarce. Oil prices worldwide are constantly rising. Average crude oil price during 2006 was over USD 65 per barrel almost 20 percent above the 2005 average. [32] Oil prices peaked above USD 78 per barrel in 2006. [33]

It is essential at this point to know the country's current standing on its energy competence, the intensity of existing energy technologies as well as the emergence of new energy technologies in the international market. With these calculations, an attempt should be made to integrate these new technologies with the old ones by assessing their improvement in individual's advancement and controlling the problem of climate change.

Looking to private sector companies (e.g. Reliance Industries Limited, etc.) for investment for these technologies or energy solutions would be crucial. In addition to ensuring an ideal connection between the government and industry, it is immensely important to understand that, India stands a potential chance in the world IT – energy collaborative knowledge. Implementing agreement provides a medium for reaching out to the rest of the world, hence India needs to participate in the energy research programmes. While, the below graphical representation negates India's participation in the International Energy Technology Collaboration programme implementing agreements.

Gathering and analyzing data quickly and effectively in a controlled laboratory environment can be difficult and in an environment such as the North Sea or Sub - Saharan Africa, this can be a monumental task. And while new technology has shown the promise of great things to come, integrating this innovative technology with existing systems, new tools and a global network of diverse business partners can be an exhausting exercise.

To attract and retain available skilled professionals, companies need to change their corporate cultures. This requires implementing innovative business processes to help people take full advantage of real - time oil field data and turn it into usable information that can help to increase recovery and production, reduce operating costs and ultimately increase profits. To accomplish this, companies should think globally and act locally. This involves establishing new responsibilities, while giving due recognition, hence a new work method is needed in order to build an atmosphere that facilitates more well - versed decision making and efficient execution of projects, where an increasing energy demand and the geopolitical forces that have created a volatile marketplace. Technology such as SCADA, DOFF and Enhanced Oil Recovery (EOR) would certainly arrest the natural decline of oil and gas reserves, resulting in a net gain in production.

Global competition for natural resources continues to drive the need to improve oil and gas reservoir performance. Although new technologies have shown promise for meeting industry needs, it is laborious to integrate new technologies with existing processes, organizations and global network with diverse business stakeholders. The convergence of forces, threats and technologies creates a ripe environment for the intelligent oil field solution. A solution that integrates people and technology to improve oil field durability, by leveraging frequently captured data that is delivered, converted into usable knowledge and responded upon in time. Successfully implementing the IT knowledge to take full advantage of all available data requires a sophisticated programming of projects designed to integrate man and machine.

Conclusion

Information Technology has emerged as a driving force in energy operations, strategies, structures, ownership and performance. IT innovations and applications have not only brought significant change into the nature of energy economics, but also have altered the energy space and its performance. Data sharing, distribution of knowledge and work without regard for geographic distance and language has become a possibility.

There is a need for a new approach which will transform the huge data volumes which are 100 to 1,000 times greater than volumes gathered by conventional technology, making the data management difficult. For example [35]: "With fewer skilled human resource and the increased demand of the world has asked for more global collaboration to leverage available skills, facilitate work in appropriate physical environments and support the use of common applications and IT infrastructures. Staff members need secure Web access anytime, anywhere and from multiple devices and more electronic documents (for example, the electronic - Well file). Non-technical (back-office) activities must be reduced to maximize limited technical (engineering and geosciences) resources; this includes functions such as linking payment transactions for complex services to real-time payment. External threats to security are increasing; these factors are being addressed by data - gathering and analysis technologies such as intelligent supervisory control and data acquisition (SCADA), early event warning, smart alarming and automated controls".

New technologies have paved the way for oil and gas companies (National Oil Companies (NOCs)/International Oil Companies (IOCs)) to collect an increasing amount of information about their assets. However, data must be catalogued, stored, backed - up and should be easily accessible, which means many companies engaged in India will have to re - evaluate their core infrastructure, including hard drive space, processor capacity, security applications and even disaster recovery plans. Energy plants need reliable communications for onshore and offshore locations and even the smallest delay can potentially cost a millions of dollar loss. Today, there are wireless broadband networks that can immediately transfer critical data and establish secure communications with teams at remote areas.

Whether the motives are to improve the operational efficiency, lower costs of upstream and downstream activities or to unlock new revenue streams, advanced broadband satellite networks are and will continue to be an essential and integral technology in the hydrocarbon industry. The opportunities are immense and the returns on investment are likely to come quickly and in a big way. In adopting these technologies rather than the aged technical solutions, the global oil and gas market will consequently, reap the benefits.

Notes

1. "IP Access International" Accessed on 22 September, 2007. Available at http://www.ipinternational.net/oil.php.
2. Slekys, Arunas G. (February 2005), *Broadband Satellite Technology: Unlocking the Benefits for the Oil and Gas Industry*, Reprinted from CIS Oil and Gas Report, p. 2. Accessed 23rd September 2007. See, http://www.hughes.com/HUGHES/Doc/0/TDMEOR4EHK74F710LJVG8J9717/Oil&Gaspercent20CIS percent20Article_11-2-04.pdf
3. Ibid.
4. National Communications System (October 2004), "Supervisory Control And Data Acquisition (SCADA) Systems", Arlington, Accessed on 23 September, 2007. Available at, http://www.ncs.gov/library/tech_bulletins/2004/tib_04-1.pdf
5. "HP Solutions for the Oil and Gas Industry" Accessed on 21 September 2007. Available at, http://h71028.www7.hp.com/erc/downloads/5982-8562en.pdf
6. Oracle, "Upstream oil and gas applications", Accessed on 21 September, 2007. Available at, http://www.oracle.com/industries/energy/oil-gas-upstream.html
7. The Public Petroleum Data Model (PPDM) is widely used in the E & P industry. By uploading your information into a PPDM data model, as is assessed by the dbMap (dbMap allows to create 2D and 3D integrated Geographic Information Systems (GIS) module, you immediately add value. DbMap extends the value of PPDM model with a built in data dictionary that provides on – line documentation of the data and its interrelations.
8. Customer Relationship Management (CRM) is the need of any organization to analyse the customer portfolio at a macro level for exceptions, trends and problems. The IT helps in managing all the competing priorities of customers without letting any failures.
9. Oracle, "Downstream oil and gas application", Accessed on 21 September 2007. Available at, http://www.oracle.com/industries/energy/oil-gas-downstream.html

10. "Oracle: Committed to the Oil and Gas Industry" Accessed on 21 September 2007, Available at, http://www.oracle.com/industries/energy/brief_oilgas_press.pdf
11. Ibid.
12. Ibid.
13. IBM Business Consulting Service, "Meeting the Challenges of Today's Oil and Gas Exploration and Production Industry", P. 1. Accessed on 23 September, 2007. Available at, http://www-935.ibm.com/services/us/gbs/bus/pdf/g510-3882-meeting-challenges-oil-gas-exploration.pdf
14. Ibid, pp. 2.
15. Clark, Kevin, "Innovate for the future: Intelligent Oil Fields", Accessed on 23 September, 2007. Available at, http://www.03.ibm.com/industries/chemicalspetroleum/doc/content/bin/ibm_intelligent_fields.pdf
16. IBM, "Intelligent Oil fields for exploration and production from IBM", Accessed on 23 September. Available at, http://www-03.ibm.com/industries/chemicalspetroleum/doc/content/solution/444490120.html
17. IBM Global Business Services (2006), "The intelligent oilfield: meeting the Industry challenges of today's oil and gas exploration and production industry", Executive Brief, Accessed on 21 September 2007. Available at, http://www- 935.ibm.com/services/us/imc/pdf/ge510-3882-oil-gas-challenges.pdf
18. IBM Global Business Services (2006), "The intelligent oilfield: meeting the industry challenges of today's oil and gas exploration and production industry", Executive Brief, Accessed on 21 September 2007. Available at, http://www-935.ibm.com/services/us/imc/pdf/ge510-3882-oil-gas-challenges.pdf
19. Lyle, Don and Leslie Haines (April 2004), "The Digital Oil Field Today", *The Digital Oil Field, a Supplement to Oil and Gas Investor,* p.6. Accessed on 23 September 2007. Available at, http://www.oilandgasinvestor.com/pdf/DOF0404.pdf
20. Decou, Reynold (April 2004), "How El Paso Production Went Digital", *The Digital Oil Field, a Supplement to Oil and Gas Investor,* p.21. Accessed on 23 September 2007. Available at, http://www.oilandgasinvestor.com/pdf/DOF0404.pdf
21. Anthony, Crooks and John Dunn (May 2006), "The Role of Information Technology in Fuel Ethanol Industry", U.S Department of Agriculture, Rural Development, pp. 9. Accessed on 21 September 2007. Available at, http://www.rurdev.usda.gov/rbs/pub/rr209.pdf
22. Lindqvist, Ulf (2006), "Securing Control Systems in the Oil and Gas Infrastructure", Business Briefing: Oil and Gas Processing Review, A Report. Accessed on 21 September 2007. Available at, www.touchbriefings.com/download.cfm?fileID=7840&action=downloadFile
23. SAP has partnership with Accenture, Logica CMG, Meridium, NRX, QUORUM, HP and IBM.
24. "Oil and Gas Industry: Strategies for Profitable Growth", Accessed on 21 September, 2007. Available at http://www.sap.com/industries/oil-gas/pdf/BWP_Strategies_Growth_Oil_Gas.pdf
25. Planning Commission of India (December 2005), *Draft Report of the Expert Committee on Integrated Energy Policy,* New Delhi: Government of India, p. 80, [Online: Web] Accessed on 27 November 2006. Available at, http://planningcommission.nic.in/reports/genrep/intengpol.pdf

26. The Economist Intelligence Unit Limited (2006), "Country Profile: India", *The Economist*, p. 30.
27. IEA, *World Energy Outlook*, 2002, Accessed on 5 September, 2007, pp. 292. Available at, http://www.iea.org/textbase/nppdf/free/2000/weo2002.pdf
28. The Economist Intelligence Unit Limited (2006), "Country Profile: India", *The Economist*, p.30.
29. IEA (2005), "Energy Technologies at the Cutting Edge" International Energy Technology Collaboration IEA implementing agreements, pp. 9. Accessed on 19 September, 2007. Available at, http://www.iea.org/Textbase/nppdf/free/2005/IAH2005mep_Full_Final_WEB.pdf
30. ITER - India along with six other partners (EU, Russia, China, Japan, South Korea and US) signed an international agreement (ITER) in Brussels. ITER, a project to demonstrate the potential of fusion as an energy source. The project ITER, is an international collaborative research project, which will reproduce the physical reaction – fusion that occurs in sun & stars.
31. Less than 5 percent of the skilled nationals of the powerhouses of the developing world, like India, China, Indonesia and Brazil, live abroad in OECD countries. In, Dugger, W.Celia (October 26, 2005), "'Brain drain' is damaging World's poorest countries, study shows", *International Herald Tribune,* Accessed on 19 September, 2007. Available at, http://www.iht.com/articles/2005/10/25/news/drain.php
32. The Economic Times (23 September, 2007), "ONGC: An energy force to reckon with", New Delhi. Accessed on 22 September, 2007. Available at, http://economictimes.indiatimes.com/articleshow/2386640.cms
33. Ibid.
34. IEA (2005), "Energy Technologies at the Cutting Edge" International Energy Technology Collaboration IEA implementing agreements, Accessed on 19 September, 2007. Available at, http://www.iea.org/Textbase/nppdf/free/2005/IAH2005mep_Full_Final_WEB.pdf
35. IBM Global Business Services (2006), "The intelligent oilfield: meeting the industry challenges of today's oil and gas exploration and production industry", Executive Brief, Accessed on 21 September 2007. Available at, http://www- 935.ibm.com/services/us/imc/pdf/ge510-3882-oil-gas-challenges.pdf

13

IT-enabled Good Governance in India

Harinam Singh

Introduction

Information technology based governance, popularly known as e-governance, is now universally accepted strategy to realize good governance. It is argued that application of information technology results are given below: - (i) simple, (ii) moral, (iii) accountable, (iv) responsive and (v) transparent, or in short SMART government. A number of advantages of an information technology based public administration system have been cited. For instance, the application of information technology makes administrative and management processes efficient, transparent, unbiased and promotes involvement and participation of all stakeholders in administrative and management processes. It helps in intuitive linkages and sharing of information resulting in comprehensive monitoring and assessment of government performance and promotes interaction and exchange of ideas, suggestions and innovations. It is also suggested that the application of information technology helps in setting in place an intelligent agent at the local level that may facilitate time production of information in a simple manner.

In this era of information technology, the key to good governance is known as E-Governance. E-Governance refers to Information and Communication Technology (ICT) enabled route to better governance. It aims at providing connectivity, networking, technology up-gradation, selective delivery systems for information and services and package of software solutions. It refers to the use of information and communication technology to promote more efficient and effective government services, allow greater public access to information and make government more accountable to citizens. It involves new styles of leadership, new ways of debating and deciding policy and investment, new ways of accessing education, new ways of organising and delivering information and services. The primary condition of implementing the E-Governance is that the government should be democratic or of partially democratic in nature. The other important requirement is that the economy should be of free nature or the mixed ones as it gives opportunity as the individuals to think big and 'out of the box' and show their enterprising nature. The system should be guided by the principles of transparency, efficiency, accountability and welfare of the society. It can help more in approaching and attaining them more than the conventional system of governance. Due to growing awareness, people's expectations in India have increased. They expect the public sector and public services to be at par with, if not better than, the private sector. The comparison with standards and norms of the developed countries are increasing. The government, must therefore, be seen performing and delivering results. E-Governance is certainly among the better tools for this purpose. This can lead to competitiveness to innovations that are so essential for reforms. Moreover, the information between the government and the citizens, thereby ruling out face-to-face interactions and provides a standard level of service. Various aspects and aspects of the government in the country are being converted into electronic or networked reforms.

Government of India has launched on ambitious programme for e-governance. The goal is to improve efficiency, convenience, accessibility and transparency in Government functions and take Government services to the common citizen. It increases the allocation for e-governance from Rs. 395 crore in 2006-07 to Rs. 719 crore in 2007-08. The Central Government supports e-governance action plans at state levels and it also increases the allocation for such support from Rs. 300 crore in 2006-07 to Rs. 500 crore in 2007-08. It also provides Rs. 33 crore for a new scheme of man power development for the software export industry (Union Budget 2007-08).

The Indian software industry has grown at a phenomenal rate over the second –half of the 1990s. Earning around 80 percent of its revenue from exports, the annual growth rate of software exports has been more than 45 percent, except for the years 1999-2000, and the succeeding years from 2001-02 to 2003-04. The slow down in exports activity due to the worldwide recession and technological downturn has now abated, as the IT industry maintains its growth momentum and consolidates its partnership with overseas customers.

Policy Initiatives for IT Sector

The policy initiatives for IT sectors are:

- The Information Technology Amendment Bill has been introduced in the Parliament on December 15, 2006. This proposes to put in place technology applications, security practices and procedures relating to such applications. It also addresses the issue of technological neutrality in IT laws as recommended by UNCITRAL Model Law on Electronic Signature.
- A proposal for IT Hardware Manufacturing Policy is also under consideration.

- In order to enhance that the benefits of IT reach the common man, Government has initiated a more to make available tools and fonts in various Indian languages freely to the general public. Tamil, Hindi, and Telugu software tools and fonts have already been released. All Indian languages are expected to be converted in the next one year.
- Government to ensure penetrations of the IT and IT related sector in the rural areas, has formulated a proposal to establish 1,00,000 Common Services Centres (CSCs) in rural areas. The scheme will be implemented through Public Private Partnership (PPP).

Governance

Improving the efficiency and effectiveness of the public administration system is a major challenge being faced by the governments all over the world. Governments are trying to meet this challenge by transforming the public administration system through the introduction of modern management concepts and modernizing administrative practices. This transformation agenda has concentrated mainly on three broad areas: finance, access and governance. In the realm of finance, a major structural adjustment programme has been launched by most of the countries. The programme sets targets for both economic growth and deficit reduction and seeks to promote job creation by creating a low-inflation, well-managed, market-driven economy. On the other hand, in an attempt towards improving access to basic development and welfare services, governments are focusing on issues related to redressal and redistribution of available resources. In the area of governance, the emphasis is on good governance which, in the ancient Indian wisdom, has been conceptualized as Ram Rajya. The concept of Ram Rajya continues to endure the Indian social and political psyche as good governance is expected to contribute significantly in improving the quality of life of the people. Good governance is also viewed as essential for efficient and effective administration in a democratic set up which is now universally accepted as a basic human right. Governance may be defined as the manner in which the society works and is organized. Good governance, therefore, is the one which power and authority of the government is strategically used for the welfare of the people and for public benefit. Good governance should be differentiated from the bad governance which means that the authority and power of the government is used indiscriminately for private gains. Information and how information is used is critical to governance. The manner in which information is used in governance processes and procedures decided whether it is a good governance or a bad governance. Selective use of information in governance results in hierarchical structures on which power and authority of the government is distributed unequally.

This skewness in power distribution at each hierarchical level is proportion to information residing at each of these levels. In case of good governance, the proportion distribution of information is uniform across the public administration system. In case of bad governance, more and more information becomes out of reach to the common man resulting in a concentration of information only to privileged groups who often use this power of information in a discriminate manner for their own interest's benefits and gains. One of the bases of good governance is informed-decision making. Well-informed decision making is critically dependent upon the quality and time liness of information. Unlimited access to information of reliable quality at right time is essential for well-informed decision making and hence for good governance. Poor quality and out-dated information leads to poor decision-making and end up in bad governance. On the other hand, restrictions on access to information lead to manipulations. Managing information is, therefore, central to the realization and institutionalization of good governance in the public administration system.

IT-enabled Structure of Government

In the traditional structure of Government, government departments provide the interface between Information Systems on the one hand and citizens on the other. Ideally, however, it should be the other way round and Information Systems should provide the means of interface between the Government and the Citizens. These citizens include businessmen and also non-government organizations. Naturally this interface will be non-discretionary and gradually available from any time to anytime and anywhere.

Concept of E-governance

The concept of e-governance in the Indian context is articulated as giving the citizens the choice of when and where they access government information and services. It takes a delivery channel view and it also means using more and more of Electronics & Information technology in many of the government functions. Proper maintenance of records, rules and procedures documentation and tracing precedents for decision making etc., are regarded as time consuming and consequently, it leads to corruption and related practices. By application of Information technology (IT) citizens can have access to information and render the process, transparent. The key concept adopted and advocated in this direction is SMART governance (Simple Moral Accountable Responsive and Transparent). One of the aspects that is crucial in this governance system is how does it work or likely to work at the grassroots level. In order to make it work at the grassroots level what initiatives are needed and how do the State, market and the NGO sector respond to make it work is also a key concern. One of the often stated advantages of e-governance is eliminating corruption.

We are still at initial stage of E-Governance. Though lot of work is under progress in G to G application, G to C applications is yet to gain momentum (G: Government, C: Citizens). E-Governance is a tool for Good Governance not an object in itself. Hence, while declaring agenda for E-Governance, it must be defined what kind of good governance will be achieved through E-Governance. In short role of 'E' is very small or like a catalyst in E-Governance policy and 'Governance' has a prime importance.

To reach at better level of Governance, the system should be effective and result oriented. Because only efficient Governance may be helpful with Government to Government or Government to Employee application but service to citizen may not have precedence in efficiency. Minimum discretion in policy implementation as well as in service delivery is another essentiality of Good Governance. To achieve this it will require high degree of coordination and exchange of information. This coordination would be not only horizontally across the departments but also should include crisis-cross coordination model. This will facilitate exchanging as well as making data available to all levels of department seamlessly and also speedily. This can happen when departmental walls as well as protocol and procedures created for pasting and receiving information should be either torn apart or those should be so transparent that co-ordination, exchange of data becomes effortless.

When review is taken of all successful information systems in Government, we observe that usage and benefit of the system is restricted to purpose specific user. Slightest deviation in the requirement is not even entertained by the system. It is, therefore, necessary to have one common application system spread horizontally across between various Government departments and members of public.

Aims of E-governance

The aims of e-governance are:

- To strengthen and upgrade the present infrastructure of information technology in the industry.
- To bring in transparency in the functioning by placing the information, to the extent possible, in the public domain.

- To interlink the functioning of different offices to bring about synergy in the aspirations and enable effective and timely decision-making.
- To develop public grievances and complaint system and eliminate the need of personal visits to offices by the public.
- To make available forms and information on the Internet and the facility of filling the applications from remote rural areas-Web Enabled application system.

E-governance in India

The initiatives towards IT at various stages of development/implementation of central government includes: India Portal, National Institute of e-governance, central repository of data, dissemination of information relating to best practices/innovations in e-governance, awards for best websites and innovative use of IT in the delivery of public services. In addition citizen service centers were set up for one stop and non-stop delivery of services to the public. India Portal is a user-friendly portal of all government websites for providing information and delivery of services.

Several state governments have taken various innovative steps to promote e-governance. One of the most popular as well as significant measures has been Chief Minister's Information System (e.g. Andhra Pradesh, Madhya Pradesh and Rajasthan), which monitors a range of activities from developmental programmes to redress public grievances. The Andhra Pradesh Development Monitoring System (based on a multipurpose household survey) has a database with spatial as well as non-spatial parameters of the entire population (75 million) of the state. Similarly, Vikas Darpan (mirror of development) of Rajasthan envisages GIS-based planning and decision support system. Andhra Pradesh has also introduced APSWAN (Andhra Pradesh State Wide Area Network), a state wide network for voice, data and video communication, which is the basic information highway for improving government-citizen and government-industry interface. The Secretariat Knowledge Information Management System (SKIMS) of Andhra Pradesh efficiently manages information in the secretariat. The Disaster Management System in Gujarat maintains communication during natural disasters. In Karnataka, computerization of treasuries captures every single transaction at all district and taluk (sub-district) treasuries.

In Gujarat, VIDYUTNET, India's first VSAT-based communication network supports real time data applications for power generation and distribution. The Government of Kerala has introduced the RD Net project (Information Kerala Mission) to connect all the 152 block offices in the state with a view to transform local bodies into genuine institutions of self-governance. Office of Controller of Entrance Examinations has been automated to bring about transparency in allocation of colleges to successful students in Kerala. In Maharashtra, the Connectivity Project has networked 3000 offices.

The National Informatics Center (NIC) introduced the information technology in the rural sector with the active support of the department of rural development, Government of India and consequently Computerized Rural Information System Project (CRISP) was launched. It aimed at facilitating the District Rural Development Agenda (DRDA) in the monitoring of exercise of poverty-alleviation programmes through Computer Based Information System (CBIS). Till date four versions of CRISP application software packages have been developed. The fourth version was renamed Rural Soft. Rural Soft 2000 was developed using a centralized-decentralized approach. It was decided to develop a core module, representing the commonalities which existed across the layers of administrative set up for the poverty alleviation programmes throughout the country.

The Government of India approved the National E-Governance Action plan for implementation during the period of 2003-07 which had been presented the Prime Minister on 6 November, 2003.

The National E-Governance Action Plan provides an insight into the action plan for e-governance that needs to be implemented in India. It focuses about the policies, infrastructure, human resources, technology, awareness campaigns, infrastructure and various projects that will be required for success of E-Governance in India. This paper goes beyond the strategy and talks about the actions required for e-Governance which are: (i) Core Policies, (ii) Core Infrastructure, (iii) Human Resources Development, (iv) Technical Assistance, (v) Research and Development, (vi) Service Delivery, (vii) Awareness and Assessments, (viii) Organization Structure, (ix) Core E-Projects and (x) Integrated Service Projects.

The National E-Governance Action Plan has suggested a list of core policies which are:

- Vision, Mission, Strategies, Policies, guidelines, frameworks, roadmaps.
- E-Governance Technology Architecture, framework, guidelines.
- Funding Strategy.
- Human Resources Strategy for project leadership, implementation, operations.
- Security Policy, framework, guidelines.

- Policy on front ends (kiosks, integrated service centres, etc. - ownership, funding/subsidy, types of services, charges etc.).
- Policy on middleware (portals, vans, datacentres, etc. - ownership, exclusivity, interconnect, charges, etc.), Policy on backends (department automation - role of private sector, role of government, safeguards, right to information, privacy, etc.).
- Policy on integrated services (ownership, responsibility etc.).
- Process Re-engineering - policies, guidelines.
- Policy on GIS–guidelines.
- Policies and guidelines on electronic payment mechanisms.
- Policies on employment generation through Employment Guarantee Schemes.
- Prioritization policy.
- Policy on interoperability.
- Policy on spread of access.
- Policies, strategies and guidelines for outsourcing of Employment Guarantee projects.
- Policies and guidelines on managing content.

The National E-Governance Action Plan has given a list of core infrastructure which are:

- National e-Government Intranet (NICNET, ERNET, other service providers).
- State-wide Intranets.
- National E-Government Data Center.
- State Data Centers.
- Security Infrastructure & Resource Centre for E-Governance (PKI, etc.).
- GIS - National Spatial Data Infrastructure, and
- Language Resource Centre.

Human resource development has following action plan:

- Training for EG Policy Makers.
- Training for CIOs.
- Training for specific projects.
- General IT skills.
- Special training programmes for specialists: Security, Languages etc.
- Advanced courses on HLD, architecture, language technologies, etc.
- Equipping National/State Institutions of Public Administration for E-governance training.

Technical Assistance

Under this technical assistance to the states will be given to the states for various areas where the DIT or one of its agencies has developed expertise and can support the development of the same in the states.

Research and Development

The National E-Governance Action Plan has suggested the research and development strategies which are: - (i) Architecture, (ii) Integration strategies, (iii) Electronic payment systems, (iv) Other areas related to Electronic Governance.

Service Delivery

The National E-Governance Action Plan has Service Delivery Infrastructure at State, District, Block and Village level, and E-Postal Services like electronic money orders.

Awareness and Assessments

The National E-Governance Action Plan has following awareness and assessments which are: (i) E-Readiness Assessment of various States/ Departments, (ii) Virtual E-Governance Forum, (iii) Assessment of E- Projects, (iv) Best Practices for E-Governance, (v) E-Governance National Resource Database, (vi) Newsletters on E-Governance, (vii) Workshops/ Seminars/ Conferences, (viii) Competitions and Awards, (ix) Electronics Governance forum for NGOs, Private. Sector, Academic, Training Institutions, and (x) Publicity and Awareness (media, films, etc.).

Organization Structure

The National E-Governance Action Plan has suggested the following organization structure: (i) National E-Governance Council (NEGC), (ii) National Institute of SMART Governance (NISG), (iii) State E-Governance Council (State EGCs), (iv) E-Governance Standards Institution.

Core E-Projects

The National E-Governance Action Plan has suggested a list of Core E-Projects which are: - (i) Land Records, (ii) Transport departments (Vehicle Registration/ Road Tax / Driving Licence), (iii) Registration, (iv) Courts, Coourtnet, (v) Municipalities, (vi) Gram Panchayats, (vii) Civil Supplies Distribution System, (viii) Citizen Data Bases, (ix) Commercial/Sales Tax/VAT9Value Added Taxes), (x) Excise, (xi) Income Tax, (xii) Company

Affairs, (xiii) Directorate General of Foreign Trade (DGFT), (xiv) Passports, (xv) E-Tourism, (xvi) Industries departments., approvals, registration, (xvii) Treasuries, Integrated Financial Management System, (xviii) Government. managed Financial Services (Insurance, PF, EGIS etc.), (xix) Police Net, (xx) Edu-net, (xxi) E-Employment, (xxii) Agri net, (xxiii) Health net, (xxiv) Social net, (xxv) Human Resource Management Systems (HRMS), (xxvi) Public Grievance Redressal System, (xxvii) Chief Minister Information System, (xxviii) Secretariat Knowledge Information System (Computerization of Secretariat), (xxix) Disaster Management, (xxx) E-Democracy (Use of ICT in Electoral Process).

Integrated Services Projects

Integrated services projects are given below: - (i) India Portal, (ii) State Portals, (iii) EDI, (iv) E-Biz, (v. E-Procurement, and (vi) Country Gateway.

SWAN

State Wide Network Area project aims at providing high speed, high connectivity networks, connecting offices at block level for faster access to government services. The Government of India declared it on 27th October, 2004. It has been proposed to provide senders like land and property records and transactions, agricultural information, credit related services, payment of bills, issue of birth, death, caste, income, certificates, pension services, grievance redressal, licenses etc. the network is to be proposed to be connected through appropriate interfaces in the form of gateways.

Implementation of E-Governance at Local (Panchayath) Level

In this section we discuss the implementation of e–governance. We take implementation at local level as an example .Before we discuss implementation of e-governance at local level we will briefly explain Panchayath Raj System in India with special reference to 73rd and 74th Constitutional Amendments.

Panchayati Raj System in India

Panchayats have been the backbone of the Indian villages since the beginning of recorded history. Gandhiji, the father of the nation, in 1946 had aptly remarked that the Indian Independence must begin at the bottom and every village ought to be a Republic or Panchayat having powers. Gandhiji's dream has been translated into reality with the introduction of the three-tier Panchayati Raj system to ensure people's participation in rural reconstruction.

73rd Amendment Act, 1992

The passage of the Constitution (73rd Amendment) Act, 1992 marks a new era in the federal democratic set up of the country and provides constitutional status to the Panchayati Raj Institutions (PRIs). Consequent upon the enactment of the Act, almost all the States/UTs (Union Territories), except J&K, National Capital Territory (NCT) of Delhi and Uttaranchal have enacted their legislation. Moreover, all the States/UTs except Arunachal Pradesh, NCT of Delhi and Pondicherry, all other States/UTs have held elections. As a result, 2,32,278 Panchayats at village level, 6,022 Panchayats at intermediate level and 535 Panchayats at district level have been constituted in the country. These Panchayats are being manned by about 29.2 lakh elected representatives of Panchayats at all levels (Planning Commission, 2001). This is the broadest representative base that exists in any country of the world developed or underdeveloped. The main features of the Act are: (i) a 3-tier system of Panchayati Raj for all States having population of over 20 lakh; (ii) Panchayat elections regularly every 5 years; (iii) reservation of seats for Scheduled Castes, Scheduled Tribes and Women (not less than one-third of seats); (iv) appointment of State Finance Commission to make recommendations as regards the financial powers of the Panchayats and (v) constitution of District Planning Committees to prepare development plans for the district. vi) constitution of an independent State Election Commission to hold PRIs elections on a regular basis; (vi) legal status to Gram Sabhas; and (vii) addition of Eleventh Schedule to the Constitution listing. 29 Subjects within the jurisdiction of PRIs. Further,the Panchayati Raj Institutions have been endowed with such powers and authority as may be necessary to function as institutions of self government and contains provisions of devolution of powers and responsibilities upon Panchayats at the appropriate level with reference to (a) the preparation of plans for economic development and social justice; and (b) the implementation of such schemes for economic development and social justice as may be entrusted to them.

Gram Panchayats

E-Governance at the village level is catching up with the Award winning *gyandoot project*. The various services which can be included (currently being enabled by gyandoot model) are: (i) Rate of Agricultural Commodities (Mandi Bhav) (ii) Online Registration of Application (Avedan Patra) (iii) Online Grievance Redressal (Samasyain) (iv) Land records (Bhu Abhilekh) (v) Net Supported Health Services (Swasthya Sewa), (vi) Rural E-mail (Gram Daak), (vii) Rural E-auction (Gram Haat), (viii) Matrimonial Alliances (Vaivahiki), (ix) Information Regarding Government Programmes (Suvidha), (x) Ask

Savaliram (Savaliram se Puchhiye), (xi) Ask the Experts (Salahkar), (xii) Intimation on Social Issues (Samaj Seva), (xiii) Online Employment Exchange (Rozgardoot), (xiv) Application Forms (Avedan), (xv) Local weather Report (Meghdoot), (xvi) Transparency in Government, (xvii) Functioning (Pardarshita), (xviii) Emergency Services (Teevradoot), (xix) Time-tables (Samay Sarini) (xx) E-agenda (e-Prajatantra), (xxi) E-newspaper (e-Samachar), (xxii) Rural Innovators (Avishkarak), and (xxiii) E-education (Gyanmitra).

E-Panchayat

E-Panchayat is a comprehensive suite of Panchayat applications designed and developed by NIC to effectively solve the information management problems at the village level. It benefits the citizens, the elected representatives, the Gram Panchayat and other village level officials, the Administrators and Planners at district and state level. For the first time in the country e-Panchayat has been introduced in Andhra Pradesh by National Informatic Centre (NIC). As a pilot in Ramachandrapuram Gram Panchayat in Medak District near Hyderabad, all the functions of the Panchayat are computerised and web enabled. Internet based services for Birth and Death Registrations, House Tax Assessment Collections, Trade Licenses, Old Age Pensions, Works Monitoring, Financial Accounting, Management and Information System (MIS) for Panchayat Administration are being provided as part of e-Panchayat system. Additional services such as market prices and agricultural extension advice are also being provided to the citizens of the village through e-Panchayat. This project is also being implemented in some more places such as Denduluru, Pedapadu in West Godavari District, Anantapur Rural Village Panchayat. It is going to be extended to all villages in Andhra Pradesh in a phased manner and can also be extended to other States by appropriate adaptation. The National Panchayat Portal, accessible through http://panchayat.nic.in, will eventually act as the front end to deliver e-Panchayat services at a later stage.

Conclusion

E-Governance can use IT as a vehicle to create a knowledge society leading to the improvement in life of a common man. The only handicap in the e-governance system is that in India, there are a large number of illiterates and even a large number of computer illiterates. Technology is a tool in this effort. In the modern age, the latest technology and scientific knowledge has become imperative for maximizing social welfare of the people. Undoubtedly, a lot of progress has been made within the last decade or so in the realm of innovations and their applications. However, much is left to be done in our rural areas. We are not losing sight of the fact that unless life in the rural areas is improved with the help of modern technology, it would not be correct to say that India has been progressing in maximizing its social welfare.

Information and communication technologies (ICTs) are playing increasingly important roles in the day-to-day lives of people, revolutionizing their work and leisure and changing rules of doing businesses. Development of ICT leads up to E-governance, which has now become the most talked word around the world.

References

Baweja, Yogesh (2005), "Does E-Governance means Good governance", 'Yojana' Vol. 49, Ministry of Information and Broadcasting, New Delhi, August.

Chaurasia, Alok Ranjan (2003), "Information Technology in Governance – Potentials and Prospects", Centre for Excellence in Information Technology, RCVP Noronha Academy of Administration and Management, Bhopal, September

Government of India (2007), "Economic Survey (2006-07)", Ministry of Finance, Economic Division, Govt. of India Press, New Delhi.

Kaushiva, Anamika (2005), "E-Governance for Good Governance-The National E-Governance Action Plan", Kurukshetra Volume 53, No. 5, March.

Khitoliya, R. K., and Tulsi Khitoliya, (2003), "Role of Information Technology in E-Governance", Punjab Engineering College, Chandigarh.

Mani, Sunil (2007), "Revolution in India's Telecommunications Industry", Economic and Political Weekly, Vol.XLII, No.7, PP 578-580, February 17-23.

Misra, D.C., and N. Vijayaditya, (2006); "Development Informatics – Reaching the Rural India-Role of NIC" Article 2 , Volume 16, No. 1, 'Information Technology in Developing Countries' International Federation for Information Processing (IFIP), Working Group 9.4, Centre for Electronics Governance, Indian Institute of Management, Ahmedabad, February.

Murthy, Narayana N.R. (2007); "The Path Long Travelled and The Journey Ahead", 'Yojana', Vol. 51, Ministry of Information and Broadcasting, New Delhi, August.

Planning Commission (2001), "Report of Task Force on Panchayati Raj Institutions (PRIs)", December, New Delhi.

Report of the Twelfth Finance Commission, (2005-2010), Government of India, New Delhi.

Sanjay, B.P. (2000), "E-Governance: A Perspective on State, Market and the NGO Sector", Seminar Paper presented to "the 3rd AMIC- India National Conference on E-Governance and Democracy in the New Millenium: Challenges and Opportunities", Chennai- December 11-12.

Sachdeva, Sameer (2002), "E-Governance Action Plan For India", Paper of Dream Government Online, December.

Soni, Sunil (2003), "Good Governance or E-Governance", Directorate of Information Technology, Municipal Administration, Maharashtra State, E-Library October 14. (Available on the website: http://www.maharashtra.gov.in/english/eLibrary/index.php).

Taganas, Rey A.L. and Vijay Kumar Kaul (2006), "Innovation Systems in India's IT Industry-An Empirical Investigation", Economic and Political Weekly, Vol.- XLI, No.39, PP No. 4178-4185, September 30-October-6.

The Economic Times (2007), "Union Budget-2007-08" Lucknow Edition., March 01. retrieved from the website: www.economictimes.com.

Vijayanand, K. and Ananda Sagar K. (2003), "Information and Communication Technology: An Indian Perspective", Paper of GIS Development. (Available on website: http://www.gisdevelopment.net/proceedings/mapasia/2003/papers/i4d/i4d004.htm).

14

E-governance in Vietnam

Pham Van Bo

Introduction

In today's fast changing and globalizing world, Information and Communication Technology (ICT) has become a powerful instrument for boosting the competitiveness of nations and in facilitating economic growth. Technology offers the hope of helping countries use the economies of scale generated by information advances to cut costs while also improving productivity and service delivery. Governments who use ICT effectively are able to lower their unit cost of operations and produce the same (or higher) outputs with less money as well as less time. [1]

This study examines the current ICT and E-governance environment in Vietnam. It also examines the investment and commitments the Vietnamese government has made in recent years and explores Vietnamese policy framework for e-governance and public administration reform (PAR). Besides we assess Vietnam's e-readiness based on results of differing methodologies from United Nations Online Network in Public Administration and Finance (UNPAN), the Economist Intelligence Unit (EIU), and other indicators. The paper also compares Vietnam's e-readiness with other countries' e-readiness by e-readiness ranking.

Development of Vietnam's E-governance

Policy Framework for E-governance: The Vietnamese government has long recognized the important role of ICT in faster economic development of the country. Since 2000, a number of strategies, laws, decrees and decisions have been approved and a number of government and party committees and boards have been established to implement these plans. In e-governance, besides the 112 projects [2] and its steering board, following two other groups has also influenced e-governance development efforts (Table 14.1):

- The IT Steering Committee, established by the Communist Party of Vietnam (CPV); and
- The National Steering Board of Information Technology, of which the administrative office is under the supervision of the Ministry of Post and Telematics.

Officially launched in 2001, Vietnam's E-governance development has focused on three main tasks:

- Computerizing government administration, especially at the national and provincial levels, equipping government offices with networked computers and connection to the Internet and training IT skills for government officials.
- Building government websites;
- Implementing pilot e-government projects in customs, banking, and taxation.

Table 14.1: Milestones of E-governance Development

Document, Issuer	Date	Core Issues
Decree 58 – CT/TW, CPV Politburo	October 17, 2000	Urging the government to speed up application and development of Information technology to serve industrialization and modernization
E-ASEAN Framework Agreement, The fourth ASEAN informal Summit (in Singapore)	November 25, 2000	Setting up the framework for cooperation amongst ASEAN's member countries in embracing the benefits of the IT revolution, aiming to (a) Facilitate the growth of electronic commerce in ASEAN (b) Promote and facilitate the liberalization of trade and investment in production of IT products and services; (c) Develop an e-society in ASEAN and capacity building; (d) Promote e-government
Decision 112/2001/QD-TTg, Prime Minister	July 25, 2001	Approving the project "State Administrative Management Computerization, Period 2001-2005 "
Decision 136/2001/QD-TTg, Prime Minister	September 17, 2001	Approving the master plan on state administrative management reform period 2001-2010
Decision 137/2001/QD-TTg, Prime Minister	September 17, 2001	Establishing the executive board of the project "State Administrative Management Computerization period 2001 - 2005"

Source: Compiled by author from various documents

CT: Decree, TW: Central Committee, CPV: Vietnamese Communist Party

QD: Decision, TTg: Prime Minister

In reference to this issue, Mai Liem Truc, Deputy Minister of Post and Telecommunication said. "If administrative systems are not improved and administrative procedures still overlap between different government bodies, we cannot effectively implement e-governance." In order to carry out e-governance in Vietnam, he said, it is necessary to first carry out administrative reform. The following section explores about public administration reform.

Public Administration Reform

In 2001, the Prime Minister approved the Public Administration Reform (PAR) master plan 2001-2010. In order to achieve the goals of the PAR master plan and develop a modern administrative service, seven action programs were identified:

- Improvements of legal documents;
- Reform of roles and functions of administrative agencies (Organizational Restructuring);
- Resizing of civil service;
- Training and retraining of civil servants;
- Reform of compensation of civil servants;
- Reform of state financial management;
- Modernization of administration system, including computerization and e-governance initiatives;

Public Administration Reform is an essential element for e-governance success and will also impact project 112. The PAR master plan also contained a program aimed at reforming and modernizing the state administrative system. Public administration reform in general has been slow in all areas. Vietnamese administration across ministries and agencies is still highly bureaucratic with very untidy systems, which are highly paper based. There has not been any significant reform of these systems, which makes it very difficult to develop e-governed solutions for business and citizens and also for the bureaucracy themselves. Both state administrative management computerization (SAMcom) or project 112 and PAR are mainly supported to buy hardware, equipment and reduce transaction cost. The major objectives are network connection to use emailing and internet surfing for government civil servants, with few using the new hardware or systems to deliver services to the public.

E-readiness for E-governance

We use three major e-readiness assessments: UNPAN e-readiness assessment, 2005; EIU e-readiness rankings, June 2007; and key economic and IT indicators.

Each of these assessments looks at e-governance in different ways, and is based on differing methodologies. We have found Vietnam to be lagging behind not only its Asian neighbors but also much of the world.

UNPAN Global-Government Readiness Report: Web measure is evaluated through an assessment of website content (one of five e-governance stages centering on the degree of citizen services and feature). The telecommunications index assessed telecommunications infrastructure (PCs/1000 people, Internet users/1000 people, telephone lines/1000 people, mobile phones/1000 people, and online population). Finally, the human capital index measured countries human resource endowment (adult literacy and student enrollments). Table 14.2 shows position of Vietnam and selected countries in Asia.

Table 14.2: UNPAN E-readiness Assessment Results in Asia, 2005

Rank out of 191 countries		Country	Web Measure Index	Infrastructure Index	Human Capital Index	E-government Readiness Index	
2005	2004					2005	2004
5	5	Republic of Korea	0.9769	0.6713	0.9700	0.8727	0.8570
7	8	Singapore	0.9962	0.6448	0.9100	0.8503	0.8575
14	18	Japan	0.8154	0.5850	0.9400	0.7801	0.7260
41	47	Philippines	0.7423	0.0840	0.8900	0.5721	0.5260
43	42	Malaysia	0.5769	0.3048	0.8300	0.5706	0.5409
46	50	Thailand	0.6654	0.1299	0.8600	0.5518	0.5096
57	67	China	0.5692	0.1241	0.8300	0.5078	0.4356
73	63	Brunei	0.2462	0.2264	0.8700	0.4475	0.4360
87	**86**	**India**	**0.5827**	**0.0277**	**0.5900**	**0.4001**	**0.3879**
94	96	Sri Lanka	0.3192	0.0359	0.8300	0.3950	0.3748
96	85	Indonesia	0.2962	0.0494	0.8000	0.3819	0.3909
105	**112**	**Vietnam**	**0.2231**	**0.0489**	**0.8200**	**0.3640**	**0.3378**
128	129	Cambodia	0.2308	0.0060	0.6600	0.2989	0.2860
129	123	Myanmar	0.1538	0.0040	0.7300	0.2959	0.3030
147	144	Laos	0.0788	0.0074	0.6400	0.2421	0.2330
162	159	Bangladesh	0.0731	0.0055	0.4500	0.1762	0.1788

Source: UNPAN 2004 & 2005

In 2005 Vietnam had increased its ranking in the UNPAN survey from 112th /191 in 2004 to 105th / 191 in 2005. In comparison to its Asian neighbours, Vietnam remained at the lower end of the table, 12th out of 16 Asian countries ahead of Cambodia, Myanmar, Laos and Bangladesh but substantial distance behind Thailand, Malaysia and the Philippines. South Korea remains the leading Asian government in e-government readiness. India ranked higher position than Vietnam in 2004 and 2005.

Economist Intelligence Unit (EIU): Nearly 100 quantitative and qualitative criteria, organized into six distinct categories, feed into the e-readiness rankings. The six categories (and their weight in the model) are connectivity and technology infrastructure (20 percent); business environment (15 percent; the nine criteria used here summaries the 70 indicators covered by the Economist Intelligence Unit's business environment rankings); social and cultural environment (15 percent); legal and policy environment (10 percent); government policy and vision (15 percent); and consumer and business adoption (25 percent).

The data used in the rankings are sourced from the Economist Intelligence Unit, Pyramid Research, the World Bank and other organizations. Qualitative criteria are assessed by the Economist Intelligence Unit's extensive network of country experts, and their assessments are reviewed by their top economists. See table 3 for Economic Intelligence Unit e-readiness ranking (Table 14.3).

Table 14.3: Economist Intelligence Unit e-readiness Rankings, 2007

2007 e-readiness (of 69)	2006 e-readiness (of 69)	Country	2007 e-readiness score (of 10)	2006 e-readiness score (of 10)
4	10	Hong Kong	8.88	9.00
6	13	Singapore	8.60	8.24
16	18	South Korea	8.08	7.90
18	21	Japan	8.01	7.77
36	37	Malaysia	5.97	5.60
49	47	Thailand	4.91	4.63
54	**56**	**India**	**4.66**	**4.04**
54	50	Philippines	4.66	4.41
56	57	China	4.43	4.02
61	59	Sri Lanka	3.93	3.75
65	**66**	**Vietnam**	**3.73**	**3.12**
67	62	Indonesia	3.39	3.39

*New to the annual rankings in 2007
Source: Economist Intelligence Unit.

From Table 14.3, we find that in the 2006 and 2007 EIU study, Vietnam ranked 66th and 65th respectively out of 69 countries rated. This was low position ranking of any Asian country in the study. Out of total 10 points, Vietnam received only 3.12 and 3.73 points respectively. The highest ranking Asian nations were Hong Kong (4th in 2007, 10th in 2006), Singapore (6th in 2007, 13th in 2006), South Korea (16th in 2007, 18th in 2006), and Japan (18th in 2007, 21st in 2006). India ranked 56 in 2006 and increased its ranking to 54 in 2007 and obtained 4.04 and 4.66 respectively. Vietnam is ahead of Indonesia in 2007.

Vietnam's ICT Infrastructure: Population of Vietnam is over eighty million; average annual GDP growth rate is 7 percent that can make success in innovation and creativeness of technologies. But Vietnam has reformed public administration system slowly, so e-governance development is limited and faulty. Tables 14.4 and 14.5 give a panoramic view of the Vietnam's ICT infrastructure.

Table 14.4: Key Economic and IT Indicators: Vietnam vs. Asian Countries, 2003

Country	Population (million)	GDP per Capita	IT Spending		ICT Penetration per 1,000 Inhabitants			
			US$ per Capita	As percent of GDP	Main line telephone	Mobile phone	PC	Internet User
Japan	128.0	46,108	2,489	7,4	472	679	408	667
Singapore	4.3	26,992	2,254	10.5	450	852	761	602
Hong Kong	6.8	26,247	1,921	8.4	559	1,079	461	707
South Korea	47.9	14,623	842	6.7	538	701	58	633
Malaysia	24.8	4,965	289	6.9	182	442	67	379
Thailand	62.0	3,184	82	3.5	105	394	48	128
Philippines	81.5	1,240	57	5.8	41	270	35	93
Indonesia	215.0	1,091	33	3.4	39	87	13	70
China	1,290.0	1,028	58	5.3	209	215	40	79
Sri Lanka	19.2	940	54	5.7	49	73	19	13
India	**1,060.0**	**528**	**21**	**3.7**	**46**	**25**	**9**	**36**
Vietnam	**81.3**	**438**	**35**	**7.3**	**54**	**34**	**11**	**64**
Bangladesh	138.0	410	10	2.7	5	10	8	2

Source: World Bank Development, 2004; WITSA Digital Planet (2004); Internet World Stats, June 2005.

The quality of ICT infrastructure is a crucial ingredient in the long-term success of e-governance initiatives. In looking at the potential for electronic government in Vietnam, it is useful to compare Asian countries on key economic and ICT indicators. As a percentage of GDP, Vietnam's ICT spending is competitive with many other Asian countries. Vietnam spent 7.3 percent of its GDP on ICT (the development of information technology and media applications from 2006 to 2010, Vietnam plans to invest a total of VND (Vietnames dong) 6.5

trillion (US$411.1 million) to develop its e-governance systems), while Japan spent 7.4 percent, Hong Kong 8.4 percent, and South Korea 6.7 percent. Indeed, with the exception of Singapore (10.5 percent), Hong Kong and Japan, Vietnam devoted a higher share of GDP to technology than any other Asian nation. India spent on ICT only 3.7 percent of GDP.

Table 14.5: Internet Development: Key Indicators for Vietnam, 2003-2007

Indicator	June 2003	June 2004	June 2005	June 2006	June 2007	Growth (times)			
						2004 /2003	2005 /2004	2006 /2005	2007 /2006
Number of Equivalent Subscribers*	0.46	1.26	2.85	3.01	4.60	2.7	2.3	1.1	1.5
Number of Internet Users*	1.90	5.11	7.71	11.09	16.51	2.7	1.5	1.4	1.5
Internet Penetration (as percent of Population)	2.4	6.3	9.4	13.36	19.87	2.6	1.5	1.4	1.5
Total International bandwidth (Mbps)	349	1,096	2,690	3,615	8,703	3.1	2.5	1.3	2.4
Total Number of Sites with. Vn	2,907	7,376	11,032	20,020	43,575	2.5	1.5	1.8	2.2

* Figures in millions
Source: Internet World Stats

Despite this investment Vietnam still lags behind most countries in Asia in terms of ICT penetration with 54 main line telephone users per 1,000 people, 34 mobile phone users, 11 personal computer users, and 64 internet users. This is much lower than the comparable figures for South Korea (538 main line telephones, 701 mobile phones, 558 personal computers and 633 Internet users) and Malaysia (182 main line telephones, 442 mobile phones, 167 personal computers, and 379 Internet users).

However, internet users in Vietnam are increasing year by year. Table 14.5 shows that Vietnam had 1,903,640 internet users in June 2003, but this figure reached 16,511,849 internet users in June 2007. As a percentage of the population, its Internet penetration level increased from 2.4 percent in June 2003 to 19.87 percent in June 2007.

Number of equivalent subscribers have also increased annually (2004/2003: 2.7 folds, 2005/2004: 2.3 folds, 2006/2005: 1.1 folds, 2007/2006: 1.5 folds). Vietnam had only 465, 710 subscribers in 2003 and reached 4,603,296 subscribers in 2007. Vietnam's international bandwidth has also risen dramatically, from 349 mbps (million of bits per seconds) in June 2003 to 8,703 mbps in June 2007. Total number of web sites *with .vn* in the address has risen from 2,907 in June 2003 to 43,575 in June 2007.

Therefore, Vietnam has gained progress of ICT development for current year. However, it is difficult for business sector to find information, guidance or analysis in Vietnam's websites as it is found these in foreign websites. These websites are not clear in structure, and not appropriate to serve business and citizens.

Conclusion and Suggestions

Vietnam's e-governance development has been poorly designed, not properly coordinated and not well managed leading to substantial losses of resources, time and creditability for E-government Vietnam. E-governance initiatives have been too focused on modernization and purchase of hardware for government agencies. Public administration reforms are carried out slowly. Untidy complex administration system has inhibited development of even basic e-government. All initiatives (112 project and Public Administration Reform initiative) only focused on hardware, did not rely on real demand of organizations, business units and citizens.

Investment in Information and Communication Technology was about 7.3 percent of GDP to promote potential of competition in the region. Vietnam had highest growth rate of internet penetration. Number of websites .vn has risen excellently.

E-government and PAR did not integrate in sight not leading to gains effective e-government.

Taking the above-mentioned findings, we proposes following suggestions:

- Establishing of an e-governance steering committee in charge of the project 112, PAR initiatives.
- Investment in hardware and modernization takes place when reform of public administration is carried out and system has been reformed. Two programs must assist each other.

- The service needs of citizens and businesses are addressed and e-governance initiatives and PAR meet these needs rather than taking a supply side approach to projects
- Strong IT leadership, IT human resources development plays a very important role in implementing e-government.
- Strategic investment in e-government programs
- Gaining knowledge of e-government best practices and initiatives across the world
- Pilot projects (quick wins) on e-government development are important to assure the success of each e-government program to be invested in full-scale.
- Consultations, seeking feedback, cooperation, and engagement of the private sector and international partners in e-government development efforts.
- Benchmarking measures and e-government readiness ranking indices should be developed and widely used on a regular basis to monitor and promote e-development at both national and local level.

Notes

1. Heeks (2001), "Understanding E-Government for Development" i-development paper No. II, IDPM, University Manchester, U.K.
2. Project 112: Project "State Administrative Management Computerisation, Period 2001-2005".
3. The term E-governance and E-government are interchangeably used in this paper.

References

Heeks, R.(2001), "Understanding E-Government for Development", i-development paper No. II, IDPM, University of Manchester,U.K.

Tuan, D. (2005), "Shaky Step Forward", Vietnam Economic Times, No.136 [2005-06-01]

Vietnam Communist Party, 2000, Decree 58-CT/TW of the Vietnamese Communist Party, Hanoi, Vietnam.

Vietnam Government, 2001, Decision 112/2001/QD-TTg "State Administrative Management Computerization Period 2001-2005", Hanoi, Vietnam.

Vietnam Government, 2001, Decree 136/2001/QD-TTg, "Master Plan for State Administrative Management reform for the period of 2001-2010", Hanoi, Vietnam.

Vietnam Government, 2001, Decree 136/2001/QD-TTg, "State Administrative Management Computerization, Period 2001-2005"

VNCI, 2006, E-Government and Business Competitiveness –A Policy Review, Hanoi, Vietnam.

World Bank, 2005, Project Appraisal Document for an ICT Development, Hanoi, Vietnam.

Internet World Stats http://www.internetworldstats.com

UNPAN (United Nations Online Network in Public Administration and Finance) 2004,2005 available at http://www.unpan.org/egovernment.asp

EIU (Economist Intelligence Unit) http://www.eiu.com/index.asp?rf=0

Part III: Social Aspects of IT Sector

15

Parental Reactions to Daughter's Unconventional Work Hours at Call Centres in India

Shelly Tara and P. Vigneswara Ilavarasan

Introduction

India's large English-speaking and low wage labour pool has helped to establish it as one of the desired destination for business processing outsourcing (BPO). The rise of Indian call center business, also called as business process outsourcing industry, has been impressive from US$6.3 billion in 2005-06 to US$8.4 billion by 2006-07. The ITES-BPO employee base has grown from 4,15,000 in 2006 to 5,53,000 in 2007 (Nasscom, 2007). Employment opportunities provided by this sector has been exploited by the educated middle class people in India, especially women. Indian call center work is transnational in nature where workers physically located at Indian locations serve the Western clients virtually. This means working at night hours to be present at the day time of the Western customers. In order to achieve better service delivery, Indian workers are trained to understand the Western culture which also includes adoption of a Western pseudo name in work, (for instance, Nitesh become Nick), and receiving holidays for the Western festivals. Higher wages, an one year old call center worker earns Rs. 18000 per month, availability of exhaustive consumer goods after liberalization, and low family responsibilities have provided enough space to the call center workers to experiment with the Western habits in the Indian setting (Updhya and Vasavi, 2006).

Given high participation of women, 40 percent, as claimed by the industry (NG and Mitter, 2005), it is important to look at the reaction of the parents of women working in call centers.

Indian society is predominantly patriarchal and offers restricted roles to women in the economic domain (Uberoi, 2003). The absence of women from their homes during night time had always been linked to questionable moral values, especially in the middle class families of India. But call center jobs have introduced a new work culture for women in India, where night shifts have become synonymous of working in call centers. Extant research on call centers does not offer much insights to parental reactions to their daughters' unconventional working hours (night shifts) at the call centers and this study attempts to fill this gap. The paper also tries to understand whether the relationship between daughter and other family members are altered.

Earlier Research

The literature on the work and family dynamics of working women in the sector other than call centers reveals that women are supposed to make more compromises if they are working.

Long working hours and inflexibility of work schedule have been some of the major concerns of the working women and their families as their interaction with their family members depend on these factors (Kim and Ling, 2001).. The gender perspective (Milkie and Peltola, 1999) suggests that women would report more work family imbalance and women would make more trade-off than men especially in work adjustments due to family responsibilities. A woman's immediate and extended families along with domestic cultural expectations were perceived to be the biggest stumbling block in combining work and family life. The families do not appreciate the demands of the managerial or professional occupation faced by the working women (Rana et al. 1998).

Greenhuas and Beutell (1985) have narrowed down some of the factors leading to pressure in performing work and family roles. Time based conflicts, work schedules, work orientation, marriage, children and spouse employment patterns may all produce pressure to participate extensively in the work role or family role (Greenhuas and Beutell, 1985).

The literature on working women in non call center sector suggests that the families of the working women react negatively when the work demands more time and the work schedule is inflexible. Women are supposed to do more adjustments to balance their work and family life. This literature does not talk much about night shifts; however, the literature hovering around women employees of call centers gives some insight about parental responses towards the night shifts and the impact on the family life of the women employees.

Patel (2006) highlights negative reactions by family members on night shifts as the presence of women in India's urban nightscape is often linked to prostitution and questionable moral values. She also argued that family concern for young women working the night shift is less about physical safety and more about how a woman's presence in the urban nightscape will negatively impact a family's reputation, since in middle-class morality, women's bodies are sites of family purity and honour.

Singh and Pandey (2005) provide a contrary picture about the responses of the families on the call center jobs. According to them the families did not consider it objectionable to allow women into such professions. In lower income families, such jobs were actually required for sustenance and families were very supportive of girls in such jobs even if it meant doing night duties. Also, among higher income groups, such jobs are acceptable where women are working in the call centers as a pastime (Singh and Pandey, 2005). Singh and Pandey (2005) conclude that with such jobs being available to women, certain changes in India's patriarchal system becoming evident. Call center employers recognize the value of parental acceptance of women's work. In annual get-togethers, family members and relatives are invited to clear their misconceptions about working in the night shift (Ng and Mitter, 2005).

Ramesh (2004) mentions that the leave structure has an impact on the family life of the women employees. The workers are allowed to avail leave only with prior consent of the team leader/manager, to ensure that the work is not disrupted. Availing leaves without prior consent is treated as unauthorized absence, which is a sufficient reason to terminate the services of the workers. Even if the worker is sick, it is mandatory to get the consent of the team leader at least four to six hours before the shift, failing which the leave would be considered as an 'unscheduled leave'. All leaves (except casual leave) are not applicable for probationers. Further, as the call centers are functioning according to the time-schedule of the clients of the firm, the employees are not eligible for national holidays (Ramesh, 2004). Thus, there is a possibility of workers, including women missing the enjoyment of holidays with their family members. Double salary offered by the call center for working on festivals reduces the chances of applying for leave during festivals (Mcmillin, 2006). This further adds in the lesser interaction between a call center worker and his family.

Also, by working the night shift, and inevitably sleeping thru the day, it is possible that women become further excluded from social and economic opportunities within the larger community (Patel, 2006).

The literature related to call centers, though, talks about fathers having negative reactions and gradual adjustment with night shift jobs of their daughters but it does not deal much with the reasons of such adjustments. It also does not discuss that between father and mother who is more against the night shift job or whether they have different perceptions. Regarding the work and family dynamics the literature tries to relate the work culture of call center and the interaction with other family members but it does not explain the reactions of the family members as well as the daughters towards lesser or altered interactions with each other. This paper examines the above issues through a qualitative study.

Methodology

Empirical qualitative data has been collected through semi-structured in-depth interviews with the women who are either working or have worked in call centers located in the National Capital Region and Jaipur. All the respondents are selected by using snow ball sampling technique, [1] as the sector is known for strict restrictions against external studies. In terms of time dimension the study was cross-sectional, as the subjects were approached at one point of time only. Present paper discusses the findings based on eighteen in-depth interviews. Such a small size sample is not uncommon among studies in this emerging sector. For instance, Fuller and Narsimhan's study (2007) has only nineteen respondents.

In the interviews, the following dimensions are covered to understand the work-family dynamics of women workers in call centers and the relationship between daughters (who are working in call centers) and other family members:

- What were the initial responses of their parents?
- What is the impact of their job on the interaction with their family members?
- What is the impact on their marriage prospects and their parent's concerns regarding the same?
- What are changes in the expectations of the parents from their daughters?

Out of total 18 respondents, nine respondents are from National Capital Region, and nine are from Jaipur (Table 15.1). Jaipur is selected for comparative purposes as it is more traditional and not typically metropolitan city but is one of the most successful tier II cities for call center industry.

Table 15.1: Sample Characteristics

		Working	Ex-workers	Total
Delhi	Married	1	2	3
	Unmarried	4	2	6
Jaipur	Married	1	0	1
	Unmarried	7	1	8
Total		13	5	18

Out of 13 working women one is working as voice and accent trainer and one is working as Assistant Manager. Both of them are unmarried and earlier used to work as customer care executive. Rest of the11 respondents are working as customer care executive. The age group of the working women is 21-26, except one lady who is in early 50's. The views given by this lady are provided separately in a case study. All of them have been working in call center for at least more than a year and all the married respondents had started working in call center before their marriage (that lady being the exceptional). The purpose of taking ex-workers is to understand the differences experienced by them in the interaction with family member after leaving call center job. Most of the interviews were conducted at their home. In order to elicit valid and reliable information, the respondents were promised to protect their identities.

Findings

This section starts with the parental response towards call center job which is followed by explaining the different dimensions related to the positive response of the parents and how it brings changes in the work and family dynamics vis-à-vis the relation between parents and daughters.

None of the respondents' father had any problem with the night shifts of call center. All the respondents said that their fathers were "cool" about the night shifts. The reason for such response was the interaction done by fathers with their colleagues and friends at work place. This interaction convinced them as their colleagues were also allowing their daughters to work in call center. Nine respondents said that their mothers were really upset with the night shifts as they were housewives. One respondent, who really had to persuade her mother to work in call center said, "Though my elder sister was already working in call center but my mother was unhappy when I too decided to join call center. She wanted me to do 9 to 5 job". According to all of the nine respondents, the security of their daughter was the main concern of mothers. Later on mothers got adjusted to the night shifts when they observed the transport system of call centers and realized that it was safe to work in call center during night. Four respondents very gladly told that none of their parents had any problem with night shifts as their brothers were working in the same sector and they told their parents that the call center environment is very safe even for girls. According to four other respondents their cousins were already working in call centers, hence, the parents were aware of its night culture. Therefore, none of the parents was displeased with call center job of their daughter. One of them said, "My mother forced me to go for call center job even though it was in night shift as my cousins told her about the facilities provided by call centers and she got very impressed with that".

Respondents from Jaipur explained that Jaipur has been a very conventional place but still parents were able to send their daughters out in the night because of the reputation developed by the organization there. The organization was able to build up a positive reputation for providing safe work environment to its female employees consistently. The responses about the parental reaction to night shifts reveal that night shift was not a big problem for parents, only mothers resisted initially but latter on they also got adjusted. The feedback from their son, friends, and relatives about night work culture of call centers really helped parents to comfortably send their daughter out in the night to work.

After the initial reactions from the parents the next unique outcome of a call center job is lesser interaction with the family members. Almost every respondents addressed to this aspect of call center job. According to them, they hardly get chance to interact with their parents and spend time with them during week days. With this study the reaction of the parents as well as of the respondents were taken into account towards the lesser interaction with each other. Almost every respondent said that they do not complain about the lesser interaction with their parents as they have got adjusted to it and now it is a part of their life.

According to one girl, "I thought it is a demand of my job and I need to get adjusted to it". One respondent who stays away from her parents said, "Whenever I get time after my shift I talk to my parents and I do not feel there is any communication gap".

Respondents were very satisfied with the two days off given in call center job. They said two days offs are the best part about their job. During these two days they could spend time with their parents and could help in household work.

Even on festivals these girls stay away from their families and work in their organization but they take it as yet another demand of their job. One respondent from Jaipur said, “We have to be professional with our work. We can not crib all the time about not being with our families”. Respondents revealed that they have very nice festival celebrations at their call centers. One girl from Delhi said, “Call centers celebrate festivals in a better way than we celebrate at home. We have Rangoli competition, we burst fire crackers. In fact people do not like to take holiday on that day. We get compensatory off and double salary for working on that day”.

When asked about the parent’s response towards the lesser interaction with their daughters then they said that parents understand their work and are o.k. with lesser interaction due to odd timings. One respondent was really grateful to her parents and said, "If they would have not understood my work conditions then I would not have been working here. It has become possible because of their support only. When you work in call center then parents take care of your other daily needs as you do not get time to do all that". According to ex-workers though their present job is giving them relatively more time to spend with their family members but when they had joined call center the lesser interaction was never a problem with the parents. One of the ex-worker said, "When I used to come back from my shift, I used to feel so exhausted that I never felt like talking to my parents. But they were very understanding". Some of the respondents, whose mothers were upset with night shifts, admitted that initially it was little problematic for the mother when the interaction got reduced with their daughters but soon they understood the demands of the job and got normal with that.

Respondents from Jaipur explained that being from a conventional place, daughters have always been expected to help mothers in household work like cooking or going on shopping with mother but now such expectations have shifted to weekends as they sleep in day time during working days.

According to all the respondents the call centers organize lots of parties for employees. These parties are conducted in pubs, discos or in some hotels. Parents allow their daughters to attend these parties as they are conducted by the organization. Regarding the night outs planned with colleagues and friends; the respondents in Jaipur denied about such plans but were happy that parents allow them to go to night parties of their organization. Some of the respondents from Delhi said that they manipulate their parents to go out in the night with their colleagues. But some respondents confidently told that their parents know about their night out with colleagues as well. One of the respondents from Delhi said, "We do have parties at disco and pubs by our organization and our parents know about that. Those parties are conducted in night only and our parents allow us. They do not mind our night outs with colleagues as well because our night parties and enjoyment is not unusual for them".

Parents have not only changed their attitude towards their daughter's night shift work but towards attending late night parties by them as well. Even in Jaipur, where unlike Delhi night life has never been so active for youngsters, parents are not against the night parties attended by their daughters.

The findings reveal one more aspect related to daughter's call center job. Parents face difficulty to find a suitable match for their daughters especially in arrange marriage. Respondents themselves admitted this fact that the groom's family starts giving a second thought as soon as they find that the girl is or used to work in call center. According to one girl who left her call center job as she was looking for some permanent settlement in life with marriage, " if people get to know that the girl used to work in night shifts then they think that she must be involved in so many relationships. Society has a very negative impression about the call centers". One call center worker said, "In arrange marriage when you tell that I work in call center, so they think that she works in night shifts, she might be meeting different kind of people or might be indulged into bad practices. Most of the girls smoke in call center, so these things hit people some where". Only one girl from Jaipur gave a different view about this issue who said, "I do not think people have any bad impression about call center. They take it as

MNC(Multi National Company) job and ask the girl to leave the job as in Jaipur people want a housewife".

While discussing about the same issue the respondents commented that one can have a good married life in call center job if your spouse is also from the same sector. So most of the people go for love marriages within the same sector and this is something which has become very common in call centers. One of the respondents from Delhi said," people have affairs here only and they get married without telling there parents because they are not from Delhi". Another respondent from Delhi said, "Having a boyfriend is a very important thing in call center you have to be very strong if you do not have a boyfriend as you feel left out. People try to find out someone whom they can confide into because they are cut from their families; they have no time with their family so they start searching their families outside. They need somebody who can listen to them". One girl from Jaipur said, "call center job can be compatible with marriage if your spouse work in the same organization or atleast in the same sector. Then you understand each other's job demands and spouse cooperates with you".

The marriage of the daughter which has been a most important concern of the parents is taking different shape with call center jobs. Though parents are aware about aversion in the society towards girls working in call centers but it does not seem to be a big problem. There is a way out of marrying someone from the same sector but here parents have lesser role to play.

Discussion

The findings of the study reveal that the parental, especially father's, reaction to the night shift work of their daughters is very supportive. Though in some cases mothers were reluctant initially about night shifts of call center but after some time even they got convinced. Even in II tier city, Jaipur, none of the respondent said that their father had any problem sending their daughter out in night to work in call center. The reputation established by the company and the feedback given by other female employees about safe work environment in night shifts of call centers, played a substantial role to make parents comfortable.

The lesser interaction between parents and daughters as an outcome of night shifts of call center work has also not been posed as a problem for any of them. Female workers accept it as part of their job and are happy to spend time with their parents on weekends. On the other side even parents understand the demands of their daughter's job and do not complain when their daughters do not spend time with them or help their mothers in household work during weekdays. The absence of daughters during festivals has also been accepted as part of their job. Such findings make respondents from Jaipur to accept that Jaipur is taking a leap from a conventional place where the expectations from daughters used to be different at home.

Night parties of call center have also been accepted by the parents as a part of the professional life of their daughter. In some cases parents permit their daughter to go for night outs with colleagues as well. Night parties are new phenomena in the middle class families, in fact in Jaipur it used to be absent. But parents do not react negatively towards it and allow their daughters to attend night parties even thought they are conducted in pubs or discos.

The negative perception in the society about call center job does not bother parents even it may create some problem at the time of their daughter's arrange marriage. Call center opens new avenues of marring some one from the same sector. All the respondents admit that it is very important to carry on call center job after marriage that your spouse is from the same sector and parents do not have problem with that. They think it is important for a good married and professional life of their daughter.

This study is an attempt to bring a new insight about parental responses towards night shifts of call centers. It reveals that the parents have started accepting the work culture of call centers and they allow their daughter to work even in night shifts. The sense of stigmatization about their daughters being out during the night hours is getting lost among parents. The expectations from daughters to take care of the household work are also getting reduced since they sleep during the day time and their mothers take care of their food and other needs.

However, the socio-economic background of the family for contributing in the responses of the parents should be covered in further studies. Though parents are "cool" about their daughter's night shift duties but whether these parents would be "cool" enough if their daughter-in-law does night shift duties in call center. The perception of the Indian families on having such daughter-in-laws who work in call centers with night shifts is also needed to be studied further.

Pratibha's Case Study (pseudo name)

A respondent from Jaipur, Pratibha, who is 54 years of age, was born, brought up and got married in Jaipur and has been working in call center for one year. Her views seem important to be discussed separately because, these days, even older people are joining call centers, hence, her views provide an insight in understanding about the response of the family members when some woman decides to join call center at such age.

Pratibha decided to join call center because she liked the ambience of the organization and found learning and growth opportunities there. Her daughter used to work in the same call center before her marriage and Pratibha got a chance to see her work place when she was invited in a get together organized by her daughter's organization.

Regarding the response of the family members when Pratibha decided to go for call center job, she said, "My daughter used to work in a call center earlier and she was dead against when I decided to join call center, even my husband resisted. They were not against the night shifts but they thought that I would not be able to cope up with that young environment and stress level. Then, they led me to have a try as they were sure that I will quit but did not. Gradually they saw that I have got adapted to my job so my daughter got normal and husband got supportive". Her daughter mentally prepared her about the work culture of call center, so according to Pratibha she did not get any shock when she saw girls smoking there.

When asked about her interaction with the family members, she replied, "I spend four hours with my family when I am awake and look forward to weekends. I hardly attend any family get-together but I feel it is part of my job". Pratibha attends parties organized by call center if the venue is not very far away and works on festivals also. When asked about her family's reaction towards her job's demands, she gladly said, "My family now feels that I need this job to be active and busy. Now they convince me not to leave this job as they think that I am meeting so many people regularly and it keeps me busy".

Being a native of Jaipur her views about the impact of call centers on the culture of Jaipur and on the perception of people about their daughter's work sound relevant. She said, "I feel good when girls come to me and tell me that they have saved so much of money out of their salary in few months. It never used to happen earlier with girls in a conventional place like Jaipur".

About marriage prospects of a female employee of call center Pratibha told that some people have bad impression about call center in Jaipur but she says the term "call center" gives wrong perception to people. According to her, "though people allow their daughter to work in call center as it can be a stepping stone in her career but they do not expect their daughter-in-law to work in call center". She confirms that marriage and call center job can go together if spouse is also working in the same sector. Apart from that she reveals that many marriages take place within organization only, because the social circle of the employees revolves around their organization.

She explained that call centers are certainly having their impact on the perception of people in Jaipur about their daughter's work as people have got liberal here to send their daughters out in night. Even her case study shows that her family was supportive towards her job and never had any problem with night shifts. Rather they think that her job keeps her busy and active.

Note

1. A non-probability sampling method often employed in field research whereby each person interviewed may be asked to suggest additional people for interviewing (Babbie, 2004).

References

Babbie, E. (2004),"*The Practice of Social Research.*", Belmont CA: Wadsworth/Thomson Learning.

Fuller, C.J., and Narasimhan, H. (2007), "Information Technology Professionals and New Middle Class in Chennai (Madras)", Modern Asian Studies, 41(1), 121-150.

Greenhaus, J.H., and Beutell, N.J. (January 1985). "Sources of Conflict Between Work and Family Roles", The Academy of Management Review, 10(1), 76-88.

Kim, J.L.S., and Ling, C.S. (2001). "Work and Family Conflict of Women Entrepreneurs in Singapore, Women in Management Review, 16(5), 204-221.

McMillin, C. D. (2006 January). "Outsourcing Identities, Call Centre and Culture Transformation in India"., Economic Political Weekly. Retrieved on February 9, 2007, from http://www.epw.org.in/bin/categorywisearch.php

Milkie, M.A., and Peltola, P. (May 1999), "Playing All the Roles: Gender and the Work-Family Balancing Act",. Journal of Marriage and the Family, 61(2), 476-490.

Nasscom. (2007),. "Indian ITES-BPO Factsheet",. National Association for Software and Service Companies, Retrieved on October 15, 2007, from http://www.nasscom.in/Nasscom/templates/NormalPage.aspx?id=2374.

Ng, C. and Mitter, S. (2005), "Valuing Women's Voices: Call Centre Workers in Malaysia and India", Gender Technology and Development, 9(2), 209-233.

Patel, R. (2006), "Working the Night Shifts: Gender and the Global Economy" An international E- journal for critical geographies, 5(1), 9-27, from http://www.acme-journal.org/vol5/patel.pdf

Ramesh, P.B. (2004 January),"Cyber Coolies in BPO: Insecurities and Vulnerabilities of non Standard Work". Economic and Political Weekly, Retrieved on February 9, 2007, from http://www.epw.org.in/bin/categorywisearch.php

Rana, B.K., Kagan, C., Lewis, S., and Rout, U. (1998), "British South Asian Women Managers and Professionals: Experience of Work and Family", 13(6), pp- 221-232.

Singh, P.,and Pandey, A. (2005 February), "Women in Call Centre". Economic and Political Weekly, Retrieved on March 15, 2007 from http://www.epw.org.in/bin/categorywisearch.php

Uberoi,P. (2003), "The Family in India: Beyond the Nuclear versus oint Debate". In Das.V (Ed), *The Oxford india Companion to Sociology and social Anthropology*, (pp.1061-1103). New Delhi: oxford.

Upathdya, C., and Vasavi, R.A. (2006,August), "Work, Culture, and Sociality in the Indian IT Industry: A Sociological Study",Final Report (submitted to Indo-Dutch Programme for Alternatives in Development), Bangalore.

16

Groupware Research and Technology Issues in Software Process Management

M. Y. Khan

Introduction

The discipline of social cybernetics concerns the nature of the relationship between groups and technology. In particular, the concept of computer-supported cooperative work (CSCW) specifically emphasizes the role of computer-based technologies in enhancing socially organized work. The development of CSCW-related technologies depends on the identification of specific human enterprises for which these new technologies might prove useful and on the elaboration of workable plans for the design of systems to support these enterprises.

We anticipate a continuing trend toward an information based society of highly specialized and dispersed experts that need to cooperate via computer assistance to solve the complex problems of the future – the Strategic Defense Initiative and the mapping of the human genome, for example. Thus, we expect to see significant demand for CSCW technologies. The domain for development of such systems, however, has not been clearly explored. Critical issues for investigation are addressed in emerging research questions that lead to the investigation of complex interactions between technology and social dynamics of organizations, as elaborated at the first conference on compute-supported collaborative work, CSCW'86 (Krasner, 1986). These questions include the following:

- What is the proper balance of control between human and machines?
- What are the effects of current technology on group structures?
- How will technological innovation affect the way large scale group activity is organized in the future?
- What research methods are appropriate for understanding/studying these phenomena?

Given the current state of the CSCW field and of the technology to support it, which is sometimes called "groupware" we believe that these issues are best addressed in the context of specific applications. Current technology for facilitating the interactions of groups tends to be either general or specific in its application focus. Systems suited to general audiences include groupware for distributed electronic communication, such as MIT's Object Lens (Lai and Malon, 1988) for interaction management such as MCC's graphical issue-based information system (GIBIS) (Begeman and Conklin 1988) and for editing of documents by teams, such as Bellcore's Quilt (Leland et al 1988). Other types of systems are aimed to more specific kinds of applications. Groupware for biomedical work, for example, includes Baylor College Hospital's COSTAR (Gorry et al 1988) and its Finnish modification FINSTAR (Engestrom et al. 1988) and the University of Oslo's Florence project (Bjerknes and Bratteteig 1988). An example of groupware for software engineering application is Bellcore's ICICLE (Intelligent Code Inspection in a C Language Environment) system for code inspection (Brothers et al 1988).

The application emphasized in this paper is software process management. By software process we mean the teamwork, cooperation, coordination, and communication that occur within and among groups throughout the life of a software project. This includes those processes that occur under the broad categories of proposal writing, software engineering, development and maintenance. The problem of developing automated group support tools for software process management is especially interesting for two reasons. First, the application is sufficiently specific: supplying a focus that can drive technology forward. Second, the application is sufficiently broad: providing a wide selection of potential sub applications with corresponding integration challenges. Traditional methods for software project management are quantitative; they focus on cost, schedule, variances and so forth, as described by Boehm (1981). In a software process management environment, on the other hand, the focus is on qualitative support. Methods and technologies are designed to support and manage the group processes of designing and developing software, as described by Krasner and Pore (1989).

Previous empirical studies of the process of software design and development have investigated the causes of performance difficulties encountered on large software projects. Many of the difficulties have been attributed to organizational behavioral problems, as described by Bostrom and Heinen (1977). One factor specifically identified by Curtis, Krasner, and Iscoe (1988) as a source of performance problems is that the technical design, management and formulation of user needs are not visible to the various project stakeholders (customers, potential users, contract managers, project managers, software engineers and so forth) to the degree needed to effectively coordinate such diverse efforts.

The identification and description of actual coordination processes can provide a basis for more effective project coordination, including the use of automated tools to support the software engineering process. Our research represents an attempt to define cooperative work in the context of software process management and to identify those aspects that enable it to occur effectively. An analysis of issues in terms of effectiveness (or productivity) and of aspects of software process management may provide a workable set of requirements for automated tools that support interactive coordination processes. Specially, our research addresses the development of a process framework for studying the interactions (potential and realized) between technology and the social dynamics involved in the software development task. Aspects that facilitate cooperative work in this context include answers to questions such as the following:

- In what situations is a shared information space (model) necessary?
- What are interesting modes of cooperation (for example, design)?
- How important is negotiation and conflict resolution in specific situation?

Groupware Defined

11th International Conference on Software Engineering in May 1989, prompts engineers to reflect on yesterday, and look to tomorrow (McLaughlin, 1989). This characterization of software helps distinguish current examples of automated tools aimed at collaborating groups from our notion of groupware, which is oriented toward the "Tomorrow" column of the table.

Some systems intended for managing and/or supporting working groups exist today, as reported by Kraemer and King (1986) for project management and electronic mail for broad based communication. However, most of these systems do not support groups as such, but rather individuals working in groups. These systems can filter or recover data and generate and route reports, but they cannot flexibly orchestrate the overall group process. Such systems are essentially passive in nature, reflecting the current status of groupware technology.

Groupware is computer technology that actively facilitates groups of collaborating users. This notion of active support distinguishes the systems of today from the groupware of tomorrow. The goal of groupware is to improve the group process, bringing about more effective performance of the group task. Groupware is concerned with facilitating interactions within collaborative groups, and it must provide capabilities for orchestrating the overall process. For example, some way of coordinating the actions of users may be needed, such as matching responses with questions. This leads to the following definition of groupware.

Groupware is computer-based technology that: (i) actively facilitates two or more users working on a common task, possibly simultaneously, using a shared environment and (ii) provides synergistic mechanisms for coordinating each user's actions with respect to the rest of the group and the system.

The emphasis on active facilitation distinguishes our definition from those of others such as Ellis (1989) and Johansen (1988).

Dimensions of Groupware

Groupware systems can be characterized by the way they deal with the set of key parameters of the conceptual model of task-related interaction that is, time, place, task, group context, group composition, human-machine allocation of work, and artifact or process focus. These terms, as they pertain to groupware, are defined in the following paragraphs.

- Time is the period over which task-related interactions of interest occur. Task-related interactions may be synchronous (occurring at the same time) or asynchronous (occurring at different times). These interactions may occur over a short period or last for an extended period.
- Place refers to the physical location(s) of the participants and whether they can communicate face-to-face or are geographically dispersed.
- Task refers to the group's objective and the general nature of the task(s) involved: setting policy or deliberating on a jury, for example.
- Group context includes the group's environment, the incentives, and the social conventions present in the cultural infrastructure of the working group, including the group's physical surroundings, its adopted style (pecking order), the level of formality in its communication, its willingness to try new things and so forth.
- Group composition includes the characteristics of the individuals who make up the group and the relationships among those characteristics, for example, the distribution of task-related expertise among group members.
- Human-machine allocation of work refers to the extent to which both the task to be performed and the coordination of the work are performed by humans versus the extent to which they are automated.
- Artifact or process focus is the extent to which the groupware focuses on products of a process versus the extent to which it focuses on processes. For example, groupware for document production tends to focus on products. On the other hand, groupware for issue management, such as MCC's GIBIS (Begeman and Conklin, 1988) focus on structuring and capturing discussion processes.

Other authors, notably DeSanctis and Gallupe (1985) and Johansen (1988,1989) have discussed the role of time and place at length, observing that most groupware is developed for one of four conceptual quadrants, according to whether the team members' work occurs at the same time or at different times and in the same place or in different places.

Johansen (1988,1989) points out that although much groupware has supported groups working at the same time and in the same place, there is a trend toward emphasizing groupware to facilitate work occurring at different times and in different places. He asserts that it is important for future groupware to overcome barriers between group members' working at the same different times and between their working in the same and in different places, that is, providing smooth transitions between synchronous and asynchronous work and between collocated and dispersed work. (See also Turoff, 1989). This emphasis is important because of related societal trends, such as job sharing and employees working at home.

Other authors, notably Huber (1982, 1984) have addressed the significance of the task and of group processes. Study carried out by our team members further identified that the group context (group norms and organizational culture) is important, along with the groups composition in terms of role, skill, and so forth. (See also Turoff 1989). The allocation of work between human and machine components is also a crucial dimension, as is the dimension of focus on processes versus focus on artifacts produced by processes.

The parameters of the conceptual model must be considered within the context of a specific application and driven by the application's needs and requirements, in this case, software process management. The development of models of the overall process that an application is intended to support is central to the identification of the characteristics of any proposed groupware application.

Research Issues

A number of interesting research issues relating to aspects of large-scale software coordination process models have emerged. The issues listed here help identify rich potential areas for groupware development.

- Synchronous and asynchronous interaction;
- Individual and shared work spaces;
- Facilitating social protocols;
- The initiative paradigm: active and passive aspects;
- Embedded knowledge;
- Support for changing roles and group composition;

- The capture and analysis of the interaction process;
- Integration – seamless transitions among groupware and related applications.

In the following section, these issues are discussed in terms of the parameters of the conceptual model of interaction described previously. We also identify some of the difficult underlying concerns with respect to technology design, and we discuss their implications for the building of groupware applications. While these issues are highly interrelated, they are described here separately for the sake of clarity.

Synchronous and/or Asynchronous Interaction: Groupware applications are typically intended to support either synchronous interaction, where people interact in real time, or asynchronous interaction, where members contribute at different times. Very few of today's systems are designed to handle both because the support needed for interaction activity over time may need to be very different for synchronous and asynchronous interaction. For example, in collaborative writing, when two or more individuals are editing the same document at the same time, it may be important to show the placement of other's cursors and to identify ownership at different times, it may be important to indicate what has changed since a user's last access by using change bars, for example. Sometimes interactions shift between synchronous and asynchronous models, such as when an attempt at real time interaction results in the leaving of an asynchronous message. As noted by Johansen (1989), it is important to provide appropriate, seamless transitions between these models.

In both kinds of interactions, notification of others' actions is important. The term notification refers to informing the audience of an action of the initiator. The granularity and timing of the notification are important design issues. Notification can be synchronous or asynchronous. Systems providing synchronous notification are sometimes called real time groupware. Examples include highly interactive multi user games, such as Xtrek (Guthrie, 1986). In the contrast, electronic mail and bulletin board systems are not real-time groupware because the notification they typically provide is asynchronous.

Individual and Shared Work Spaces: Task-related interactions may occur synchronously or asynchronously within the context of an individual work space or a shared work space. A shared work space is sometimes referred to as a shared work surface. Shared work surfaces (data, documents, etc.,) have been shown to be powerful in specific situations (Stefik et al. 1987). Groupware interfaces are inherently more complex than single user interfaces because of the added more burden of informing each user of the actions of other users. While managing multiple (and possibly concurrent) accesses to single work spaces, such interfaces must support synchronous and asynchronous use as well as remote and local interactions.

A shared work space is a means for multidirectional, long term conversation based communications, and it allows several members of a groups to engage in an ongoing conversation on-line. Communications can be remote as well as local, and conversations can be persistent and ongoing. Conversations also can be stored and retrieved by various group members, and items can be added to conversations incrementally. Each item can be tagged to indicate what type of them it is, who contributed it, who has read it, and so on. Some of the novel issues related to group interfaces include the following:

- Operations on shared data and the visual representation of authorization information, for example, colours to indicate that something is being modified, has been touched, is not writable, or is private.
- Presentation and retrieval: when to use query-oriented interaction and when to use network-browsing-oriented interaction, an issue raised by Begeman and Conklin (1988)
- The provision of multiple visualizations of the shared data state, for example, information in a document presented in the text form or as a tree; Lantz (1989) points out the need for highly personalized views, specialist views, handicapped views, and so forth in some situations.
- Support for different views in time and space, for example, emphasis of unread messages in a shared bulletin board.
- The coupling of windows shared by group members, along with the issue of which window operations require notification of other users.
- The possible use of auxiliary means of coordination in addition to computer workstation-based support, such as video technology, face-to-face communication, or telephone conversations for example, annotating material to indicate, "Don't touch that until I have finished").
- Distinctions between issues related to the human-computer interface and those related to underlying data: decoupling of these issues is difficult in many groupware situations; object-oriented systems tend to blur this distinction.

Supporting Social Protocols: Social protocols are accepted rules or policies that govern interaction (for example, taking turns conversationally). Thus, social protocols are a critical element of the group context. When and how technological mechanisms are implemented to support particular social protocols of interaction is a key issue in a groupware system. Experience has shown that some policies are best determined by a groups' social process, as opposed to embedding social protocols in technology a priori, as in the experience with the Coordinator. The Coordinator, as CSCW tool, focuses on the use of language in human cooperative activity. In particular, the first version made rigid use of the "language/action" paradigm developed by Winograd and Flores (1986) and Winograd (1988). In a study of the use of this tool, Carasik and Grantham (1988) found that the tool forced violation of some established norms of interaction, which led to serious breakdowns in communications.

We believe that social protocol and policy determine the amount of computer support (the technological mechanism) required. We would like to develop technology that is sufficiently flexible so that different policies can be provided at different times. A groups' size and organizational structure may influence the degree of social structure that is appropriately imposed by the software.

An example of the issue of support (or a lack of support) for a particular social protocol occurs in satellite telephone technology in which only half-duplex conversations are permitted; that is, only one person can talk and be heard at a time. Via a voice-activated switch, the person currently speaking prevents the other person from interrupting as long as he or she is making sounds. This technology does not support normal informal conversation. In which one person support normal informal conversation, in which one person can interrupt another. In our use of this technology, it seems awkward and unnatural.

Cultural and infrastructure underlies social protocols. Therefore, another important aspect of the groups context for groupware applications (as with any multi-user system) deals with incentives for using the system, specifically with the disparity between who does the work and who gets the benefit, as observed by Grudin (1988). Problems may arise if an application requires that some people do additional work but do not benefit from the use of the system. This aspect has eluded many groupware builders, notably in group calendar systems. Such systems can also remain unused because they contain objectionable functionality, allowing someone to schedule an appointment on a calendar without permission from the calendar owner, for example. It is best if groupware technology is specifically designed for the group context. This may involve support for the informal group culture, support for a specific type of culture (for example, democratic or consensus based), or support that can be specifically tailored for multiple cultures. Empirical evaluation and adaptation are crucial.

There is a critical need for collaboration between technology builders and social scientists in realizing effective and useful groupware systems, especially with respect to social protocols. This is the primary thrust of the CSCW research community (Krasner, 1986; Greif, 1989). Technology builders need social scientists to help identify system requirements and to help evaluate prototype technologies. Builders of technology must remember that they are not always typical users.

Initiative Paradigm: Active and Passive Aspects: Data-base-oriented information systems of the last decade are considered to be primarily passive in the sense that users needing information have to initiate the action necessary to retrieve that information. This may be through query, browsing, or program. In some cases, the group process may be facilitated if the groupware system itself becomes an active participant in the interaction process and is capable of providing information to users based on anticipated needs, as in some artificial intelligence (AI) systems. Johansen (1988) strikes a balance between the utility and the hyperbole of AI as follows:

Against a backdrop of marketing hype and frustrated expectations, it is easy to overlook the fact that AI has made considerable progress. This progress is often hidden because from the perspective of AI professionals, compared to unmet research goals, it seems trivial. It is precisely the undramati , overlooked understandings...that we are expecting in the groupware field.

In particular, Johansen sees "agents: 'intelligent' systems capable of acting on behalf of human beings" as "very relevant". As artificial intelligence technology matures and as coordination processes become more precisely defined, we expect groupware systems to contain more intelligent autonomous agents; however, most interaction processes will be guided by the expertise of the individuals involved. For the near future, therefore, we see a mixed-initiative paradigm in which the group delegates contain responsibilities for taking initiative to the system.

Embedded Knowledge: The knowledge represented within groupware for software process management should include expertise from a variety of domains. Specifically, such a system ought to include knowledge about principles of coordination (in general) about the organization (its structure, policies, etc.) and about software process management (both in general and as the concepts relate to a specific organization or project). It is reasonable to expect that the requisite knowledge for such a system will not be represented in a single expert. Thus, multiple experts must be involved in the process of embedding knowledge into groupware. Relevant expertise must be identified from multiple sources and integrated into a comprehensive representation. The knowledge support system KSS (Shaw and Gaines, 1988) provides a method for accomplishing this.

Support for Changing Roles and Group Composition: The roles assumed by individuals vary according to the task or subtask at hand and the characteristics of the individuals involved in the task (group composition). In many work situations, individuals assume several interaction roles concurrently or serially throughout a working period; therefore, support for transition between multiple roles (some of which are individual and some, team oriented) becomes a major issue. For example, in a software development situation, an individual might assume roles in several tasks concurrently: as a member of a design review team, as an individual programmer, or as a member of a process evaluation task force. Context-sensitive window objects for different roles have been postulated to support such situations, as in the Tebasco Project (Evangelist et al. 1987) at MCC.

Situations where the size and composition of the group vary over time levy additional requirements on groupware systems. The system must properly handle new users and clean up after departing users. Also, groups frequently split into subgroups into the larger group. This pattern occurs in large software projects; in other application contexts, it may not be important to provide subgroup support within a groupware system.

Capture and Analysis of Interaction Process: In many situations, the historical record of the interaction process can serve as important input to the task at hand. This is the case in software design where process rationale becomes critical during reengineering activities. Precisely what types of data to capture depend on the types of analysis envisioned. In the design of systems for process management software, this data-capturing and data-analysis function can be critical for elicitation of original requirements, as well as for reengineering activities. In software process management, product quality measurement and statistical process control techniques are used to pinpoint potential trouble spots. Development of empirically based groupware to support software engineering processes requires such capture and analysis. We recommend that groupware be instrumented to collect required data in relatively unobtrusive ways.

Integration: Seamless Transitions among Groupware and Related Applications: It is helpful if groupware technology is integrated so that it is easy for different members of a group to make transitions easily as different needs arise. Integrated groupware technology can be developed within a flexible and seamless groupware support environment, as illustrated by the following examples:

- Elements of a conversation between two persons can be made available to other individuals.
- Annotations to a document can be made part of the document (and vice versa).

- An object type can be changed; for example, an "issue" in a system such as MCC's graphical issue-based information system (Begeman and Conklin, 1988) if appropriate.
- A person's role can change.
- Items in a relatively passive system can be mailed electronically to members of a group. (Electronic mail can be integrated with the rest of the groupware, thereby making the system less passive).

Software Process Models and Enabling Technology

According to Steiner (1972), the quality of a group's performance on a task depends on three characteristics: the requirements of the task, the resources that the group brings to the task, and the process (steps) in which the group engages to complete the task. The notion of process in this context involves all the intra and interpersonal acts and behaviours by which individuals transform their resources into a product. This includes the intellectual and communicative behaviors through which individual group members assemble their resources and partition the task among group members. Unlike task requirements and resources, the actual process in which a group engages cannot be measured or evaluated before work begins. Prescriptions for guiding and managing groups do, however, exist.

New large-scale models of software engineering processes are emerging to meet the new emphases on exploratory (Krasner, 1989). Creating process models of the teamwork, cooperation, coordination, and communications aspects of large software projects is especially challenging because it is groundbreaking applied research. Database and language approaches have been considered. Passive, integrated data bases are currently used to deal with process coordination problems, but the assumption that simply providing a shared information base in a software engineering environment produces coordination is overly optimistic. We have found that more active communication is necessary, such as through electronic mail. On the other hand, languages that allow representation of coordination activities in organizations are emerging from the computer sciences area of concurrent programming languages. However, these languages (see Harel, 1984) are designed primarily for systems of cooperating computer programs, not people. There is need to explore representation techniques for cooperative process modeling as well as integration with other process model aspects.

Current technology for software process management (SPM) operates on the simple model of passive synchronization of programming-level tasks through the check-out/check-in policies of centralized configuration management. Although workable for small collocated teams where the real coordination is done "over the partitions," it is clearly not workable for large projects with several teams that may be geographically distributed. Current software engineering environment technology, notably database locking mechanisms, can actually inhibit certain types of close cooperation, such as cooperative authoring of a document or on-line team design reviews, as pointed out by Rein (1990). New varieties of groupware that can be applied directly to software engineering include the following:

- On-line discussion groups, ranging from electronic bulletin boards to structured tools such as The Coordinator, based on Winograd's language/action paradigm (Winograd, 1988; Winograd and Flores 1986); MCC's GIBIS (Begeman and Conklin 1988); Computer-controlled video conferences, including the Crusier at Bellcore.
- Multimedia groupware, such as rapport at AT&T and Project Andrew [1] at Carnegie-Mellon University.
- Groupware using hypermedia, such as Quilt at Bellcore.
- Context-sensitive mail or intelligent routers and filters, such as in the Object lens at the MIT Sloan School of Business.
- WYSIWES (What You See Is What Everyone Sees) group interfaces, such as in Xerox PARC's Colab (Lantz 1989).
- Group decision support systems, such as at the University of Arizona Planning Lab (Nunamaker et al 1990).
- Groupware based on methods such as described by Mason and Mitroff (1981) for surfacing strategic assumptions.

For such technology to be most effective in software development environments, groupware for software process management (SPM groupware) has to capture and use critical coordination information not included in today's software engineering environments. This will enable to environment to serve as an active coordination agent using the groupware medium. When such technology is coupled with artificial intelligence applied to project management, we expect capabilities to emerge that can help analyze project behavior, support to emerge that can help analyze project behavior, support managerial decision making, provide real-time understanding of project status and problems, and ensure complete and consistent project data from start to finish. Software process models provide a means for describing and making inferences (dynamically) about processes of interest. Specific issues about software engineering processes can be both generated and investigated from a theoretical basis. Descriptive studies can help characterize the overall process and can lead to more normative investigations, which attempt to identify better processes or characteristics of processes. Specific process-model-based research can be seen in the 'Callisto' (Sathi et al. 1986) and 'Tebasco' (Evangelist et al. 1987). The majority of these modeling projects and others like them are based on extensions of either Petri net technology [2] or knowledge-based technology from artificial intelligence.

Software process models that can be supported by empirical investigations (e.g. see Guindon et al. 1988) will be valuable resources for the determination of groupware requirements. Some of these requirements will be formally expressible within groupware tools. The development and use of the tools themselves can be provided additional input concerning the validity or suitability of the underlying process models.

Applying Groupware to Software Process Management

As pointed out previously, the information requirements for SPM groupware include the storage of several new types of information. Some of these responsibilities; commitments; expertise; issues (open, resolved, dissolved), positions, and arguments; decisions; assumptions; goals and objectives; and shared models of product and process. Also, new hypermedia technology must be developed to handle the different types of data (voice, video and pictorial) in an integrated fashion.

While groupware requires effective networking and database technology to allow forms of close cooperation, a more critical need, pointed out by Krasner (1989), is for an organization to be explicit about its coordination processes, specifically, the process model building and analysis activities that do not normally occur in today's project environments. Although the basic concepts exist for modeling cooperative processes, very little research outside the domain of office automation has occurred to date. This presents a significant challenge.

Empirical studies can provide insights throughout the process of designing SPM groupware. Initial process models and analysis can provide a foundation for development of groupware. The complexity of the operational domain of such systems requires, however, that empirical evaluation be conducted throughout the stages of development and use. As previously discussed, a system for CSCW that is based on formally rigorous theories and models can affect the work environment in unforeseen ways.

Conclusion

We assert that active, intelligent, adaptive, orchestrated, groupware that manages multiple and possibly concurred accesses to shared work spaces in the desirable goal for future technology, for example, a groupware system for software process management that plays an active role (takes some initiative) suggesting assignments for members of group, in accordance with the project's needs and schedule and based on the experience and availability of project staff.

When considering groupware for software process management, however, we believe that the system must contain one important quality: flexibility. The groupware must be flexible enough to allow the dynamics of the group and the culture of the organization to influence its use. One group, for instance, may need to use a tool in an entirely different manner from another group, perhaps even employing it in a manner not foreseen by the tool designer. In an authoritarian group, a manager may use an on-line group calendar facility to call meetings or even to assign particular times for employees to perform certain tasks. In a more egalitarian group, members may use the information on the group calendar facility to request for meetings with each other.

Our ideal, then, is groupware that is flexible as well as active, intelligent, adaptive, and orchestrated. It will be challenging to develop such groupware, but such a combination of factors will be necessary for groupware, to gain acceptance.

Notes

1. AT&T is the name of Company of US who conduct research and product development in the area of IT and Project Andrew was developed in Carnegie Mellon University
2. Petri net – A graphical mathematical modeling tool. It can be used as visual communication aid similar to flow charts, block diagram and network.

References

Begeman M. L. and J. Conklin (1988), "The Right Tool for the Job", Byte, Vol. 13, no. 10, October, pp. 255-266.

Brothers 1., V. Sembugamoorthy, and M. Muller (1988), "ICICLE: Groupware for Code Inspection," in CSCW 90: Proc. Conf. Computer-Supported Cooperative Work, Sept. 1988, pp. 169-181.

Beohm B., (1981), "Software Engineering Economics". Englewood Cliffs, NJ: Prentice-Hall.

Bjerknes G. and T. Bratteteig (1988), "The Memoirs of two Survivors or Evaluation of a Computer System for Cooperative Work", in CSCW 88 : Proc. Conf. Computer-Supported Cooperative Work, September, pp. 167-177.

Bimson K., and L. Burris (1989), "Assisting Managers in Project Definition: Foundations for Intelligent Software Management, IEEE Expert, pp. 66-76, Summer.

Bostrom R. and J. S. Heinen (1977), "MIS problems and Failures: A Socio-technical Perspective," MIS Quart, pp. 17-32, September.

Carasik R., and C. Grantham (1988), "A Sase Study of CSCW in a Dispersed Organization," in proc. CHI '88: Human Factors in Computing Szyst., May, pp. 61-65.

Curtis B., H. Krasner, and N. Iscoe (1988), "A Field Study of the Software Design Process for Large Systems," Commun. ACM. vol. 31, no. 11, pp.1268-1287, November.

DeSanctis G., and B Gallupe (1985), "Group Decision Support Systems: A New Frontier," Data Base, Winter, pp. 3-10.

Ellis C., (1989), "remarks" made in the IFIP technical workshops report "User Technologies of the Future: Groupware Technology," at the XI World Computer Congress: IFIP Congress'89, San Francisco, CA, August.

Engstrom Y., R. Engestro, and O. Saarelma, (1988), "Computerized Medical Records, Production Pressure, and Compartmentalization in the Work Activity of Health Center Physicians," in CSCW 88: Proc. Conf. Computer-Supported Cooperative Work, September, pp. 65-84.

Evangelist M., I..Forman, and G. Bruns (1987), "Tebasco: A Team-based Coordination System," MCC Tech. Rep. ITM-87-07, August.

Gorry G. A., A. M. Burger, R. J. Chaney, K. B. Long and C. M. Tausk (1988) "Computer Support for Biomedical Work Groups," in CSCW 88: Proc. Conf. Computer Supported Cooperative Work, September, pp. 39-51.

Greif I.,(1989), Ed., "Computer-Supported Cooperative Work: A Book of Readings". San Mateo, CA: Morgan Kaufmann.

Guindon R., H. Krasner, and B. Curtis (1988), "Control of Cognitive Processes during Software Design: What Tools are Needed?" in Proc. CHI '88': Human Factors in Computing Systems, May.

Grudin J., (1988), "Why CSCW Applications Fail: Problems in the Design and Evaluation of Organizational Interfaces," in CSCW 88: Proc. Conf. Computer-Supported Cooperative Work, September, pp. 85-93.

Guthrie C., (1986), Xtrek, a Multiplayer, Space Shoot-;em-up Game," XII Version by J Weinstein.

Harel D., (1984), "Statecharts: A Visual Approach to Complex Systems," Rehovot, Israel: Dept. Appl. Mathematics, The Weizmann Institute of Science.

Huber G., (1982), "Group decision support systems as aids in the use of structured group management techniques," DSS-82 Conf. Proc., pp. 96-108.

Huber G., (1984), "Issues in the Design of Group Decision Support Systems," MIS Quarterly, September, pp. 195-204.

Johansen R., (1988), Groupware: Computer Support for Business Teams. New York: The Free Press (Macmillan), pp. 44-99.

Johansen R.,(1989) "Groupware Future Directions and Wild Cards," presented at the Conference of Organizational Computing, Coordination, and Collaboration: Theories and Technologies for Computer-Supported Work, Austin, TX, November.

Krasner H. and M. Pore (1989), "A Software Process Management Approach to Productivity and Quality," in Proc. 7th Northwest Software Quality Conf. Portland, OR, September.

Kraemer K. 1 ., and J. L. King (1986), "Computer-based Systems for Cooperative Work and Group Decision-making: Status of Use and Problems in Development," Proc. CSCW'86: Conf. Computer-Supported Cooperative Work, Austin, TX, December, pp. 353-375.

Krasner H., (1989), "Requirements Dynamics in Large Software Projects: A Perspective on New Directions in the Software Engineering Process," on Proc. 11th World Computer, Congress (IFIP89), August, pp. 211-216.

Lantx K.,(1989)," remarks" made at the IFIP Groupware Technology Workshop, Palo Alto, CA, August.

Leland M..D..P., R..S. Fish, and R.E. Kraut (1988), "Collaborative Document Production Using Quilt," in CSCW 88: Proc. Conf. Computer-supported Cooperative Work, September, pp. 206-215.

Lai K., and T .W. Malone (1988), "Object lens: A 'Spreadsheet' for Cooperative Work" in CSCW 88: Proc. Conf. Computer-supported Cooperative Work, September, pp. 115-124.

Mason R. and I. Mitroff (1981), "Challenging Strategic Planning Assumptions: Theory, Cases, and Techniques". New York: Wiley.

Malone T. W., K.R Grant, K.Y. Lai, R. Rao and D. Rosenblitt (1987), "Semi-Structured Messages are Surprisingly Useful for Computer-supported Coordination," ACM Trans. Office Inform, Syst., vol. 5 no. 2, pp. 115-131, April.

McLaughlin F., (1989), "ICSE 11 Prompts Engineers to Reflect on Yesterday, look to Tomorrow," computer, vol. 22, no. 7, pp. 110-112, July.

Nunamaker J., D. Chappel, D. Vogel and C. McGoff (1990), "Electronic Meeting System Tutorial," Tuscon: MIS Dept. Karl Eller Graduate School of Business, University of Arizona.

Rein G., (1990) in "How Can We Make Groupware Practical?" (panel led by B Ensor), in Proc. CHI '90': Human Factors in Computing Syst., April, p. 89.

Sathi A., T. Morton, and S. Roth (1986), "Callisto: An Intelligent Project Management System," AI Magazine., Winter, pp. 34-51.

Shaw M. L. G., and B R Gaines (1988), "A Methodology for Recognizing Consensus, Correspondence, Conflict, and Contrast in a Knowledge Acquisition System," paper presented at the Workshop in Knowledge-Based Systems, Banff, AB, Canada, November.

Stefik M., G. Foster, D. G. Bobrow, K. Kahn, S. Lanning, and L Suchman, (1987) "Beyond the Chalkboard: Computer Support for Collaboration and Problem Solving in Meetigns," Communication. ACM, vol. 30, no. 1. pp 32-47, January.

Steiner I. D., (1972), "Group Process and Productivity", Academic, New York.

Turoff M.,(1989), "Computer-Mediated Communication Requirements for Group Support," presented at the Conference on Organizational Computing, Coordination, and Collaboration: Theories and Technologies for Computer-Supported Work, Austin, TX, November.

Walz D., (1989), "A Longitudinal Study of Group Design of Computer Systems", Ph.D. dissertation, Dept. Management Science. Information. System., University of Texas, published as Microelec. And Comput. Technol. Corp., Austin, TX, Tech. Rep. STP-110-89, March.

Winograd T. (1988), "A Language/action Perspective on the Design of Cooperative Work," in Proc. CHI '88: Human Factors in Computing System., May.

Winograd T. and F Flores (1986), 'Understanding Computers and Cognition". Norwood, NJ: Ablex.

Appendices

Appendix 1

Telecommunications in India

Importance of Telecommunications

The telecommunications sector has undergone a total transformation throughout the world over the last two decades. Technological advances have revolutionised the quality and range of services available. Moreover, developments in the sphere of information technology, satellite-based television broadcasting, new forms of communication such as data communication through e-mail and associated services through the internet, are transforming the way people communicate and conduct business.

Clearly, a country's ability to benefit from this revolution depends heavily on the modernity of its telecommunications network. Countries that can acquire and access information on demand and then integrate them usefully into their industrial structure through a modern telecommunications network are most likely to experience high rates of growth. Large-scale use of information and telecommunications technologies directly influences productivity, cost effectiveness and competitiveness in industries with high levels of product differentiation and low levels of unit prices. In these industries, prompt availability of information about demand trends or price movements can boost competitive advantage. International competitiveness of the textile, garment, toy and consumer electronics industries in Taiwan and the Philippines have been found to be effective due to advanced telecommunications services that allow tight links between commercial distribution in American and European markets and local production.

An advanced telecommunications system is equally important for service industries like banking, trading, retailing, transportation, maintenance and insurance where information and real-time communication are vital to the production process. A reduction in the costs of these services will indirectly enhance international competitiveness within the entire economic system, since lower marketing costs mean lower costs for manufacturing firms exposed to international trade. The extraordinary efficiency of financial markets of Hong Kong and Singapore is based on the extensive use of such advanced telecommunications services.

Thus, improved communication is vital for productivity in all spheres of activity: agriculture, industry, trade and commerce. In addition, with the increasing movement of people within and outside the country, easy communication is essential for enhancing human welfare so that families, friends and acquaintances can overcome the barriers of distance. For a developing economy like India, it is thus important that policies and programmes be initiated, and resources committed to bring about rapid growth in the sector.

History of Telecommunications in India

In operation since 1851, Indian telecommunications is among the oldest such systems in the world. Telephone services in the country were originally operated by private companies which were taken over by the Government of India in 1943. Since then, telecommunications in India has been a public utility owned by the Central Government with its operation and maintenance regulated by the Indian Telegraph Act of 1885 (ITA) and the Wireless Telegraphy Act of 1933.

At the time of independence, the government took charge of a network that connected only 82,000 subscribers in 321 exchanges. In the ensuing centralised planning process, resource allocation for the telecom sector was low and priority was given to sectors such as roads, power, sanitation, education, health etc. Till 1985, resource deployment in the telecom sector was around 2 to 3 percent of the total budget allocation. With no access to any other sources of funds, this sector registered low to moderate growth.

At the beginning of the Seventh Five Year Plan in 1985, the Government decided to focus on improving this sector and constituted the Telecommunications Board and the Department of Telecommunications (DoT) within the Ministry of Communications to oversee operations, maintenance and development of telecom services (telecom was till then under the purview of the Posts and Telegraph Department within the Ministry of Communications). A new emphasis was also laid on the funding of the telecom sector. Whereas the plan budget of the Department of Telecom was still regulated by the central planning process and the Ministry of Finance, it was agreed in principle that additional resource mobilisation through market borrowing will be permitted in respect of the telecom sector. The most striking success of liberalization policy is visible in telecommunications. Teledensity has risen dramatically, an entire industry with private players has come about, privatisation has begun with VSNL, and new technologies have come about. Looking forward, the telecom sector will feature lively competition between multiple private firms, with a strong role for TRAI to establish pro-competitive policies.

New Telecom Policy (NTP), 1999

The NTP announced in 1999 provides the basic framework for the future development and growth of the telecom sector in the country. The NTP took into account the far-reaching technological developments taking place in the telecom sector globally and stressed on implementing the Governments' resolve to make India a global IT superpower. The objectives of the NTP are to

1. Make available affordable and effective communications for the citizens.
2. Strive to provide a balance between the provision of universal service to all uncovered areas, including the rural areas and the provision of high-level services capable of meeting the needs of the country's economy.
3. Encourage the development of telecommunication facilities in remote, hilly and tribal areas of the country.
4. Create a modern and efficient telecommunications infrastructure taking into account the convergence of IT, media, telecom and consumer electronics and thereby propel India into becoming an IT superpower.
5. Convert Public Call Offices (PCOs), wherever justified, into Public Teleinfo Centres offering multimedia services like Integrated Services Digital Network (ISDN) services, remote database access, government and community information system etc.
6. Transform in a time bound manner, the telecommunications sector to a greater competitive environment in both urban and rural areas providing equal opportunities and level playing field for all players.
7. Strengthen research and development (R&D) efforts in the country and provide an impetus to build world-class manufacturing capabilities.
8. Achieve efficiency and transparency in spectrum management.
9. Protect the defence and security interests of the country.
10. Enable Indian telecom companies to become truly global players.

Under NTP, the government no longer engages in telecom service provision which is now the task of private and public sector companies operating in a competitive market. The Telecom Regulatory Authority of India (TRAI) now performs the regulatory functions of the sector. The Department of Telecom and the Telecom Commission have re-focused on the functions of policy formulation, licensing, wireless spectrum management, administrative monitoring of PSUs, research and development and standardization/validation of equipment etc.

Some major initiatives in telecom policy in recent years include the following:

1. Opening up of international long distance tariff.
2. Permitting internet telephony.
3. Setting up a mechanism for Universal Social Obligation (USO).
4. Permitting a fourth operator in the cellular mobile segment.
5. Commissioning of National Internet Backbone (NIB).

The NTP, 1999 has envisaged that India should emerge as a major manufacturing base and major exporter of telecom equipment. The manufacturing capacity of the indigenous industry is small in relation to the other major operators in the world and exports constitute a very small proportion of the total production of telecom equipment in the country. Promoting exports as a thrust area and development of India multinationals should be among the major goals in this sector. To achieve this objective, the Tenth Plan aims at removing the various constraints relating to transfer of the latest technology, access to cheap international finance, joint ventures with foreign companies, rationalization of custom and import duties on inputs and development of a strong industry-sponsored R&D base. Due to fast changing technology in this sector and the existing technological gap in relation to the developed countries, India will have to rely on transfer of technology in a big way in the immediate future. To achieve this, strategic alliance with leading international companies will have to be an integral part of the overall exercise of technology transfer.

Bharat Sanchar Nigam Limited (BSNL) and Mahanagar Telephone Nigam Limited (MTNL)

The telecommunications network of these two public sector undertakings is one of the largest telecom networks in Asia. BSNL provides a cost-effective alternative to mobile subscribers with low entry charges, no security deposit for local/ISD connection, no CLIP charges, no additional airtime charges while roaming in BSNL network, free zonal roaming for pre-paid subscribers, no monthly charges for national roaming under Plan 325, free SMS services for both incoming and outgoing to all prepaid customers and additional grace-period of 90 days for SIM-activation without activation charges.

In order to supplement the efforts of public sector providers and to ensure greater competition in providing basic telephone services, private companies are being licensed to offer basic services.

MTNL, the telecom operating company for the cities of Delhi and Mumbai was created in 1986. It was decided that MTNL be authorised to raise funds as a frontal agency for the DoT. The growth of the telecom sector underwent a qualitative change after resource mobilisation was allowed through MTNL.

Telecommunications Regulatory Authority of India (TRAI)

TRAI has taken the following measures in recent years with a view to obtain a sound framework for competition between multiple technologies and vendors:

- TRAI has given two recommendations on VSAT. First relates to issuance of licenses for VSAT services such as 'Receive only VSAT' because it has great potential to open new vistas of innovation in information sharing and spread of education and in facilitating the efforts of the government in bridging the Digital Divide. The second recommendation relates to facilitating VSAT operations by permitting Higher Data Rate, Reduction in Licence fee for Captive VSAT Network and reducing minimum antennae size. In order to promote internet in the country, TRAI has made a number of recommendations including setting up of National Internet Exchange of India (NIXI). In order to ensure for the consumers equal ease of access to different domestic and international long distance service providers, TRAI has issued a directive to all access providers and national/ international long distance operators regarding implementation of carrier selection in their respective network. A regulatory principle was maintained to prevent any anti-competitive or predatory behaviour on part of dominant operators.

- The quality of services of all service providers is being monitored by TRAI on the basis of quarterly reports submitted by them. TRAI is also conducting an objective survey of Quality of Service (QoS) parameters of basic and cellular service in different circles through an independent agency. The agency has also been assigned the task of making an assessment of customer satisfaction through a survey. TRAI has also issued a regulation on quality of service for Voice over Internet Protocol (VOIP) based International Long Distance Service.
- It has recommended deregulating regulatory intervention in cellular tariffs. This implies that operators no longer need to have prior approval of the regulator for implementing tariff plans except under certain conditions. Guidelines were also issued to service providers in respect of pre-paid cards which would enable increased consumer awareness and transparency for the subscriber.
- It has finalized the System of Accounting Separation (SITUATION), providing a detailed accounting and financial system maintained by telecom service providers and has forwarded the same to DoT for notification under Section 35 (2) (d) of the TRAI Act. The proposed system, based on the principles of accounting separation, would help monitor and measure financial performance of individual telecom products/ network services and help disinvestment aggregate costs to the level of network elements. It would also help in identification of cross-subsidisation practices in the industry, wherever these exist. By providing for maintenance of detailed cost records right up to the level of network elements, the system will help generate accurate information on costs, which is necessary for tariff and interconnected

regulations. TRAI's document on accounting separation would go a long way in promoting healthy competition with adequate mechanisms built in to detect unfair competitive practices. While the system is mainly aimed at enabling better monitoring and control by TRAI, the new system is also expected to help operators establish a more comprehensive system for arriving at product/network service wise costs and revenues and thus enhance their operating efficiency.

The National and International Long Distance markets in telecom have been opened up to competition. These policy measures have resulted in significant reduction in long distance tariffs due to competitive pressures. This has drastically reduced the margin available to fund the Access Deficit incurred. In a multiple operator environment, it is important to specify an International Usage Charge (IUC) regime which gives greater certainty to the Inter-operator settlements and facilitates inter-connection agreements. Therefore, TRAI notified its Telecommunication Inter-connection Usage Charges (IUC) Regulation, 2003 in January 2003 which covered arrangements among service providers for payment of Inter-connection Usage Charges for telecommunication services, covering basic services which include WLL (M) service, Cellular Mobile Service providers and Long Distance Operators throughout the territory of India. This regulation has specified the IUC for origination, transit and termination in a multi-operator environment.

FDI in Telecommunications

In basic services including, cellular mobile, paging and value added service, and global mobile personnel communications by satellite, FDI is limited to 49 percent subject to grant of licence from the Department of Telecommunications (DOT) and adherence by the companies (who are investing and the companies in which investment is being made) to the licence condition for foreign equity cap and lock-in-period for transfer and addition of equity and other licence provisions.

FDI up to 100 percent is allowed for the following activities in the telecom sector:

1. ISPs not providing gateways (both for satellite and submarine cables).
2. Infrastructure providers providing dark fibre (IP category).
3. Electronic Mail.
4. Voice Mail.

Up to 100 percent FDI in telecom manufacturing activities on automatic approval basis is also allowed.

Automatic approval for foreign equity in software and almost all areas of electronics is permitted. Automatic approval is accorded for foreign technology agreements in all areas of electronics except aerospace and defence, subject to specified conditions. 100 percent foreign investment is permitted in units set up exclusively for exports. Such units can be set up under any one of the following schemes; EHTPs, STPs, Free Trade Zones/EPZs, and 100 percent EOUs.

Since the initiation of the economic liberalisation process in 1991, sectors such as automobiles, chemicals, food processing, oil and natural gas, petro-chemicals, power, services, and telecommunications have attracted considerable investments. Today, in the changed investment climate, India offers exciting business opportunities in virtually every sector of the economy. Telecommunications and computer software/hardware are among the leading sectors attracting FDI

Telecommunication Services in Rural Areas

The availability of modern telecom services has raised a controversy about a *digital divide*, where rural areas are disadvantaged by having inadequate telecom services. In order to enhance teledensity in rural and remote areas, guidelines for implementing Universal Service Obligation (USO) were issued which became effective from April 1, 2002. To meet the investment requirement, a 5 percent Universal Levy has been imposed on the Adjusted Gross Revenue on all telecom operators, excluding pure value added service providers. The quantum of levy can be increased in the event of an increase in the scope of USO. The Administrator for implementation of USO was also appointed on June 1, 2002. The support from Universal Service Fund (USF) covers both public access through public or community telephones in villages as well as individual household telephones in identified Net High Cost rural/remote areas. In short, main features of the Universal Service Support Policy (USSP) are as under:

1. USSP came into effect from April 1, 2002. Guidelines for USSP were issued on March 27, 2002. Administrator USF was appointed on June 1, 2002.
2. Universal Levy (USL) is fixed at 5 percent of the Adjusted Gross Revenue (AGR) of all telecom carriers or operators, excluding pure value added service providers such as ISPs, e-mail, voice mail service providers.
3. The amount of USL will remain within the present cap of licence fee.
4. Support from Universal Service Fund (USF) shall cover both public access through public or community telephones in villages as well as individual household telephones in identified net high cost rural/remote areas.
5. Disbursement of funds to the service providers will be through a multi-layered bidding process on the basis of least quoted subsidy, after necessary verification of the targets achieved.

Recent Policy Initiatives in Telecom Sector

These include the following:

1. BSNL announced 33 percent reduction in call charges for all the countries for international calls.
2. Annual licence fee for National Long Distance (NLD) as well as International Long Distance (ILD) licences reduced to 6 percent of Adjusted Gross Revenue (AGR) with effect from January 1, 2006.
3. Entry fee for NLD licences reduced to Rs. 2.5 crore from Rs. 100 crore prospectively, i.e. date of issue of amendment to the existing guidelines to that effect.
4. Entry fee for ILD reduced to Rs. 2.5 crore from Rs. 25 crore.
5. NLD service provider shall be permitted to carry intra-circle traffic with mutual agreement of originating service provider. Agreement with terminating service provider shall not be required.
6. Mandatory roll out obligations for future NLD licensees as well as existing NLD licensees waived off.
7. No more mandatory roll out obligations for ILD service licensees except for having at least one switch in India. Roll out obligations for existing ILD service licensee stands waive off from the date of issue of orders.
8. Net worth and paid up capital of the applicant company for NLD as well ILD service licence shall be Rs. 2.5 crore only and while counting the net worth, the net worth of promoters shall not be counted.
9. NLD service providers can access the subscribers directly for provision of leased circuits/closed user groups and can provide last mile connectivity. The ILD service providers can access the subscriber directly only for provision of leased circuits/closed user groups.
10. Access service provider can provide internet telephony, internet services and broadband services. If required, access service provider can use the network of NLD/ILD service licensee.
11. No more IP-II and IP VPN licences to be issued with immediate effect as these licensees are allowed to migrate to NLD/ILD service licence.
12. Internet service provider (ISP) with internet telephony (restricted) to be charged licence fee at 6 percent of AGR with effect from January 1, 2006.
13. Annual licence fee in respect of VSAT commercial to be charged at 6 percent of AGR with effect from January 1, 2006.
14. FDI Ceilings raised from 49 percent to 74 percent. 100 percent FDI is permitted in the area of telecom equipment manufacturing and provision of IT enabled services.

Agenda for Reforms

Telecommunications should continue to be treated as a major element of infrastructure for growth and development of the Indian economy. For rapid and balanced growth of the sector, the Government should not view it as an opportunity area for resource mobilisation. This needs to be reflected in its licensing and taxation policies.

Inter-circle long-distance services should be opened to the private sector. The process leading to this action should incorporate appropriate lessons from the experience of privatization of the telecom services in India and abroad.

There is a definite cross subsidisation when the international and domestic long distance tariffs subsidise local tariffs. A more cost-oriented tariff structure is needed. This should be initiated with a major tariff study to understand the level of cross-subsidisation existing today. The study could provide vital inputs to formulate a sound tariff policy aimed at providing a tariff structure which is reasonable and affordable for telecom users, and an implementation plan that would induce competition, without ignoring government's social welfare plans.

As the Indian telecom network is already the 14th largest in the world and is expected to become a global player, it is important that the country starts investing in developing equipment manufacturing capabilities. This will help India establish itself as a manufacturing base that can take care of the future telecom requirements of the country as well as export sales. Rationalisation of import tariff should be extended to components and inputs required by manufacturers of such products so as to encourage domestic production. This is in compliance with the National Telecom Policy objective of ensuring that India emerges as a major manufacturing base and an exporter of telecom equipment.

Within a couple of years, there may be as many as 100 or more service providers who would come under the telecom regulatory net. It must be emphasised here that telecom regulation will necessarily be technically complex and hence the TRAI must be appropriately equipped to deal with many issues that are likely to arise. The success or failure of deregulation will largely depend on how well the TRAI functions. One such important issue is to do with interconnections between networks of different operators. The TRAI should ensure that the inter-connectivity is seamless as far as consumers are concerned and the agreements legally commercially and financially sound so as to support private sector entry.

Regulation in the area of tariffs and across charges will also become increasingly complex as the number of service providers increase. There will be a plethora of tariff regime which will require regulation along with appropriate consumer protection.

The Government will be receiving large amounts of funds from the telecom sector in the form of licence fees, duties and taxes. It is important that this money be used as seed money for developing infrastructure facilities in the country including telecom. An infrastructure fund should be created and all telecom licence fees be transferred to this Fund. The Fund should provide debt and participate in equity of various infrastructure projects. The telecom sector could be given the first option to utilise a certain portion of this fund. The portion earmarked for the telecom sector can be utilised as follows:

1. Providing equity and long term debt funds to all the players in the sector. This fund could be administered by the Infrastructure Finance Development Corporation (IFDC) with transparent government guidelines regarding fund allocation.
2. Providing resources for evaluation of the new emerging technologies and their relevance in the Indian context.
3. Funds needed for addressing the needs of technical manpower training of both existing employees of the DoT as well as of future personnel.
4. Funds required for the initial establishment of the TRAI and specialised training for its personnel.
5. Funding initial investment required to create an information superhighway in the country.

A major development in the recent past has been the convergence of telecom, broadcasting and information technology. Primarily driven by emerging technologies, the information era is fast demolishing the traditional walls between telecom, broadcasting and computer technology in the marketplace. Government policy-makers have to take cognisance of this happening and shape future policies accordingly. A broader regulatory body will then be called for to address issues like airwaves regulation which does not fall under the current ambit of TRAI.

Presently, the Ministry of Information & Broadcasting, the Department of Telecom and the Department of Electronics are responsible for the three sub-segments. A highest level inter ministerial body should be constituted that can consider an integrated national policy on information technology which is binding on the three ministries/departments.

There are substantial implementation issues which still remain to be resolved till it can be ensured that adequate funds flow into the sector. A long term implementation Monitoring Group should be created to ensure that the above recommendations which are accepted by the government are effectively implemented.

The twin objectives of the telecom sector are (a) delivering low-cost voice telephony to the largest possible number of individuals and (b) delivering low-cost high speed computer networking to the largest number of firms.

With the introduction of competition in the market, the focus of planning needs to shift from the overall expansion of DELs and network to providing requisite policy framework for the sector/market to grow as required and consistent with the overall policy objectives.

Telecommunications is one of the fastest growing sectors in India. However, viewed in the context of global growth patterns and indicators, the sector is still in the early stages of development. The sector also needs, especially in terms of broad-band, to expand at substantially higher rates to meet the needs of related sectors like IT, I&B and other sectors of the economy. Keeping this perspective in view, the sector needs to be treated essentially as an infrastructure sector for the next decade or so. Once the required teledensity is achieved and the necessary support network has been created, the sector could be treated as service sector.

With a view to ensuring optimum growth in the coming years, Government's broad policy of taxes and regulation for the telecom sector has to be a promotional one. Mopping up of resources or revenue generation by the Government should not be a determinant of the policy governing the sector. The incidence of licence fees in the form of revenue sharing and spectrum charges has to be guided by this principle. Keeping in line with the policy adopted by most of the progressive administrations in the world, the licence fee needs to be aligned with the cost of regulation and administration of Universal Service Obligations (USO). As part of the promotional policy, there is need for the TRAI to work out afresh, the revenue shares and USO regime.

The presence of multiple operators in various sectors implies a need to focus on the conditions that will enable these operators to function smoothly. Specific planning would be required to prepare the grounds for a multi-operator system to develop and the subscriber base to expand without impediments. Ensuring fair and timely interconnection is a major part of such an endeavour. At present, a major obstacle to interconnection arises if the incumbent has to increase capacity and the demand for interconnection takes up only a small portion of the installed capacity. In such a situation, the incumbent finds it difficult to provide adequate funds to install the required capacity since it would not generate adequate payment to cover costs. Such a situation would merit Government intervention and it would be appropriate to establish a fund that will help meet the cost of installing the capacity in such situations of inadequate demand. The Government may recover the amount funded over a period of time.

The existence of an independent and effective regulatory body is crucial for ensuring optimum growth and free and fair competition. The independence of the regulator depends, to a large extent, on the funding mechanism, the constitution of the regulatory body and the principles guiding its functioning. Against the present system of funding from the budget support, a mechanism for making the regulatory body self-financing needs to be put in place. One such option could be provision of necessary funds out of levies/fees collected from telecom operators. Ensuring stability and clarity of regulatory principles is equally important for making the regulatory mechanism effective.

To ensure efficient functioning of the regulatory body, two crucial inputs required are, continuous upgradation of the skills of the staff and regular flow of all relevant information, including the one relating to latest technological developments. A structured approach needs to be developed to achieve these objectives. Besides ensuring that the competent personnel are attracted by the regulator, adequate training facilities and a framework of research and information service needs to be developed. A team of research institutes could be identified for specific tasks in this regard. The Planning Commission itself could be a part of such a network.

Private investment is also expected to play a leading role in the expansion of telecom services. In the area of value added services, the private sector would continue to play a dominant role. The quantum of investment by the private operators would basically get determined by the rate of return on such investments—both basic as well as value added services. Foreign Direct Investment (FDI) also has a major role to play in supplementing the resources of the domestic private sector as the scale of investment envisaged is large. To boost private sector investment, appropriate policy initiatives should to be undertaken.

Eleventh Five Year Plan (2007-12) on Telecom and Connectivity

According to the Approach Paper on Eleventh Plan released by the Government in November 2006, "India has witnessed an impressive telecom revolution during the last decade especially during the last two years; this sector has added 75 million telecom subscribers. The total number of subscribers as on August 2006 stands at more than 165 million with an average addition of about 5 million per month during the year 2006. The sector is poised to grow further and by the end of 11th Plan it proposed to reach a target of 575 million subscribers. Presently, this sector contributes about Rs. 17,000 crore to the national exchequer which will reach a figure of more than Rs. 60,000 crore by the end of 11th Plan period.

From one of the highest telecom tariffs in the world, India has moved to one of the lowest telecom tariffs without compromising on the quality of service that has in fact generated substantial customer satisfaction. With the introduction of

One-India call at Re.1 per minute telecommunication has now become well within the reach of every citizen. The substantial growth in the telecom sector can be attributed to the proactive policies of the Government, thereby creating a level playing field for all the operators.

In addition to telephone connectivity, which is predominantly used for voice, we also need to establish internet connectivity for data. Broadband—"an always-on network capable of providing interactive voice, data and video services on public network"—is fast becoming a prerequisite for rapid economic growth, maintain competitiveness and social transformation. Access to the internet in urban areas increased significantly due to its affordability but it is still low and in rural areas it is negligible. We need to make special intervention to connect rural areas and develop a strategic plan to carry text, data and video to make rural telephony a viable proposition. A Universal Service Obligation (USO) Fund would encourage sharing of critical infrastructure amongst the service providers.

The new wireless technologies (such as 3G-GSM or CDMA, Wi Fi and Wi Max) now provide various last mile connectivity options for different range of distances. It is, therefore, imperative that the traditional fixed line voice telephone connectivity is augmented or supplemented using latest wireless technologies for last mile connectivity. Government of India through the State Wide Area Network (SWAN) Programme has provided connectivity to 6,000 block headquarters on fibre. Wireless technology options could be used for providing last mile connectivity to every village from the block headquarters. Andhra Pradesh proposes to connect every village on fibre and provide gigabit connectivity. These options need to be examined to arrive at the best option for the country.

The backbone for the wireless communication infrastructure is availability of adequate spectrum and its judicious allocation. A considered effort needs to be taken to identify unused spectrum which is available with various agencies such as Defence & Space and reposition the same for the growth of this sector.

This vibrant sector needs to be adequately backed by strong R&D support. The Government should facilitate establishing *Centres of Excellence* in premier educational institutions in the country in public private partnership mode. The centres can focus on the emerging areas such as 3G and beyond, Next Generation Networks, IP Telephony and IPv 6 Networks etc. In addition to this, the Government needs to make a consolidated effort to put in place a system for addressing the critical aspects related to telecom network security.

Telecom Target for the 11th Plan

- A telecom subscriber base of 575 million with one phone for two rural households by 2010, i.e. to reach a rural teledensity of 10 percent from the present 1.9 percent.

- On demand telephone connection across the country at an affordable price.
- 20 million broadband connection and 40 million internet connections by 2010 and broadband connection on demand across the country by 2012.
- 3G services in all cities/towns with more than 1 lakh population.
- Broadband connectivity to every school, health centre and Gram Panchayat.
- To make India a hub for telecom manufacturing by facilitating establishment of telecom specific SEZs." [1]

Vision for the Future

According to *Economic Survey*, 2006-07, "By the end of 2012, a total of 650 million telephone connections (including 66 million wired and 584 million wireless connections) are expected to be achieved. Concurrently, there is also a vision of providing 200 million rural telephone connections, which translates into a rural teledensity of 25 per cent. Broadband connectivity would be made available on demand, without limiting the speed. Each village would have at least one broad-band enabled kiosk. Broad-band connection would be provided to schools, health centres and panchayat offices. It is also envisaged that internet and broad-band subscribers will increase to 40 million and 20 million, respectively, by 2010." [2]

India is the preferred location for software vendors for its quality and cost. India has strong Unix base which provides opportunity for the development of products for internet based applications. Further, India has global connectivity with international dialling facility from over 13220 locations, leased/switched high-speed data links from major centres through STPs and VSNL. Internet connectivity is provided through several networks. Abundant investment opportunities exist in India in the following thrust areas:

1. Communication infrastructure.
2. Optic fibre cable.
3. Gateways.
4. Satellite-based communication wireless.
5. Software development.
6. IT-enabled services.
7. IT-education.
8. IT-enabled education.
9. Data Centres and server farms.

The future policies and programmes for the telecom sector should be guided by the basic goal of creating a world-class telecom infrastructure in order to meet the requirements of IT based sector and needs of a modernizing economy on the least cost basis. Ensuring value for money to the consumers and easy and affordable access to basic telecom services to every one and every where, should be the other goal of policies. The major objectives for the future should include the following:

1. Affordable and effective communication facilities to all citizens.
2. Provision of universal service to all uncovered areas, including rural areas.
3. Building a modern and efficient telecommunications infrastructure to meet the convergence of telecom, IT and the media.
4. Transformation of the telecommunications sector to a greater competitive environment, providing equal opportunities and level playing field for all the players.
5. Strengthening R&D efforts in the country.
6. Achieving efficiency and transparency in spectrum management.
7. Protecting the defence and security interests of the country.
8. Enabling Indian telecom companies to become truly global players.

Conclusion

Telecommunications is now universally acknowledged as one of the prime movers of a modern day economy. Hence, it is of vital importance for a developing economy like India. There is already a large unmet and unsatisfied telecom demand which needs to be addressed at the earliest.

Telecommunications is one of the prime support services needed for rapid growth and modernization of various sectors of the economy. It has become especially important in recent years because of enormous growth of Information Technology (IT) and its significant impact on the rest of the economy. India is perceived to have a special comparative advantage in IT and in IT-enabled services. However, sustaining this advantage depends critically on high quality telecommunication infrastructure. Keeping this in view, the focus has to be on the provision of world class telecommunication facilities at reasonable rates. Provision of telecom services in rural areas would be another thrust area to attain the goal of accelerated economic development and social change. Although the telecom network has grown rapidly in recent years, its growth needs to be accelerated further. It is equally important to speed up the structural changes in this sector in line with trends in other countries to ensure that telecommunication services are not only made available on the scale needed to sustain rapid growth in the economy as a whole but also that their cost are in tune with the expectations of a modernizing economy.

For a dynamic sector, reforms are a continuous process necessitated by dynamics of change including technological innovations. The telecom sector in India has been witnessing a continuous process of reforms since 1991. With the opening of international long distance services and internet telephony from April 2002, the process of liberalization and opening up the sector for competition was completed.

India's vast size, the large number of spread out settlements and its large unserved population provides a huge potential for the expansion of telecommunication services. This opportunity must be seized by

- Raising substantially the penetration ratio in order to provide access to dependable voice communication means with at least national Subscriber Trunk Dialling (STD) connectivity to a much larger cross-section of the population.
- Satisfying the more demanding audio, video and data communication needs of the business community in all major business districts in the country, and
- Making provision for easy up-gradation of the network to meet future communication needs.

All this would need a rapid expansion and upgradation of the existing network. If the telecom network in India is able to grow at even the current annual growth rate of 20 percent for the next five years, it would rank among the six largest networks in the world. India has also expressed its first commitment to make large investments in value-added services by opening up this sector. All this would place India among the leading countries in terms of equipment purchase. This highlights the importance that India would command in the global telecom business in the near future, a fact which should be used for strategic leveraging. Concerted efforts need to be made to attract foreign companies to set up manufacturing bases in India for supporting domestic as well as export sales.

Substantial investments will be required to ensure that India acquires the status of a global player in telecom. This cannot be achieved through governmental efforts or through a monopoly state-owned operator alone. The Government of India has realised the ground realities and initiated the process of progressive deregulation of the telecom sector.

Apart from opening up the basic as well as value-added market segment of telecom services to the private sector, the Government has taken major initiatives in offering a level playing field to operators by constituting the Telecom Regulatory Authority of India (TRAI). The Ministry of Communications is also examining restructuring of the Department of Telecommunications (DoT) into two separate bodies, one dealing with policy planning and the other in charge of service operations. However, a great deal needs to be done, especially in terms of the implementation strategy to that the country can enjoy the benefits of the reforms.

Notes

1. Government of India, Planning Commission, *An Approach to the Eleventh Five Year Plan*, November 2006, pp. 47-48.
2. Government of India, *Economic Survey*, 2006-07, p. 190.

Appendix 2

Information Technology in India's Financial Sector

Introduction

Information technology (IT) has transformed the functioning of businesses, the world over. It has bridged the gaps in terms of the reach and the coverage of systems and enabled better decision-making based on latest and accurate information, reduced costs and overall improvement in efficiency. In the Indian context, the financial sector, especially the banking sector, has been a major beneficiary from the inroads made by IT. Many new processes, products and services offered by banks and other financial intermediaries are now IT-centred. Effective integration of technology with sound business practices requires business process re-engineering and banks in India need to follow up on the beginnings made in this regard. Newer delivery channels to customers – Automated Teller Machines (ATMs), and the networking of ATMs in the form of Shared Payment Networks, Internet Banking – and implementation of Core Banking solutions by most banks are some examples.

The RBI has played a proactive role in the implementation of IT in the banking sector. IT based initiatives are focused on meeting the three pronged objective of better house keeping, improved customer service and overall systemic efficiency. The RBI has come out with a Financial Sector Technology Vision Document outlining the approach to be followed for IT implementation for the medium-term period of about three years. This document will help banks in finalising their IT plans in tandem with the overall approach for the banking sector, as envisioned by the RBI.

Most of the initiatives regarding technology are aimed at providing better and more efficient customer service by offering multiple options to the customer. The death of distance, which is a by-product of technology, has become a reality in the banking sector. Technology is also playing a key role in banks' strategy for gaining a competitive edge.

Many banks have commenced the process of setting up core banking solutions, which are at various stages of implementation. While such systems are already in place in respect of new private sector banks, a few old private sector banks and public sector banks are also quickly moving to set up such processes. This would benefit the customer in the banking sector. Computerisation of the business of banks has been receiving high importance. The public sector banks have already crossed the 70 percent level of computerisation of their business. The directive from the Central Vigilance Commission (CVC) to achieve 100 per cent computerisation has resulted in renewed vigour towards computerisation of branches.

Networking has been receiving focused attention by banks. This activity is also being monitored by the RBI. Most banks have their own corporate networks to facilitate inter-branch and branch-controlling office communication in an electronic mode. Inter-bank and inter-city communication takes place through the Indian Financial Network (INFINET). As part of the INFINET, the terrestrial lines have been augmented to provide for increased data transfer capabilities. All these have resulted in the dependence of banks on network-based computing which has benefited the customer.

INFINET also provides for safe and secure transmission of electronic messages with the use of Public Key Infrastructure (PKI) which has the legal backing of Information Technology Act, 2000. It also provides for messages to flow in a structured environment, using the Structured Financial Messaging Solutions (SFMS), which provides for inter-operability of messages so that straight through processing (STP) is achieved.

Another major development witnessed in recent years is the growth in multiple delivery channels to customers such as internet-based banking, mobile banking and anywhere banking. This has benefited the customers and the banks alike. While customers have now a wide variety of options to choose from, banks have been able to reduce costs which have had a positive impact on their profitability.

These developments have, however, also posed certain challenges. In a world where geographical barriers are losing significance, it is essential that security is given prime importance in a trans-national scenario where large sums of money are at stake. While the challenges relating to physical security could be confronted with relative ease, the position is much more complicated in respect of IT security. The RBI has, therefore, provided guidelines on information system (IS) security as also IS audit which banks can use for their benefit. These are generic in nature and do not have any prescriptive tones.

Financial Sector Technology Vision Document

The RBI released the draft Financial Sector Technology Vision document on May 6, 2005. It provides a broad overview of the thrust areas of the direction provided by the RBI in respect of IT for the financial sector for more than two decades and sets out a roadmap for 2005-08. The Vision document focuses on the following:

1. IT for regulation and supervision,
2. IT for the Financial Sector and
3. IT for Government related functions.

The Vision Document envisages emerging challenges in the form of implementation of standardisation across a variety of hybrid systems at different financial entities, need for decision support systems and the technology to facilitate risk based off-site supervision. It envisions common inter- operable web -based structures for transmission of data relating to regulatory functions and the use of a single centralised database for all information, apart from hiving off the operation of non- critical functions by the RBI.

The Vision Document also visualises Institute for Development and Research in Banking Technology (IDRBT) which is to be a premier research institute, concentrating on research and development for the banking and financial sector, providing educational/training facilities and hiving off business related activities.

Recognising the requirements of IT for the financial sector, the Vision Document elucidates the thrust areas of the RBI by providing generic information on various standards and approaches, IS Audit and requisite focus on business continuity plans. The Vision Document proposes that specific attention would be devoted to percolation of technology efforts to all types of banks and all sections of the customers in the banks with specific reference to the rural areas and the use of affordable technology products which can be easily used by the target clientele with inter-shareable resources.

The document also details the use of IT in the Government sector transactions (which has the largest potential to grow significantly in the years to come), with specific attention on the need for business process re-engineering, changes in rules and procedures for aligning them with e-governance in a manner so as to achieve implementable objectives.

Electronic Banking

Electronic banking (E-Banking) is a generic term encompassing internet banking, telephone banking, mobile banking etc. In other words, it is a process of delivery of banking services and products through electronic channels such as telephone, internet, cell phone etc. The concept and scope of E-Banking is still evolving.

Several initiatives taken by the Government of India as well as the Reserve Bank of India (RBI) have facilitated the development of e-banking in India. As a regulator and supervisor, the RBI has made considerable progress in consolidating the existing payment and settlement systems, and in upgrading technology with a view to establishing an efficient, integrated and secure system functioning in a real-time environment, which has further helped the development of e-banking in India. The Government of India enacted the IT Act, 2000 with effect from October 17, 2000, which provides legal recognition to electronic transactions and other means of electronic commerce.

E-Banking: Global Experience: Finland was the first country in the world to have taken a lead in e-banking. The Scandinavian countries have the largest number of Internet users, with up to one-third of bank customers in Finland and Sweden taking advantage of e-banking. Internet banking is also widespread in Austria, Korea, Singapore, Spain, Switzerland, *etc.* E-banking facilitates an effective payment and accounting system thereby enhancing the speed of delivery of banking services considerably. While the e-banking has improved efficiency and convenience, it has also posed several challenges to the regulators and supervisors.

In response to the challenges thrown by the Internet banking, regulators and supervisors from various countries have prepared their own mechanism of regulation. There is a matrix of legislation and regulations within the United States that specifically codifies the use of and rights associated with the internet and e-commerce, in general, and electronic banking and internet banking activities, in particular. The concerns of the Federal Reserve are limited to ensuring that Internet banking and other electronic banking services are implemented with proper attention to security, safety and soundness of the bank, and the protection of the banks' customers.

In the U.K, there is no specific legislation for regulating e-banking activities. The FSA is neutral on regulations of electronic banking. In Sweden, no formal guidance has been given to examiners by the Sveriges Bank on e-banking. General guidelines apply equally to Internet banking activities. The role of the Bank of Finland has been, as part of general oversight of financial markets in Finland, mainly to monitor the ongoing development of Internet banking without active participation. The Reserve Bank of New Zealand applies the same approach to the regulation of both Internet banking activities and traditional banking activities. There are however, banking regulations that apply only to Internet banking. Supervision is based on public disclosure of information rather than application of detailed prudential rules.

The Monetary Authority of Singapore (MAS) subjects Internet banking to the same prudential standards as traditional banking. The MAS drafted an 'Internet Banking Technology Risk Management Guidelines' in September 2002, which calls upon all banks providing internet banking to establish a sound and robust risk management process. The Hong Kong regulatory approach towards e-banking is less specific in nature. The Hong Kong Monetary Authority (HKMA) expects their banks to undertake a rigorous analysis of the security aspects of their system by getting it reviewed by qualified independent experts.

Like many of these countries, India does not have specific regulatory laws for e-banking. The existing regulatory framework over banks has been extended to Internet banking as well. However, certain guidelines have been issued to banks to recognise the risks arising from electronic modes and to devise control mechanisms that are needed to mitigate such risks. Banks offering the e-banking services in India need to comply with the guidelines.

E-Banking and RBI: The RBI has been gearing up to upgrading itself as a regulator and supervisor of the technologically dominated financial system. In 1998, it availed the technical assistance project of Department for International Development (DFID), UK for upgrading its supervisory system and adaptation of its supervisory functions to the computerised environment. It issued guidelines on 'risks and control in computer and telecommunication system' in February 1998 to all the banks advising them to evaluate the risks inherent in the systems and put in place adequate control mechanisms to address these risks, which can be broadly put under three heads, *viz.,* IT environment risks, IT operations risks and product risks.

The existing regulatory framework over banks has also been extended to internet banking. These guidelines cover various issues that would fall within the framework of technology, security standards and legal and regulatory issues. Virtual banks, which have no offices and function only on line are not permitted to offer e-banking services in India and that only banks licensed under the Banking Regulation Act and having a physical presence in India are allowed to offer such services. Further, banks are required to report to the RBI every breach or failure of security systems and procedures in Internet banking, while the RBI at its discretion may decide to commission special audit/inspection of such banks.

As per recent guidelines, banks no longer need any prior approval of the Reserve Bank for offering the internet banking services. Nevertheless, banks must have their internet policy and they need to ensure that it is in line with parameters as set by the Working Group on Internet Banking in India (2001). The main recommendations of this Working Group are set forth below.

Main Recommendations of the Working Group on Internet Banking (Chairman: S.R. Mittal), 2001

Reserve Bank of India constituted a Working Group to examine different issues relating to internet banking and recommend technology, security, legal standards and operational standards keeping in view the international best practices. The Group was headed by the Chief General Manager–in–Charge of the Department of Information Technology and comprised experts from the fields of banking regulation and supervision, commercial banking, law and technology. The Bank also constituted an Operational Group under its Executive Director comprising officers from different disciplines in the bank, who would guide implementation of the recommendations.

The Working Group, as its terms of reference, was to examine different aspects of Internet banking from regulatory and supervisory perspective and recommend appropriate standards for adoption in India, particularly with reference to the following.

1. Risks to the organization and banking system, associated with Internet banking and methods of adopting International best practices for managing such risks.
2. Identifying gaps in supervisory and legal framework with reference to the existing banking and financial regulations, IT regulations, tax laws, depositor protection, consumer protection, criminal laws, money laundering and other cross border issues and suggesting improvements in them.
3. Identifying international best practices on operational and internal control issues, and suggesting suitable ways for adopting the same in India.
4. Recommending minimum technology and security standards, in conformity with international standards and addressing issues like system vulnerability, digital signature, information system audit etc.
5. Clearing and settlement arrangement for electronic banking and electronic money transfer; linkages between i-banking and e-commerce
6. Any other matter, which the Working Group may think as of relevance to Internet banking in India.

Keeping in view the terms of reference, the Group made a number of recommendations. A summary of these recommendations is given below.

Technology and Security Standards: The role of the network and database administrator is pivotal in securing the information system of any organization. Some of the important functions of the administrator via-a-vis system security are to ensure that only the latest versions of the licensed software with latest patches are installed in the system, proper user groups with access privileges are created and users are assigned to appropriate groups as per their business roles, a proper system of back up of data and software is in place and is strictly adhered to, business continuity plan is in place and frequently tested and there is a robust system of keeping log of all network activity and analyzing the same.

Organizations should make explicit security plan and document it. There should be a separate Security Officer/Group dealing exclusively with information systems security. The Information Technology Division will actually implement the computer systems while the Computer Security Officer will deal with its security. The Information Systems Auditor will audit the information systems.

Access Control: Logical access controls should be implemented on data, systems, application software, utilities, telecommunication lines, libraries, system software, etc. Logical access control techniques may include user-ids, passwords, smart cards or other biometric technologies.

Firewalls: At the minimum, banks should use the proxy server type of firewall so that there is no direct connection between the Internet and the bank's system. It facilitates a high level of control and in-depth monitoring using logging and auditing tools. For sensitive systems, a stateful inspection firewall is recommended which thoroughly inspects all packets of information, and past and present transactions are compared. These generally include a real-time security alert.

Isolation of Dial up Services: All the systems supporting dial up services through modem on the same LAN as the application server should be isolated to prevent intrusions into the network as this may bypass the proxy server.

Security Infrastructure: PKI is the most favoured technology for secure Internet banking services. However, it is not yet commonly available. While PKI infrastructure is strongly recommended, during the transition period, until IDRBT or Government puts in place the PKI infrastructure, the following options are recommended.

1. Usage of SSL, which ensures server authentication and the use of client side certificates issued by the banks themselves using a Certificate Server.
2. The use of at least 128-bit SSL for securing browser to web server communications and, in addition, encryption of sensitive data like passwords in transit within the enterprise itself.

Isolation of Application Servers: It is also recommended that all unnecessary services on the application server such as ftp, telnet should be disabled. The application server should be isolated from the e-mail server.

Security Log (audit Trail): All computer accesses, including messages received, should be logged. All computer access and security violations (suspected or attempted) should be reported and follow up action taken as the organization's escalation policy.

Penetration Testing: The information security officer and the information system auditor should undertake periodic penetration tests of the system, which should include the following.

1. Attempting to guess passwords using password-cracking tools.
2. Search for back door traps in the programs.
3. Attempt to overload the system using DdoS (Distributed Denial of Service) &

DoS (Denial of Service) attacks.

4. Check if commonly known holes in the software, especially the browser and the e-mail software exist.
5. The penetration testing may also be carried out by engaging outside experts (often called 'Ethical Hackers').

Physical Access Controls: Though generally overlooked, physical access controls should be strictly enforced. The physical security should cover all the information systems and sites where they are housed both against internal and external threats.

Back up and Recovery: The bank should have a proper infrastructure and schedules for backing up data. The backed-up data should be periodically tested to ensure recovery without loss of transactions in a time frame as given out in the bank's security policy. Business continuity should be ensured by having disaster recovery sites, where backed-up data is stored. These facilities should also be tested periodically.

Monitoring against Threats: The banks should acquire tools for monitoring systems and the networks against intrusions and attacks. These tools should be used regularly to avoid security breaches.

Education and Review: The banks should review their security infrastructure and security policies regularly and optimize them in the light of their own experiences and changing technologies. They should educate on a continuous basis their security personnel and also the end-users.

Log of Messages: The banking applications run by the bank should have proper record keeping facilities for legal purposes. It may be necessary to keep all received and sent messages both in encrypted and decrypted form. (When stored in encrypted form, it should be possible to decrypt the information for legal purpose by obtaining keys with owners' consent.)

Certified Products: The banks should use only those security solutions/products which are properly certified for security and for record keeping by independent agencies (such as IDRBT).

Maintenance of Infrastructure: Security infrastructure should be properly tested before using the systems and applications for normal operations. The bank should upgrade the systems by installing patches released by developers to remove bugs and loopholes, and upgrade to newer versions which give better security and control.

Approval for I-banking: All banks having operations in India and intending to offer Internet banking services to public must obtain an approval for the same from RBI. The application for approval should clearly cover the systems and products that the bank plans to use as well as the security plans and infrastructure.

It should include sufficient details for RBI to evaluate security, reliability, availability, auditability, recoverability, and other important aspects of the services. RBI may provide model documents for Security Policy, Security Architecture, and Operations Manual.

Legal Issues: The banks providing Internet banking service, at present are only accepting the request for opening of accounts. The accounts are opened only after proper physical introduction and verification. Considering the legal position prevalent, particularly of Section 131 of the Negotiable Instruments Act, 1881 and different case laws, the Group holds the view that there is an obligation on the banks not only to establish the identity but also to make enquiries about integrity and reputation of the prospective customer. The Group, therefore, endorses the present practice but has suggested that after coming in to force of the Information Technology Act, 2000 and digital certification machinery being in place, it may be possible for the banks to rely on digital signature of the introducer. The present legal regime does not set out the parameters as to the extent to which a person can be bound in respect of an electronic instruction purported to have been issued by him. Generally authentication is achieved by security procedure, which involves methods and devices like user-id, password, personal identification number (PIN), code numbers and encryption etc., used to establish authenticity of an instruction. However, from a legal perspective a security procedure needs to be recognized by law as a substitute for signature. In India, the Information Technology Act, 2000, in Section 3(2) provides for a particular technology (viz., the asymmetric crypto system and hash function) as a means of authenticating electronic record. This has raised the doubt whether the law would recognize the existing methods used by banks as valid methods of authentication. The Group holds the view that as in case of other countries, the law should be technology neutral.

In keeping with the view that law should be technology neutral, the Group has recommended that Section 3(2) of the Information Technology Act, 2000 needs to be amended to provide that in addition to the procedure prescribed there in or that may be prescribed by the Central government, a security procedure mutually agreed to by the concerned parties should be recognized as a valid method of authentication of an electronic document/transaction during the transition period. Banks may be allowed to apply for a license to issue digital signature certificate under Section 21 of the Information Technology Act, 2000 and function as certifying authority for facilitating Internet banking. Reserve Bank of India may recommend to Central Government for notifying the business of certifying authority as an approved activity under clause (o) of Section 6(1) of the Banking Regulations Act, 1949.

Section 40A(3) of the Income Tax Act, 1961 recognizes only payments through a crossed cheque or crossed bank draft, where such payment exceeds Rs. 20,000, for the purpose of deductible expenses. Since the primary intention of the above provision, which is to prevent tax evasion by ensuring transfer of funds through identified accounts, is also satisfied in case of electronic transfer of funds between accounts, such transfers should also be recognized under the above provision. The Income Tax Act, 1961 should be amended suitably. Under the present regime there is an obligation on banks to maintain secrecy and confidentiality of customer's account. In the Internet banking scenario, the risk of banks not meeting the above obligation is high on account of several factors like customers not being careful about their passwords, PIN and other personal identification details and divulging the same to others, banks' sites being hacked despite all precautions and information accessed by inadvertent finders.

Banks offering Internet banking are taking all reasonable security measures like SSL access, 128 bit encryption, firewalls and other net security devices, etc. The Group is of the view that despite all reasonable precautions, banks will be exposed to enhanced risk of liability to customers on account of breach of secrecy, denial of service etc., because of hacking/ other technological failures. The banks should, therefore, institute adequate risk control measures to manage such risk. In Internet banking scenario there is very little scope for the banks to act on stop-payment instructions from the customers. Hence, banks should clearly notify to the customers the timeframe and the circumstances in which any stop-payment instructions could be accepted.

The banks providing Internet banking service and customers availing of the same are currently entering into agreements defining respective rights and liabilities in respect of Internet banking transactions. A standard format/minimum consent requirement to be adopted by banks may be designed by the Indian Banks' Association, which should capture all essential conditions to be fulfilled by the banks, the customers and relative rights and liabilities arising there from. This will help in standardizing documentation as also develop standard practice among bankers offering Internet banking facility.

The concern that Internet banking transactions may become a conduit for money laundering has been addressed by the Group. Such transactions are initiated and concluded between designated accounts. Further, the proposed Prevention of Money Laundering Bill 1999 imposes obligation on every banking company to maintain records of transactions for certain prescribed period. The Banking Companies (Period of Preservation of Records) Rules, 1985 also require banks to preserve certain records for a period ranging between 5 to 8 years. The Group is of the view that these legal provisions which are applicable to all banking transactions, whether Internet banking or traditional banking, will adequately take care of this concern and no specific measures for Internet banking is necessary.

The Consumer Protection Act, 1986 defines the rights of consumers in India and is applicable to banking services as well. Currently, the rights and liabilities of customers availing of Internet banking services are being determined by bilateral agreements between the banks and customers. It is open to debate whether any bilateral agreement defining customers rights and liabilities, which are adverse to consumers than what is enjoyed by them in the traditional banking scenario will be legally tenable. Considering the banking practice and rights enjoyed by customers in traditional banking, it appears the banks providing I-banking may not absolve themselves from liability to the customers on account of unauthorized transfer through hacking. Similar position may obtain in case of denial of service. Even though, The Information Technology Act, 2000 has provided for penalty for denial of access to a computer system (Section-43) and hacking (Section 66), the liability of banks in such situations is not clear. The Group was of the view that the banks providing Internet banking may assess the risk and insure themselves against such risks.

The Information Technology Act, 2000, in Section 72 has provided for penalty for breach of privacy and confidentiality. Further, Section 79 of the Act has also provided for exclusion of liability of a network service provider for data travelling through their network subject to certain conditions. Thus, the liability of banks for breach of privacy when data is travelling through network is not clear. This aspect needs detailed legal examination. The issue of ownership of transactional data stored in banks' computer systems also needs further examination.

Regulatory and Supervisory Issues: All banks, which propose to offer transactional services on the Internet should obtain approval from RBI prior to commencing these services. Bank's application for such permission should indicate its business plan, analysis of cost and benefit, operational arrangements like technology adopted, business partners and third party service providers and systems and control procedures the bank proposes to adopt for managing risks, etc. The bank should also submit a security policy covering recommendations made in chapter-6 of this report and a certificate from an independent auditor that the minimum requirements prescribed there have been met. After the initial approval the banks will be obliged to inform RBI any material changes in the services/products offered by them.

RBI may require banks to periodically obtain certificates from specialist external auditors certifying their security control and procedures. The banks will report to RBI every breach or failure of security systems and procedure and the latter, at its discretion, may decide to commission special audit / inspection of such banks.

To a large extent the supervisory concerns on Internet banking are the same as those of electronic banking in general. The guidelines issued by RBI on 'Risks and Controls in Computers and Telecommunications' will equally apply to Internet banking. The RBI as supervisor would cover the entire risks associated with electronic banking as a part of its regular inspections of banks and develop the requisite expertise for such inspections. Till such capability is built up, RBI may outsource this function to qualified EDP auditors.

Record maintenance and their availability for inspection and audit is a major supervisory focus. RBI's guidelines on 'Preservation and Record Maintenance' will need to be updated to include risks heightened by banking on the net. The enhancements will include access to electronic record only by authorized officials, regular archiving of data, a sufficiently senior officer to be in charge of archived data with well defined responsibilities, use of proper software platform and tools to prevent unauthorized alteration of archived data, availability of data on-line, etc. If not available on-line, the system should be capable of making available the data for the same financial year within 24 hours and past data within a period of maximum 48 hours.

Banks should develop outsourcing guidelines to manage effectively, risks arising out of third party service providers such as risks of disruption in service, defective services and personnel of service providers gaining intimate knowledge of banks' systems and mis-utilizing the same, etc. Alternatively, IBA or IDBRT may develop broad guidelines for use of the banking community.

With the increasing popularity of e-commerce, i.e., buying and selling over the Internet, it has become imperative to set up 'Inter-bank Payment Gateways' for settlement of such transactions. The Group have suggested a protocol for transactions between the customer, the bank and the portal and have recommended a framework for setting up of payment gateways. In their capacity as regulator of banks and payment systems of the country, the RBI should formulate norms for eligibility of an institution to set up a payment gateway and the eligible institution should seek RBI's approval for setting up the same.

Only institutions who are members of the cheque clearing system in the country may be permitted to participate in Inter-bank payment gateways for Internet payment. Each gateway must nominate a bank as the clearing bank to settle all transactions. Only direct debits and credits to accounts maintained with the participating banks by parties to an e-commerce transaction may be routed through a payment gateway. Payments effected using credit cards, payments arising out of cross border e-commerce transactions and all intra-bank payments (i.e. transactions involving only one bank) should be excluded for settlement through an inter-bank payment gateway.

Inter-bank payment gateways must have capabilities for both net and gross settlement. All settlement should be intra-day and as far as possible, in real time. It must be obligatory for payment gateways to maintain complete trace of any payment transaction covering such details like date and time of origin of transaction, payee, payer and a unique transaction reference number (TRN).

Connectivity between the gateway and the computer system of the member bank should be achieved using a leased line network (not through Internet) with appropriate data encryption standard. All transactions must be authenticated using user-id and password. Once, the regulatory framework is in place, the transactions should be digitally certified by any licensed certifying agency. SSL/128 bit encryption must be used as minimum level of security. Adequate firewalls and related security measures must be taken to ensure privacy to the participating institutions in a payment gateway. Internationally accepted standards such as ISO8583 must be used for transmitting payment and settlement messages over the network.

The RBI may have a panel of auditors who will be required to certify the security of the entire infrastructure both at the payment gateway end and the participating institutions end prior to making the facility available for customers use. The credit risk associated with each payment transaction will be on the payee bank. The legal basis for such transactions and settlement will be the bilateral contracts between the payee and payee's bank, the participating banks and service provider and the banks themselves. The rights and obligations of each party must be clearly stated in the mandate and should be valid in a court of law. It will be necessary to make customers aware of risks inherent in doing business over the Internet. This requirement will be met by making mandatory disclosures of risks, responsibilities and liabilities to the customers through a disclosure template. The banks should also provide their latest published financial results over the net.

Hyperlinks from banks' websites often raise the issue of reputational risk. Such links should not mislead the customers in to believing that they sponsor any particular product or any business unrelated to banking. Hence, hyperlinks from a banks' websites should be confined to only those portals with which they have a payment arrangement or sites of their subsidiaries or principals. Hyperlinks to banks' website from different portals are normally meant to pass information pertaining to purchases made by banks' customers in the portal. Banks must follow the minimum recommended security precautions while dealing with such request, which includes customer authentication through user-id and password, independent confirmation of transaction by the customer and authorizing payment, use of SSL and 128 bit encryption for all communication both with the portal and customer browser terminal, etc.

On the question of additional capital charge on banks, which undertake Internet banking, the group held the view that standards have not yet been developed for measuring additional capital charge for operational risk. However, this requirement could be covered as the RBI moves towards risk based supervision.

The applicability of various existing laws and banking practices to e-banking is not tested and is still in the process of evolving, both in India and abroad. With rapid changes in technology and innovation in the field of e-banking, there is a need for constant review of different laws relating to banking and commerce. The Group, therefore, recommends that the Reserve Bank of India may constitute a multi disciplinary high level standing committee to review the legal and technological requirements of e-banking on continual basis and recommend appropriate measures as and when necessary.

The regulatory and supervisory framework for e-banking is continuing to evolve and the regulatory authorities all over the world recognize the need for cooperative approach in this area. The Basle Committee for Banking Supervision (BCBS) has constituted an Electronic Banking Group (EBG) to develop guiding principles for the prudent risk management of e-banking activities. This Working Group, therefore, recommends that the Reserve Bank of India should maintain close contact with regulating/supervisory authorities of different countries as well as with the Electronic Banking Group of BCBS and review its regulatory framework in keeping with developments elsewhere in the world.

The Group submitted its report in June 2001 and the Reserve Bank while accepting the recommendations of the Working Group, issued guidelines on 'Internet Banking in India' for implementation by banks. It also stated that the earlier guidelines issued by the Reserve Bank on 'Risks and Controls in Computers and Telecommunications' (1998) would equally apply to Internet banking as well.

E-Banking: Challenges and Concerns

E-banking is based on technology that by its very nature is designed to expand the "virtual" geographic reach of banks and customers without necessarily requiring a similar "physical" expansion. Such market expansion can extend beyond national borders which significantly increases cross-border cooperation challenges for bank supervisors due to the following factors.

1. The potential ease and speed with which banks located anywhere in the world

can conduct activities with customers over interconnected electronic networks4 into countries where a bank is not licensed or supervised.

2. The potential ability of a bank or non-bank to use the Internet to cross borders and to seamlessly link banking activities that have typically been subject to supervision with non-banking activities that might be unsupervised by any financial market authority.
3. The practical difficulties faced by national authorities wishing to monitor or control local access to e-banking sites originating in other jurisdictions without the cooperation of home country authorities.

Banking organisations have been delivering services to consumers and businesses remotely for years. Electronic funds transfer, including small payments and corporate cash management systems, as well as publicly accessible machines for currency withdrawal and retail account management are global fixtures. However, delivering financial services over public networks such as the Internet is bringing about a fundamental shift in the financial services industry.

The changes created, and some of the technical characteristics of internet technology raise new concerns for both bankers and supervisors. Banking organisations are focusing increasingly on their e-banking activities and are globally expanding Internet banking activities, exploring the use of wireless networks and venturing into some new areas of electronic commerce.

Banks offer e-banking services to defend or expand market share or as a cost saving strategy to reduce paperwork and personnel. The Internet also provides banks with substantial opportunity to extend their customer reach beyond existing boundaries. However, the nature of the open network and the evolution of electronic commerce exposes banks to significant competition from both banking and non-banking firms. In addition, electronic delivery channels operate in an uncertain legal and regulatory environment that differs by jurisdiction.

All these factors present new challenges for financial institutions in managing security, integrity and availability of services provided while remaining sufficiently profitable.

Following are the emerging trends and issues that could impact bank risk profiles.

1. A significant increase in competition in the electronic financial services industry as both banking and non-banking firms rapidly introduce new financial products and services.
2. Rapid technological improvements in telecommunications and computer hardware and software enabling greater speed in transactions processing.
3. Bank management and staff often lack expertise in technology and e-banking risk issues.

4. Greater reliance on outsourcing to third party service providers, and a proliferation of new alliances and joint ventures with non-financial firms.
5. Greater demand for global infrastructures for technology that are scalable, flexible and interoperable, both within and across enterprises and that can ensure the security, integrity and availability of information and services.
6. Increased potential for fraud, due to the absence of standard business practices for customer verification and authentication on open networks like the Internet.
7. Legal and regulatory ambiguity and uncertainty with respect to the application and jurisdiction of current laws and regulations to evolving e-banking activities.
8. The collection, storage and frequent sharing of significant quantities of customer data can lead to customer privacy issues that potentially create prudential risks for banks (e.g. legal and reputational).
9. Questions regarding the effectiveness and efficiency of online disclosures. Lengthy or complicated online disclosures may cause customers to simply "click through" or even quit a web site; moreover, extensive disclosure reduces the speed at which web sites and pages can be downloaded.

Banks and bank supervisors generally agree that the supervisory principles that apply to traditional banking are applicable to e-banking. However, the combination of rapid changes in technology and the degree of bank dependence on technology vendors and service providers modify and sometimes magnify traditional risks. Hence, there is a need for additional supervisory guidance in selected areas to enhance the overall risk management framework for e-banking activities.

These developments in e-banking to date suggest the following.

1. The desire to benefit from the advantages of e-commerce in financial services has become widespread. The financial services industry is increasingly focused on providing technology-based financial services solutions directly to customers in order to help build and retain customer bases.
2. Speed-to-market has become a critical factor for success in e-banking. To reduce time to market, banking institutions are allying with non-banking firms to provide total financial services solutions.
3. The current trends in the formation of strategic alliances and technology outsourcing will grow.

These developments present challenges for both banks and bank supervisors. Bank management needs to re-evaluate the robustness of traditional risk management practices in light of the new risks posed by e-banking activities. Also, bank supervisors need to take a balanced approach to the introduction of new regulation and supervisory policy on e-banking, so as to ensure safe and sound operations of banks while at the same time not stifling innovation and the competitiveness of the banks relative to non-banks.

E-Banking Risks and Their Management

E-banking using the Internet as an added delivery channel may shift bank risk profiles to some degree and create new risk control challenges for banks. Accordingly, bank supervisors need to consider the implications of a bank's use of the e-banking delivery channel on its strategic risk, operational risk, reputational risk, legal risk, credit risk, liquidity risk, market risk and foreign exchange risk.

Strategic and Business Risk: Strategic risk is one of the most significant risks that e-banking activities present for banking organisations. Strategic risk differs from other risk categories in that it is more general and broad in nature. Strategic decisions to be taken by a bank's Board of Directors and executive management will have implications for all other risk categories.

Given growing customer acceptance and demand for e-banking, as well as the potential efficiencies afforded, most banks will need to develop a strategy to use the Internet delivery channel to provide informational content and/or transactional service to customers. The rapid changes in technology, the pace of competition with other banks and non-bank competitors and the nature of that strategy could expose banks to substantial risk if the planning and implementation of the strategy is flawed or otherwise not well thought through.

Some of the strategic risks involved with e-banking are directly linked with timing issues. There can be significant strategic risk associated with a management decision to be a technology pioneer, particularly if the institution becomes burdened with systems made redundant by rapid technological changes. Likewise, an overly cautious technology follower may find itself unable to adequately position itself in a saturated market or a market that is consolidating rapidly.

Prior to the Internet, banking institutions used proprietary networks within their consolidated enterprise and connected in limited ways to other banks. These proprietary networks helped provide a strategic defence against new entrants and provided individual franchise protection. However, the Internet as an open network with open access allows both banks and non-banks freedom to create and leverage existing business without the need for expanded physical presence. Consequently, competition within the financial services industry has been significantly increased and is likely to increase further.

Most bankers believe that the e-banking delivery channel will enable them to reduce operational expenses. However, many bank customers wish to maintain a traditional banking relationship, which makes it difficult for banks to abandon the existing physical infrastructure. This means that banks will - at least for the foreseeable future - need to run multiple delivery channels for sometime and e-banking will be a net additional expense. Thus, operational expense savings may occur over the long run only.

The challenge the banking industry faces in maintaining market share is complicated by the entry of new firms that are providing individual financial services via the Internet to existing bank customers. The emergence of aggregation and screen scraping technologies poses both strategic opportunities and threats to banks. Depending on the evolving relationship between the aggregator, the banks affected and the consumer, banks may get further disintermediated as aggregators potentially disrupt the traditional relationship between the customer and the bank and "limit" the direct access that banks will have on-line to retail customers. In addition, aggressive aggregation by both banks and non-banks may lead to greater commoditisation of banking products and services, thereby reducing bank profit margins and adding new security and legal risks.

In essence, bank management needs to carefully consider how its Internet strategy will help maintain the competitiveness and profitability of the institution yet not lead to unwarranted increases in its risk profile. Supervisors should expect banking institutions to carefully assess the pros and cons associated with their strategic options.

Operational Risk: Because of the reliance on technology for all facets of e-banking, operational risk is one of the more significant risks. To limit operational risk, banking organisations may want to consider implementing an integrated enterprise-wide architecture and technology infrastructure that can facilitate interoperability, ensure the security, integrity and availability of data and support the management of relationships with third-party service providers. Further, as technology is also dramatically changing business models and operating processes, banks need to ensure that they have appropriate control procedures (including change control) and audit processes.

***Technology Infrastructure*:** E-banking has brought the issue of technological systems and applications integration to the forefront. Many large banks are now faced with the task of integrating systems for e-banking activities with their existing legacy systems and with the systems of multiple service providers and partners. These banks are exposed to significant operational risks from errors in transaction processing if the systems for e-banking are not properly integrated.

Accordingly, many large banks are making significant investments in technology infrastructure in order to create improved internal controls and enhanced risk management oversight processes. The banks are also hoping to improve flexibility, scalability and interoperability of their systems and operations both within their enterprises and across outside service providers.

While these general developments by large banks are positive, in general the banking industry has much further to go towards improving its systems and risk management infrastructure to effectively support e-banking. Small to medium-sized banking organisations are particularly challenged because of budget restrictions for acquiring hardware and software, as well as attracting and keeping technical staff. Many of these banks rely significantly on third party service providers to manage the necessary technological infrastructure to support the bank's e-banking operations. In this situation, the bank still retains ultimate responsibility for ensuring that these operations are well controlled and managed, and the bank supervisor will wish to assess the ongoing ability of bank management to do so adequately.

Security: The majority of bankers surveyed by EBG members identified security risk as a primary concern relating to e-banking. External threats such as "hacking" [1], "sniffing" [2], "spoofing" [3] and "denial of service" [4] attacks expose banks to new security risks. Open electronic delivery channels create new security issues for banks with respect to confidentiality and integrity of information, non-repudiation of transactions, authentication of users and access control.

Most banks appear to be sensitive to external security threats. Among the issues identified for immediate attention is the development of more robust tools to verify the identity and authenticity of larger value transaction requests. [5] In addition, the banking industry needs to continue to work towards international best practices for encryption requirements, including the legality of electronic signature and records. Moreover, since many banks' internal networks rely on security technology similar to that used to manage their external systems, it is important that bankers also be sensitive to managing the security risk arising from their internal networks. If not managed properly, internal security exposures can also compromise the integrity and confidentiality of bank records and customer data. [6]

Poor security may create reputational or legal risks for banks, as they may be deemed to have provided inappropriate protection for customers' personal data, with consequential legal and/or reputational damage.

At the international level, bank supervisors should encourage the development of a comprehensive approach to managing risk associated with both internal and external security exposures. Given the continuing evolution of industry standards, security risk management may be an area where bank supervisors can work collaboratively with the industry to promote the development of sound practices.

Data Integrity: Data integrity is an important component of system security. Banking organisations must improve interoperability within and across enterprises to effectively manage relationships with customers, other banks and external service providers. Until industry standards are identified for electronic data management, banking organisations will continue to be challenged to establish effective control processes to ensure the accuracy and integrity of data being transmitted and received. The processes should include, at a minimum, sound policies and practices related to project management, system development life cycle, change control and quality assurance. Bank supervisors should also encourage banks to review the integrity of the data used by their risk management systems.

Given the lower cost and ubiquitous nature of the Internet, organisations are increasingly using TCP/IP [7] as a standard communications protocol to achieve this. While there are significant benefits of communicating via TCP/IP, organisations must ensure that data transmitted between bank legacy systems and systems of other parties are translated and integrated accurately. Moreover, while the introduction of middleware [8] and languages such as XML [9] are helping to facilitate this effort, the development of industry standards to support these new technologies is still in its very early stages.

System Availability: In addition to ensuring a secure internal network for their e-banking activities, effective capacity planning is critical to ensuring the ongoing availability of e-banking products and services. Transaction volumes may become increasingly volatile due to price sensitivities and greater automation. Also, competitive pressures and increased reliance on having services available 24 hours a day and 7 days a week (24x7) have raised customer expectations considerably and in turn reduced the tolerance for error. To compete effectively and avoid potentially significant reputation risk that could arise from systems outages, banks offering e-banking services must deliver the right mix of products and services securely, accurately and consistently. These factors underscore the importance of effective business continuity, recovery and incident response plans. Moreover, trends in outsourcing make it necessary for bankers to ensure that similar plans are in place at their external service providers and are periodically tested for effectiveness.

Denials of service attacks can also reduce or eliminate a bank's ability to serve its customers while under attack. These attacks have become increasingly common against high profile e-commerce providers. An added challenge is posed by banks' inability to control the availability of the Internet network itself. Thus, a bank needs to consider, as part of its contingency plans, alternative means to deliver service in the event of a major disruption to the Internet network.

***Internal Controls/Audit*:** The ability to detect and correct errors is a critical internal control component of any banking operation. Moreover, banking organisations must have sufficient controls in place to prevent fraud from both internal and external sources and safeguard the bank's information and assets.

Much of the efficiency and cost reduction in e-banking services stem from banks' ability to implement "straight-through processing". [10] While the benefits of straight-through processing are many, the reality is that e-banking modifies how internal controls, proper segregation of duties and clear audit trails are applied over broad access channels. The challenges presented by these changes are compounded by a critical shortage of skills and expertise in the industry in both the operations and audit areas. Going forward, banks will be increasingly challenged to ensure that highly automated environments provide effective control and that the controls can be independently audited.

***Outsourcing*:** Perhaps more than any other industry development relating to e-banking, banks' increased reliance on outsourcing is having a significant impact on the risk profiles of all banking organisations – both large and small. Large banks are outsourcing many activities as they are increasingly focusing on their core businesses and partnering with other organisations for solutions outside of their core competencies. Small banks usually must outsource because they often lack the necessary technical expertise and resources to build an e-banking delivery channel on their own. Additionally, a decline in the cost of "turnkey" solutions has made it more affordable for small banks to purchase e-banking applications from vendors. These developments are positive in that they increase efficiency, they allow smaller institutions to compete more effectively and they promote the introduction of "state of the art" applications within the industry. However, they can also substantially add to banks' challenges in managing operational risk.

Preliminary indications from surveys tend to indicate that financial institutions rely upon a relatively small number of third-party providers. This seems to be especially the case for small to medium-sized institutions. In some cases, these third parties happen to be new firms with a relatively short track record. This apparent reliance on a concentrated number of third parties, which the EBG will investigate further, could have systemic implications if a major problem would arise with one of these service providers.

To properly manage the risks associated with outsourcing, banks must conduct appropriate due diligence and monitoring of the ongoing viability of third party service providers. The adequacy of terms under contract and service level agreements must also be carefully reviewed for legal risk. Operations processing and the management of security, integrity and availability risks are also more complicated. Furthermore, many technology providers and partners are newly established and may lack knowledge of the controls required within a banking environment. Minor disruptions on the part of third party service providers can expose banking organisations to potential financial loss and substantial legal and reputation risk. Complexity is also added by multiple vendor/service provider relationships that often support e-banking operations. Although to date such disruptions seem to have been controlled, in the future their potential impact could be quite considerable and raises significant concerns for supervisors and industry participants.

Outsourcing can lead to additional privacy related risk exposures. Banks may not always be aware of the exact collection and usage of customer data by vendors and other third parties, and/or may not be adequately managing such activities. Moreover, the legal rights and responsibilities of service providers and vendors may not always be clear. Banks should be encouraged to address privacy issues in their contractual and ongoing relations with vendors.

Various bank supervisors around the world have developed specific guidance related to technology outsourcing. The EBG plans to conduct a review of such guidance and to explore ways to coordinate the development of sound practices in this area for both the banking industry and its supervisors.

Reputational Risk: A bank's reputation can be impacted by any adverse development that precludes the availability of their e-banking delivery channel. Banks have long based their business on a reputation of trust. The ability to provide a trusted network to support e-banking is critical, and a bank's reputation can be damaged by Internet banking services that are poorly executed or otherwise alienate customers and the public.

A bank's reputation can suffer if it fails to deliver secure, accurate and timely e-banking services on a consistent basis. A bank's reputation can also be adversely impacted if it fails to respond to inquiries posted via e-mail, does not provide proper disclosure, or violates customer privacy.

Hypertext links from a bank to third party web sites or outsourced service providers may cause customer confusion about the provider of specific products and services offered, and whether they are insured or uninsured. Customers can also be confused about whether the links from the bank reflect an implied endorsement of the third party's products or services and may well look to the bank for recourse if problems are encountered.

Further, major security breaches in a bank or a non-bank competitor's web site could undermine overall consumer or market confidence in banks' ability to appropriately manage Internet-based transactions.

Any problems that a bank might experience with regard to data and privacy protection could threaten the reputation of that bank as well as the reputation of any other banks perceived to be involved in similar activities.

To protect against adverse situations that may cause damage to their reputation, banking organisations should develop and monitor performance standards for their e-banking activities. Regular review and testing of business continuity, recovery and incident response plans, and communications strategies are also critical to protecting banks' reputations.

Legal Risk: Legal risk arising from e-banking activities represents another area of increased concern. Currently, supervisors in every jurisdiction are examining how existing legal and regulatory frameworks originally designed to address issues affecting the "physical' world of banking interact with the developing e-banking delivery channel as well as examining potential ambiguities.

A bank that develops relationships via the Internet with customers in other jurisdictions may be unfamiliar with the banking and customer protection laws and regulations specific to those countries and may consequently incur heightened legal risks. Even banks that do not intend to solicit business from consumers in foreign jurisdictions may find that their offerings on-line are considered solicitations in some countries. For example, if a bank makes its web site available in another language, regulators in any country where that language is spoken may determine that the bank is marketing services to its citizens and may find that the bank is therefore subject to its local laws and regulations.

Unauthorised use or misuse of data collected over the Internet is another potential source of legal risk. Unauthorised individuals can attack and/or try to infiltrate the "data warehouses" [11] maintained about consumers by both banks and third party vendors. For example, hackers or others might break into banks' or vendors' databases or build their own databases and use the consumers' information to perpetrate fraud. Authorised staff may also deliberately misuse data. Surveys of banks and third party vendors conducted by the EBG have showed that such attacks on "data warehouses" have already occurred, although the impact of these attacks has been minimal to date.

The enforceability of certain emerging areas of law is also uncertain. Laws related to the enforceability of electronic contracts and digital signatures are still under development and vary from jurisdiction to jurisdiction. Effective "know your customer" (KYC) practices are also becoming more critical to bankers in their attempts to prevent fraud.

Other Traditional Banking Risks: The e-banking delivery channel also has implications for other traditional banking risks such as credit risk, liquidity risk, interest rate risk, and market risks. The impact of the introduction of e-banking does not necessarily result in an increase or decrease in the risk profile of the institution, but risks can be shifted, sometimes in complex ways.

***Credit Risk*:** The credit risk of a banking institution can be affected by e-banking activities in a number of ways. The use of the Internet delivery channel may allow banks, especially small institutions, to expand very rapidly, which could lead to heightened asset quality and internal control risks. The use of the Internet also allows banks to expand their geographic reach out of their traditional area, which increases the challenge of understanding local market dynamics and risks, verifying collateral and perfecting security liens with out-of-area borrowers. In addition, the Internet also makes it more difficult to authenticate the identity and creditworthiness of a potential customer, which are essential elements to sound credit decisions. [12] Further, there has been a tendency for some Internet-only banks to pay higher rates on deposits opened over the Internet, which could lead to a higher level of sub-prime credits at these institutions in order to support these higher deposit rates. These factors underscore the importance of sound credit underwriting policies, credit monitoring and administration practices regardless of which product delivery channel is used.

***Liquidity Risk*:** The speed with which information and misinformation moves over the Internet can have implications for the liquidity risk profile of a bank. Adverse information about a bank, whether it is true or not, can be easily disseminated over the Internet through bulletin boards and news groups. This could cause depositors to withdraw their funds in mass at any time of the day, any day of the week. Also, Internet banking can increase deposit volatility to the extent that new customers brought in through this channel maintain accounts solely on the basis of interest rate or terms. Accordingly, increased monitoring of liquidity and changes in deposits and loans may be warranted depending on the volume of activity created through e-banking.

***Market Risk*:** The impact of recent growth in securities issuance and trading over the Internet on banks' market risk profiles is complex. From a market standpoint, the increased volume of securities, which are traded over the Internet, can on the one hand lead to increased volatility, but, on the other hand, it can lead to increased liquidity. From an individual bank's standpoint, banks may be exposed to increased market risk if they create or expand deposit brokering, loan sales, or securitisation programmed as a result of Internet banking activities. As with liquidity risk, the effects of increased e-banking activities on market volatility need to be monitored by banks and supervisors.

***Foreign Exchange Risk*:** A bank may be exposed to foreign exchange risk if it accepts deposits from foreign customers or create accounts denominated in currencies other than their local currency. Since the Internet allows banks the opportunity to expand their geographic range, even internationally, some banks may take on greater foreign exchange risk through e-banking activities than they have through their traditional delivery channels. Also, foreign exchange risk can be intensified by political, social or economic developments, which a bank inexperienced in cross-border banking may not appreciate fully. Supervisors should ensure that a bank initiating cross-border e-banking activities through the Internet has the appropriate risk management systems and expertise to manage these risks properly.

As the preceding discussion indicates, the basic types of risks associated with e-banking are not new. However, the specific ways in which these risks arise, as well as the potential magnitude and speed of impact on banks, may be new for bank management and supervisors alike. In addition, while assessing risk should already be dynamic, the rapid pace of technological innovation supporting e-banking, the increased degree of systems outsourcing and the reliance of some products/services on the use of open networks such as the Internet, intensifies the need for a rigorous and ongoing risk management process.

Bank supervisors should expect their banks to have comprehensive risk management processes in place that include the three basic elements of *assessing* risks, *controlling* risk exposure, and *monitoring* risks associated with e-banking. This comprehensive risk management framework should be integrated into the bank's overall risk management framework.

It is also essential that this risk management process is supported by appropriate oversight by the board of directors and senior management and is carried out by staff with the necessary knowledge and skills to deal with the technical complexities of new e-banking developments.

Similarly, bank supervisors must recognise their own critical need for supervisory staff with appropriate technology knowledge and skills to ensure that they understand the risks and challenges arising from the development of the e-banking delivery channel. Enhanced technical training of existing supervisory staff, complemented by appropriate recruitment of outside expertise, should be a high priority in order to ensure that the supervisor keeps abreast of increasingly complex technology and market developments.

Electronic Money (E-Money)

In recent years, there has been a gradual switchover from the use of paper-based payments media to those based on electronics. While the basic characteristics of these new instruments are by and large similar to those of paper-based instruments, the former present a different set of challenges to policy makers. Electronic money (e-money) is one such new product which has appeared on Indian horizon recently.

Meaning of E-Money: E-money may be broadly defined as an electronic store of monetary value on a technical device used for making payments to undertakings other than the issuer without necessarily involving bank accounts in the transaction, but acting as a prepaid bearer instrument. These products could be classified into two broad categories, viz. (a) pre-paid stored value card (sometimes called "electronic purse") and (b) pre-paid software based product that uses computer networks such as internet (sometimes referred to as "digital cash" or "network money"). The stored value card scheme typically uses a microprocessor chip embedded in a plastic card while software based scheme typically uses specialized software installed in a personal computer.

Recommendations of the Working Group on Electronic Money (Chairman: Zaire J. Camas), 2002

In January 2002, the RBI constituted a Working Group on Electronic Money which submitted its report in July 2002. The Group identified certain areas of concern from the point of view of the central bank in the context of more widespread use of e-money so that the conduct of monetary policy is not impaired and at the same time, the integrity of the instrument is also preserved. Some of the suggestions made by the Group included (a) multi-purpose e-money to be issued only by authorized banks on a credit basis, which should be strictly regulated and closely monitored, (b) ensure redeem ability in order to preserve the unit of account function of money as well as to control money supply in the economy and (c) reporting of monetary statistics for the purposes of monetary policy and protection against criminal abuse, such as money laundering.

Summary and recommendations of the Report are set forth below.

Broadly, e-money is an electronic store of monetary value on a technical device. E-money could be classified as (a) pre-paid stored value card (sometimes called "electronic purse") and (b) pre-paid software based product that uses computer networks such as internet (sometimes referred to as "digital cash" or "network money"). The stored value card could be of three types - single-purpose card, closed-system or limited-purpose card and general-purpose or multi-purpose card. The single-purpose card generally with a magnetic chip recording the amount of fund therein is designed to facilitate only one type of transaction e.g., telephone calls, public transportation, laundry, parking facilities etc. Here, the distinguishing point is that the issuer and the service provider (acceptor) are identical for such cards. The closed-system or the limited-purpose cards are generally used in a small number of well-identified points of sale within a well-identified location such as corporate/university campus. The multi-purpose card on the other can perform variety of functions with several vendors viz., credit card, debit card, stored value card, identification card, repository of personal medical information etc.

While it may not be desirable place any limit on storing monetary value in e-money, it is expected that e-money could be used to substitute central bank notes and coins at least partially. However, the importance of e-money with respect to regulatory oversight, restrictions on issuers and their implications for monetary policy is extremely critical from the point of view of the central bank.

After considering various issues, the Group recommends that multi-purpose e-money may be permitted to be issued only against payment of full value of central bank money or against credit only by the banks. The issuance of e-money on credit basis should, however, be strictly regulated and closely monitored.

It needs to be appreciated that issuers must be under obligation to offer redemption of their e-money liabilities net of service charges, if so required. From monetary policy point of view, such redemption requirement is essential in order to preserve unit of account function of money as also to control money supply in the economy.

With regard to status of issuers of e-money, it may be indicated that there are five reasons which may warrant banks as the issuers of multi-purpose e-money. These include attributes of e-money being closure to demand liabilities of the bank, implications of e-money on velocity of circulation of money and its corresponding impact on monetary statistics, the option to impose reserve requirement on e-money, the need for closure monitoring of e-money when these would be issued as credit and the technical security of e-money. For all these reasons, the Group recommends that only banks should be allowed to issue multi-purpose e-money. However, single-purpose and limited-purpose e-money should be allowed to be issued by any entity including banks.

Non-banks should not be permitted to issue multi-purpose e-money. If they are permitted, they along with banks must conform to seven minimum prudential requirements as laid down by European Central Bank (ECB) in 1998. These are (i) prudential supervision of issuers of e-money by the central bank, (ii) solid and transparent legal arrangements codifying the rights and obligations of issuers, merchants, consumers and the regulators, (iii) adequate technical, organizational and procedural safeguards to prevent and detect threats to the security of e-money, (iv) protection against criminal abuse, (v) supplying of all relevant information to the central bank for the purpose of monetary policy, (vi) legal obligations to redeem e-money against central bank money at par at the request of the holder and (vii) the right of the central bank to impose reserve requirement on issuers of e-money.

E-money could have profound impact on compilation of monetary statistics and money supply unless regulated prudently. E-money could be issued against cash (i.e., 100 per cent backed by central bank money paid upfront). Since e-money are close substitutes of central bank money, these should be explicitly accounted for in monetary statistics. If e-money is allowed to be issued only by banks, then currency would be substituted with demand/time liabilities through e-money. In that eventually, issuance of e-money would be money stock neutral and no change would be required in the definition of money stock. However, if e-money is issued by entities other than depository institutions (i.e., banks), the money creating sector as embedded in compilation of monetary statistics would need to be broadened.

There could be a situation where residents could use e-money supplied by entities outside the country for domestic transactions. In that case, monetary aggregates would lose its predictive power.

48. It is expected that the proportion of interest bearing liabilities in monetary aggregates would grow in the event of growing use of e-money which would render them more unstable, and information content of monetary aggregates would also change.

If e-money is issued on credit, there is a possibility that the issuers may assume a leveraged position. There is, therefore, a need for continuous monitoring of the behaviour of issuing authorities for balanced growth of their assets and liabilities, particularly liabilities arising out of issuance of e-money. For these reasons, the Group recommends that the central bank should regulate and closely monitor the practice of issuing e-money on credit.

If consumers prefer to use e-money vis-a-vis currency, then for a given stock of currency, the money multiplier would go up which would in turn increase the aggregate money supply in the economy more than what would have been the case without e-money. Also, with large scale use of e-money, it has been apprehended that central bank's balance sheet may shrink to such an extent relative to that of the banking sector that it may be unable to perform its liquidity absorption function on account of non-availability of adequate volume of assets. Apart from constraining its liquidity management function, relative shrinkage in balance sheet may also have serious implications regarding loss of *seigniorage* revenue for the central bank. As a counter argument, it is maintained that there should always be a lower bound below which the use of currency notes and coins should not go down so that there should also be a limit below which reserve money should not shrink relative to broad money stock.

A review of developments indicate that while considerations of potential benefits of micro-economic efficiency, extension of banking to urban poor and rural communities and facilitation of e-governance demand that there should be increasingly private provision of payment and settlement services in the economy in future, macro-economic stabilisation policy warrants that there should be a case for public regulation over such provision. Also, currency uses are characterised by network externalities in that larger the number of users, larger is the settlement value of the currency concerned implying that currency could at best be supplied oligopolistically. There are, in fact, the fundamental reasons for which the "monopoly right" in the issuance of currency should be in the hands of some public authority, preferably the central bank. Even then, there cannot be any final judgement on this issue at this point of time. In view of all these considerations, the Group, therefore, recommends that the RBI should regularly monitor closely all these developments so that integrity of the financial system is preserved.

The RBI should also periodically review issues relating to legal framework, if any, technical security and the clearing and settlement arrangements of different e-money schemes and the practices of various e-money schemes, both in India and abroad, for preserving integrity of the financial market. On the issue whether entities other than the central bank could issue independent media of exchanges, the Group feels that such a possibility is apparently remote in India at this point of time. However, RBI may continually keep track of these developments for smooth functioning of the financial market.

Ensuring e-security

No innovation is without challenges and IT is no exception to this rule. The most prominent challenge arising from these innovations relates to the concept of security. Considering the scope for fraud in the e-banking area and the possibility of contagion, the Reserve Bank as a regulator and supervisor has been proactive in addressing the risks associated with e-banking that could have otherwise undermined the credibility of the Indian banking sector. The Reserve Bank has been promptly addressing issues related to fraud with the use of electronic banking facility. Even after issuing guidelines for a secured e-banking, the Reserve Bank from time to time advises the banks on control mechanisms to combat such frauds. In a recent case of attempt of fraud by a customer while using internet banking facility, the Reserve Bank advised the concerned bank to plug the loopholes and the same was also communicated to other banks so that they remained vigilant and control the misuse of internet banking system.

In India, the legal infrastructure for promoting e-banking has not yet been put in place in a comprehensive manner. India does not have a licensed certifying authority appointed by the Controller of Certifying Authorities to issue digital signature certificates. Also, India is not yet a signatory to the International Cyber Crime Treaty, which seeks to intensify co-operation among different signatory nations for exchanging information concerning crime and cyber criminals. Further, there are unresolved legislative issues related to cyber crimes laws, clarification relating to regulatory authority over e-money products, consumer protection and privacy laws. To make the e-banking operations in India more widespread, secure and efficient, these issues need to be addressed by relevant authorities.

As the banking practices and legislations concerning e-banking are still in the process of evolution in India and abroad because of technological innovations, there is a need for a constant review of various legislations and regulatory framework relating to banking and commerce. The Reserve Bank is monitoring and reviewing the legal and other requirements of e-banking on a continuous basis to ensure that the e-banking would develop on sound lines and the e-banking related challenges would not pose a threat to financial stability.

Reviewing the technological developments in Indian commercial banks, the *Economic Survey*, 2006-07 observed, "The banking sector in India is adapting itself to rapid innovations in technology particularly on the information-based technology front to impart efficiency in providing wide range of products and services to the public at large. Computerisation of banking operations had received high importance in 2005-06. Since September 1999 to March 2006, PSBs incurred an expenditure of Rs. 10,676 crore on computerisation and development of communication networks. Out of 27 PSBs, branches of as many as 10 PSBs were 100 per cent computerised, while more than 50 per cent branches of 12 banks were computerised by end-March 2006.

The number of branches providing 'core banking solutions' (CBS) in recent years is increasing rapidly. Under CBS, a number of services are being provided such as 'anywhere banking', 'everywhere access', and quick transfer of funds in an efficient manner and at reasonable cost. New private sector banks, foreign banks and a few old private sector banks have already put in place CBS; PSBs are increasingly adopting similar system. The total number of branches of PSBs offering CBS increased from 11.0 per cent as on March 31, 2005 to 28.9 per cent as on March 31, 2006.

Total number of Automated Teller Machines (ATMs) installed by the banks were 21,147 at end-March 2006. Nationalised banks with 7,165 ATMs were the largest providers of the ATM service followed by new private banks (6,112), SBI group (5,443), old private banks (1,547) and foreign banks (880)." [13]

Notes

1. Hacking refers to the practice of breaking into a computer without authorisation, for malicious reasons, just to prove it can be done, or for other personal reasons.
2. Sniffing involves the use of a software program that is illicitly inserted somewhere on a network to capture ("sniff") user passwords as they pass through the system.
3. Spoofing refers to an attempt to gain access to a system by posing as an authorised user.
4. A denial of service attack represents an attempt to overwhelm a server with requests so that it cannot respond to legitimate traffic.
5. Identification is concerned with positively establishing the identity of the person or organisation conducting the transaction. Passwords are a common approach although evolving approaches such as biometrics may become increasingly prevalent. Authorisation is concerned with establishing the authority that an individual has to conduct a particular transaction.
6. Attacks on internal systems including those by employees are more frequent than external attacks in many organisations.
7. TCP/IP is a standardised communications protocol for transmitting data in packets via the Internet. TCP (Transfer Control Protocol) deals with the construction of the data packets, while IP (Internet Protocol) routes them from computer to computer.
8. Middleware is transaction-processing software that facilitates client/server communications over a network (e.g. TCP/IP) allowing client applications to access and update information from unlike platforms, databases and mainframes.
9. XML (Extensible Mark-up Language) is a universal, flexible, text-based data format that facilitates structured data interchange on the Web.
10. Straight-through processing refers to automatic transaction processing without any human intervention in the transaction process flow.
11. 18 A data warehouse is a large database of customer transactions, updated regularly, which is used as a source of information for data mining.
12. 19 On the other hand, the use of the Internet delivery channel can potentially lead to a reduction in a bank's concentration of credit to a single local industry, market or geographic region.
13. 20. Government of India, *Economic Survey*, 2006-07, p. 63.

Appendix 3

Information Technology in Indian Agriculture

Introduction

India has emerged over the last decade as the most preferred destination for outsourcing of IT services. The vibrant IT industry is contributing immensely by providing information about latest technology and international business practices. Various sectors of the Indian economy such as industry, finance, insurance, communications and transport have adopted information technology in a big way. However, agricultural sector of the economy is lagging behind in utilizing IT services. Some efforts have been made in this regard but they are by no means adequate.

The rapid strides that the country has registered in the IT field will remain incomplete unless IT is fully utilised to ensure more efficient and productive Indian agriculture. Practically there is no area in agriculture in which IT has no role to play. The immense scale and diversity of Indian agriculture provides the ultimate challenge to the potential of information technology.

The adoption of IT services in agriculture depends on the main functionaries involved in Indian agriculture. These functionaries can make a big contribution to the growth of agriculture with the assistance of IT. For this purpose, the following functionaries can be considered (a) farmers, (b) industries providing inputs to agriculture, (c) industries dealing with agricultural output, (d) central and state governments and (e) NGOs working for the benefit of farmers and agricultural universities and research centres.

Indian farmers, for whose welfare a huge government machinery is devoted, still suffer from the absence of right information at the required time. The farmers require timely information on weather conditions, sowing time, availability of inputs including credit, expert advice on maintaining his crop in healthy condition, information on markets and on all other areas of interest to him and his family.

In spite of the best efforts and expenditure, the conventional apparatus has not been able to deliver the goods satisfactorily. Herein lies the role of IT which can efficiently address the concerns of farmers stationed at even remote locations. Low literacy levels, cost of computers, poor communication infrastructure make it impossible for individual farmers, particularly small farmers, to directly adopt IT. This calls for institutional efforts to provide IT-based services to farmers.

In order to keep pace with the state-of-the-art technologies, NIC has been conducting various training programmes on IT application on a regular basis from time to time for the user organisations of agriculture sector.

Agricultural Informatics Division of NIC has taken up various initiatives in bringing IT led development which includes Web enabled applications, GIS based applications, Multimedia applications, Database applications and e-Governance and training etc., in the Ministry of Agriculture.

IFFCO-ISRO Cooperation

Recently, IFFCO has initiated IT-based services for farmers and cooperative societies. It has taken up a project in association with Indian Space Research Organisation (ISRO) to utilise satellite based remote sensing data and Geographical Information Systems (GIS). Developed countries have been utilising precision farming with the help of IT tools for a long time. While this will take a long time for our country due to small holdings, it is to be noted that GIS has an invaluable role to play even in the existing conditions. Remote sensing and GIS information can provide warnings on evolving crop stresses, crop vigour etc.

The IFFCO-ISRO GIS project extends support for efficient and timely availability of IFFCO's fertiliser to farmers though better logistics and efficient operations. Based on the experience gained from this project, more intensive services based on GIS are envisaged for the entire country. In addition to the GIS-based services, effort is being made to create databases that contain information of interest to the farmers. These include recommendation on package of practices for major cereals, pulses, horticulture, floriculture and animal husbandry etc. Information on various inputs such as seeds, fertilisers, credit is provided. An important service envisaged is to provide access to the nearest expert in case of stress or any other problem witnessed in the crops. Facilities are sought to be provided to encourage and share farm experiences by forging various crop forums. Many of the agricultural extension services are also proposed to be made online using multimedia facilities.

In order to encourage farmers to obtain best possible price, information regarding prices of various crops prevailing in approachable markets (mandies) is also being provided. Other areas of interest to farmers such as distance education, location specific news etc. are also planned.

Cyber Dhabas

To take IT to the door steps of the farmers, it is proposed that a large number of access points called cyber dhabas are setup. These cyber dhabas should be provided with a PC and a modem with telephone connection to connect to the databases. The farmers can take the assistance of the operator of cyber dhabas for a nominal charge.

The major bottleneck in spreading e-culture to rural areas is related to connectivity i.e. ensuring that the access points can get connected to the databases which are in selected locations. Since dialup lines are very slow, other viable options are required to be explored. Unfortunately, the alternatives are expensive and may not be feasible. One silver lining is that several private operators are connecting important cities with fiber optic which provides a very reliable and fast access. Since these will pass through rural areas, it is possible to explore the possibility of tapping this potential by laying the last mile connectivity. Once this is done, substantial segment of rural India can access the IT-based services.

E-Choupal

E-Choupal, ITC's unique web-based initiative, offers farmers the information, products and services they need to enhance productivity, improve farm-gate price realisation, and cut transaction costs. Farmers can access the latest local and global information on weather, scientific farming practices, as well as market prices at the village itself through this web portal-all in Hindi. E-Choupal also facilitates the supply of high quality farm inputs as well as the purchase of commodities at the farm.

Given the literacy and infrastructure constraints at the village level, this model is designed to provide physical service support through a choupal sanchalak, himself a lead farmer, who acts as the interface between the system and the farmers. The contents of this site in their entirety are made available only to the registered sanchalaks.

National E-governance Plan (2003-07)

The national e-governance plan (2003-07) reflects the strategic intent of the central government in the right perspective. Many projects are earmarked under this plan, and it is trying to address the digital divide.

From a political perspective, after watching the performance of some IT-savvy states in the recent elections, the system has woken up to the need to focus more on rural development. The political systems are keener to use IT to disseminate information faster to farmers, disburse loans, improve education and the health systems in villages etc. There is a clear-cut incentive to do it as 60 percent of the vote-bank still lives in rural India.

E-governance has to be supported by the will and resources of those who are in governance, be it at the central or state level. The central government has analysed and appreciated the concept by creating a separate E-governance department headed by a secretary to trigger e-governance in India. The World Bank, ADB and UN have been approached, and in response they are generously funding e-governance projects. In future, education, agriculture, state wide area networks (SWANs) and Community Information Centre projects will be rolled out backed by a strong public private participation model (PPP) to achieve long-term sustainability. Projects with PPP models in these segments can revolutionise the governance experience.

Other IT Services to Agricultural Sector

LAN at Ministry of Agriculture: A high speed Local Area Network was established in the Ministry of Agriculture spread over various buildings viz. Krishi Bhawan (~500 nodes), Shastri Bhawan (~200 nodes), Krishi Anusandhan Bhawan (~100 nodes) etc. In addition, Agricultural Informatics Division has extended its full support (round the clock operation) to the Ministry of Agriculture at the time of super Cyclone in Orissa and Gujarat Earthquake. Video Conferencing facilities were used extensively from the Krishi Bhawan studio to Orissa and Gujarat on continuous basis.

Plant Protection Informatics and Communication Network (PPIN): The PPIN envisages to link the Directorate of Plant Protection Quarantine and Storage (DPPQS), Faridabad, National Plant Protection Training Institute, Hyderabad, CIMPS, LCC, Central Pesticides Testing Laboratories, and State Pesticides Testing Laboratories and to disseminate information related to plant protection and pesticides. As a part of project, the computerization of the CIB/RC of DPPQS has been taken up by NIC.

Agricultural Marketing Information Network (AGMARKNET): NIC has taken up an ambitious project "Agricultural Marketing Information Network (AGMARKNET)" on turnkey basis for the Directorate of Marketing and Inspection (DMI), Department of Agriculture and Cooperation, Ministry of Agriculture. The project aims at improving the prevailing agricultural marketing information system by minimising the gap between generation and dissemination of market information. The major components of AGMARKNET are the following: (a) establishment of computing facilities and networking, (b) development of human resource, (c) information transmission, (d) development of database and (e) portal on market information.

NIC has tied up with Bharat Sanchar Nigam Ltd. (BSNL) to provide internet facilities at the AGMARKNET nodes.

Strengthening of Informatics in the Offices and Field Units of the

Department of Agriculture and Cooperation (DAC) – DACNET: The Central sector scheme "DACNET" has been approved by the Department of Agriculture and Co-operation to be implemented by NIC. The project will commence in this financial year. The project proposal envisages establishment of NICNET based INTRANET linking directorates, attached offices, subordinate offices, autonomous bodies and public sector undertakings and field units of the Department of Agriculture and Cooperation, Ministry of Agriculture. This project includes networking of field offices of DAC, connectivity, procurement of H/W, S/W tools, Application Software Development and training of officials.

MIS Support to Food Processing Industries: To cater to the needs of the Department, NIC has developed various information systems viz. FPO and licensing monitoring, Plan scheme monitoring, industrial approvals, processed food products, FPO on-line status, industrial entrepreneur's memoranda query information, country profile and international cooperation etc.

Agricultural Extension Information System Network: VISTARNET: Research, education, extension, and training are considered as four pillars of sustainable agriculture. Generation and transfer of technology have become very crucial the world over. As a step towards making technology reach the small holders (resource-poor-farmers), efforts are being made to establish "**VISTARNET** – NICNET based Agricultural Extension Information System Network, in India, linking extension functionaries at Central, State, and District level. The required funds for implementing VISTARNET will be provided through the National Agricultural Technology Project (NATP). As a part of VISTARNET, Informatics Development for the Directorate of Extension has been taken up by NIC.

Indian Agriculture On-line: NIC has submitted an IT-Plan for the agriculture sector with the objective of establishing NICNET based Agricultural Informatics and Communication Network **(AGRISNET)** to facilitate higher and sustainable agriculture productivity and also putting **Indian Agriculture on-line** in the Country.

Infrastructural Facilities and Services Provided by NIC to the Ministry of Agriculture: These are the following.

1. INTRANET/INTERNET
2. Telecommuting Programme
3. Video Conferencing
4. Information Kiosk
5. In-house Training facilities
6. Web site design and development
7. Web enabled applications

8. Information Bulletin
9. Intranet applications
10. Agriculture Portal
11. NICNET based Public Information and Facilitation Centre

Animal Production and Health Information Network (APHNET): APHNET envisages building up of reliable databases and network based information systems for all activities of the animal husbandry and dairying sector at district, state and national level, using NICNET facilities. In view of importance of making technology reach the small holders through IT, the plan also suggests to establish APHNET nodes at about 42,000 veterinary polyclinics, hospitals, and dispensaries and veterinary aid centres for strengthening animal disease surveillance and advisory system in the country. As a part of APHNET, National Project on Rinderpest Eradication (NPRE) which aims at capturing animal disease related information from various State Animal Husbandry Departments has been entrusted to NIC.

Agricultural Research Information System Network (ARISNET): The ARISNET, a close user group of institutions in National Agricultural Research System of India (NARS) on NICNET, links ICAR Institutions and their Regional Research Stations, Central Agricultural University, State Agricultural Universities and their colleges, Krishi Vigyan Kendras, Zonal Research Centres and Project implementing centres with the ICAR Headquarters. ARISNET has become an integral part of agricultural research, extension and education process.

Market Information System for Horticulture: This project has been taken up by NIC on a turnkey basis. As part of the project, NICNET based internet/intranet facilities have been established at National Horticulture Board (NHB), Gurgaon. Computing facilities have been created at 33 market centres of the Board located all over the country. NIC has developed and implemented the necessary software for evolving a comprehensive database of the prices and arrivals of fruits and vegetables being received by NHB headquarters from 33 market centres on a daily basis. This information is being used by the Ministry of Agriculture, its related agencies and markets through National Bulletin on fruits and vegetables.

Integrated Fertilizer Management Information System (FERMIS): The Department of Fertilizers (DOF) in collaboration with National Informatics Centre (NIC) has introduced computer-based methods for decision support with the major objective to evolve an evaluation system which ensures a uniform system of Planning and control mechanism with signalling system to highlight deviations from desired performance indicators by Plants/organisations for all the public sector fertilizer enterprises. Thus, an Integrated Fertilizer Management

Information System (IFMIS) of worth Rs. 1.99 crore has been executed through NIC and various IT based systems developed covering Planning, Movement, Import, Handling-Payment, Project Monitoring, Performance Monitoring & Evaluation aspects to strengthen Fertilizer Informatics for decision support. Information exchange for decision support has been promoted through Fertilizer Informatics Network (FERTNET) extended to Department of Fertilizers and various Fertilizer Companies viz. IFFCO, NFL, MFL, GNFC, FCI and HFCL.

Computerisation of Agricultural Census, 1995-96 and Input Survey, 1996-97

Recognizing the predominance of the Agriculture Sector in the Indian economy, Department of Agriculture and Cooperation collects and maintains agricultural statistics such as number, area, tenancy, land utilisation, cropping pattern and irrigation particulars of different classes of operational holdings regularly and make it timely accessible to the planners and policy makers for decision making. This project was entrusted to NIC and it made significant headway in almost all the States/UTs except Bihar, where Census is yet to be conducted. The salient features are the following.

1. Creation of large database of about 8000 Million Bytes at National Level and 1 GB at state level
2. Processing and tabulation of Agricultural Census and Input Survey database at District/State/National level
3. Development and implementation of information retrieval system at micro and macro level for decision making at various levels.
4. Selective information will be put on INTERNET/INTRANET for easy access of information to public to ensure transparency in the Government functioning.
5. Training programs on Software implementation will be organized for about 200 officers of State/UTs and Central Government.

Concluding Observations

Proper planning, monitoring and corrective action requires availability of timely information of right quality. The Central and State Governments require to hasten the process of computerisation of all the records and transactions. Since the government has a major say in promoting agriculture, this will have cascading effect on the entire rural areas. To begin with, it is necessary that the government agencies should accept conventional reports in the electronic format. This will bring about much desired speed and effectiveness in governance. Many State Governments are showing interest in developing proper systems for e-governance. But the implementation of the same is still requires a lot to be desired. The Central and State Governments should also make information available for ready reference by those interested. This will be not only helpful to people but also will raise interest levels in IT-based services. Similarly, research institutions and NGOs should exchange and provide information through a well conceived network. These measures will encourage end user, i.e. Farmer's participation and will provide the much needed input from them.

The face of the Indian agriculture can be transformed by a well conceived deployment of IT. The potential of IT as yet remains untapped and urgent measures are required to derive maximum benefit. The key players involved in this process such as industry, government and educational institutions and research centres are required to make contributions in this endeavour. The initiative to develop necessary IT-based agricultural services need to be developed immediately. Parallel steps to develop necessary IT communication infrastructure are to be taken up along with the utilisation of fiber optic network wherever it is passing through the rural segments.

The Karnataka Government's Bhoomi project has led to the computerisation of the centuries-old system of handwritten rural land records. Through it, the revenue department has done away with the corruption-ridden system that involved bribing village accountants to procure land records; records of right, tenancy and cultivation certificates. The project is expected to benefit seventy lakh villagers in 30,000 villages.

A farmer can walk into the nearest taluk office and ask for a computer printout of his land record certificate for Rs 15. He can also check details of land records on a touch-screen kiosk by inserting a two-rupee coin. These kiosks, installed at the taluk office, will provide the public with a convenient interface to the land records centre.

Glossary of Internet Banking

Glossary of Internet Banking

Access Products: Products that allow consumers to access traditional payment instruments electronically, generally from remote locations.

Applet: A small application programme that is designed to do a small, specific job.

Application: A computer program or set of programs that perform the processing of records for a specific function.

Asynchronous Transfer Mode (ATM): Method of transmitting bits of data one after another with a start bit and a stop bit to mark the beginning and end of each data unit.

Auditability: The degree to which transactions can be traced and audited through a system.

Authentication: The process of proving the claimed identity of an individual user, machine, software component or any other entity.

Authorization: The process of determining what types of activities are permitted. Usually, authorization is in the context of authentication: once you have authenticated a user, they may be authorized different types of access or activity.

Bandwidth: The transmission capacity of a computer channel or communications line

Bastion Host: A system that has been hardened to resist attack, and which is installed on a network in such a way that it is expected to potentially come under attack. Bastion hosts are often components of firewalls, or may be "outside" web servers or public access systems.

Biometrics: A method of verifying an individual's identity by analyzing a unique physical attribute.

Browser: A computer program that enables the user to retrieve information that has been made publicly available on the Internet; also permits multimedia (graphics) applications on the World Wide Web.

Byte: Fixed number of binary digits, often representing a single character.

CD-ROM: Compact Disk-Read on Memory.

Chip: An electronic device consisting of circuit elements on a single silicon chip. The most complex circuits are microprocessors, which are single chips that contain the complete arithmetic and logic units of computers.

Chip Card: Also known as an integrated circuit (IC) card. A card containing one or more computer chips or integrated circuits for identification, data storage or special-purpose processing used to validate personal identification numbers, authorize purchases, verify account balances and store personal records.

Client-Server Network: A method of allocating resources in a local area network so that computing power is distributed among computer workstations in the network but some shared resources are centralized in a file server.

Closed Network: A telecommunications network that is used for a specific purpose, such as a payment system, and to which access is restricted (also referred to as a private network).

Closed Stored Value System: A system in which value is issued and accepted by either a relatively small group of merchants, or in which the system is limited geographically (i.e., university programs and fare cards for mass transit systems).

Code: Computer programs, written in machine language (object code) or programming language (source code).

CPU: Central Processing Unit.

Cracker: A computer operator who breaks through a system's security. This can be legitimate activity, such as to test system security measures.

Cryptography: The principles, means, and methods for rendering information unintelligible and for restoring encrypted information to intelligible form (i.e., scrambling a message).

Cursor: Movable dot on the computer screen which indicates a particular position.

Cyber Mall: A set of electronic or digital storefronts linked through a common web site.

Database Administrator (DBA): The individual with authority to control the data base management system.

Data Mining: Data mining refers to the practice of using algorithms to find patterns in customer transaction data to market specific products and services to customers.

Data Warehouse: A data warehouse is a large database of customer transactions, updated regularly, which is used as a source of information for data mining.

Denial of Service: A denial of service attack represents an attempt to overwhelm a server with requests so that it cannot respond to legitimate traffic.

Dial-up: The ability of a remote user to access a system by using private or common carrier telephone lines.

Digital: Referring to communications processors, techniques, and equipment where information is encoded as a binary "1" or "0".

Digital Certification: A process to authenticate (or certify) a party's digital signature; carried out by trusted third parties.

Digital Signature: A mathematical encryption technique that associates a specific person with a given computer file and indicates that the file has not been altered since that person signed it; should not be confused with making an electronic representation of a written signature.

Distributed Transaction Processing: Application processing that involves multiple users requiring concurrent access to a single shared resource.

Domain Name: An alphanumeric name for a web site that includes both the online address and online name.

DOS: Disk Operating System.

Download: To transmit a file or program from a central computer to a smaller computer or a remote site.

Electronic Cash: The digital equivalent of dollars and cents (also referred to as digital cash).

Electronic Data Interchange (EDI): The transfer of information between organizations in machine-readable form.

Electronic Document: The digital or computer equivalent of paper documents.

Electronic Money: Monetary value measured in currency units stored in electronic form on an electronic device in the consumer's possession. This electronic value can be purchased and held on the device until reduced through purchase or transfer.

Electronic Purse: A stored value device that can be used to make purchases from more than one vendor.

E-mail: Messages people send to one another electronically from one computer to another.

Encryption (Cryptography): The process of scrambling data by a device or encoding principle (mathematical algorithms) so that the data cannot be read without the proper codes for unscrambling the data.

End-to-end Encryption: The protection of information passed in a telecommunications system by cryptographic means, from point of origin to point of destination.

Ethernet: A type of local area network originally developed by Xerox, communication takes place by means of radio frequency signals carried over coaxial cable.

File Transfer Protocol (FTP): A standard way of transferring files from one computer to another on the Internet.

Firewall: A system or combination of hardware and software solutions that enforces a boundary between two or more networks.

Flowchart: A programming tool to graphically present a procedure by using symbols to designate the logic of how a problem is solved.

Gateway: A computer that performs protocol conversion between different types of networks or applications.

GB: Giga bytes = 1 billion (million x 1,000) bytes.

Graphical User Interface (GUI): A way of communicating with a computer by manipulating icons (pictures) and windows with a mouse.

Groupware: Software that allows a group of people to work on the same data through a network, by facilitating file sharing and other forms of communication.

Hacker: A computer operator who breaks into a computer without authorization, either for malicious reasons or just to prove it can be done.

Hacking: It refers to the practice of breaking into a computer without authorisation, for malicious reasons, just to prove it can be done, or for other personal reasons.

HD: Hard Disk. HD is computer's storage space. HDs are measured in mega bytes/giga bytes.

HDD: Hard Disk Drive.

Home Banking: Banking services that allow a customer to interact with a financial institution from a remote location by using a telephone, television set, terminal, personal computer, or other device to access a telecommunication system which links to the institution's computer centre.

Home Page: A screen of information made available to users through the Internet or a private intranet; it is the "main page" that users are expected to read first in order to access the other pages that comprise the web site.

Host: Also known as a host computer that is the primary or controlling computer in a computer network, generally involving data communications or a local area network.

Hypertext: Electronic documents that present information that can be connected together in many different ways, instead of sequentially.

Hypertext Mark-up Language (HTML): A set of codes that can be inserted into text files to indicate special typefaces, inserted images, and links to other hypertext documents.

Hypertext Transfer Protocol (HTTP): A standard method of publishing information as hypertext in HTML format on the Internet.

Identification: It is concerned with positively establishing the identity of the person or organisation conducting the transaction. Passwords are a common approach although evolving approaches such as biometrics may become increasingly prevalent. Authorisation is concerned with establishing the authority that an individual has to conduct a particular transaction.

Incident Response Team: A team of computer experts (internal or external) organized to protect an organization's data, systems, and other assets from attack by hackers, viruses, or other compromise.

Indent: Space left at the beginning of a line of print.

Integrated Circuit Card (IC Card): A plastic card in which one or more integrated circuits are embedded (also called a chip card).

Internet: A worldwide network of computer networks (commonly referred to as the Information Superhighway).

Internet Service Provider (ISP): An entity that provides access to the Internet and related services, generally for a fee.

Interoperability: The compatibility of distinct applications, networks, or systems.

Intranet: A private network that uses the infrastructure and standards of the Internet and World Wide Web, but is cordoned off from the public Internet through firewall barriers.

Issuer: In a stored value or similar prepaid electronic money system, the entity which receives payment in exchange for value distributed in the system and which is obligated to pay or redeem transactions or balances presented to it.

KB: Kilo bytes = 1,000 bytes (characters).

Key: A secret value or code used in an encrypting algorithm known by one or both of the communicating parties.

LAN: Local Area Network (i.e. through the server).

Local Area Network (LAN): A network that connects several computers that are located nearby (in the same room or building), allowing them to share files and devices such as printers.

Lock and Key Protection System: A protection system that involves matching a key or password with a specific access requirement.

Logging: The storing of information about events that occurred on the firewall or network.

Magnetic Stripe: Used on debit, credit, and identification cards to store encoded information read by card readers; less secure than computer chip cards.

MB: Mega bytes = 1 million (10 lakh) bytes

Memory Card: An integrated circuit (IC) card capable of storing information only.

MHz: Mega hertz = Mega cycle = 1 million hertz or cycles per second.

Middleware: It is transaction-processing software that facilitates client/server communications over a network (e.g. TCP/IP) allowing client applications to access and update information from unlike platforms, databases and mainframes.

Multimedia: Audio and visual (music, pictures, graphics, animations etc.)

Navigation: Moving through a complex system of menus or help files.

Network: A group of computers connected by cables or other means and using software that enables them to share equipment and exchange information. A system of software and hardware connected in a manner to support data transmission.

Node: Any device, including servers and workstations, connected to a network. Also, the point where devices are connected.

Non-repudiable Transactions: Transactions that cannot be denied after the fact.

Off-line: Equipment or devices that are not in direct communication with the central processor of a computer system, or connected only intermittently.

On-line: Equipment or devices that communicate with a computer network. Connections can be direct (as in a LAN using dedicated connections) or indirect (as in using the Internet).

On-line Scrip: Debit accounts on the Internet or other major computer network.

On-line Service Providers (OSP): Closed network services that provide access to various computer sites or networks for a fee.

Open Network: A telecommunications network to which access is not restricted.

Open Stored Value System: A system that may be comprised of one or more electronic cash issuers of stored value that is accepted by multiple merchants or entities.

Operating System: A program that controls a computer and makes it possible for users to enter and run their own programs.

Package: A package consists of modules. The two popular modules in MS office are: (a) MS Word for word processing and (b) MS Excel for data processing.

Packet Switching: A data transmission method that routes packets along the most efficient path and allows a communication channel to be shared by multiple connections.

Password: A unique word or string of characters that a programmer, computer operator, or user must supply to satisfy security requirements before gaining access to the system or data.

Password Cracker: A software program designed to conduct an automated brute force attack on the password security controls of an information system by "guessing" user passwords.

Password Sniffer: A software program that is illicitly inserted somewhere on a network to capture user passwords as they pass through the system.

Payment System: A financial system that establishes the means for transferring money between suppliers and users of funds, usually by exchanging debits or credits between financial institutions.

Personal Identification Number (PIN): A sequence of digits used to verify the identity of a device holder.

Piggyback (Between-the-lines Entry): A means of gaining unauthorized access to a system via another user's legitimate connection.

Point of Sale (POS): A system of terminals that debits or charges a customer's account and credits or pays a merchant's account to effect payment for purchases at retail establishments.

Port: Connection for printer.

Prepaid Card: A card on which value is stored, and for which the holder has paid the issuer in advance.

Privacy: In the context of a payment system, the property that no information which might permit determination of transactions may be collected without the consent of the counterparties involved.

Protocols: A standardized set of rules that define how computers communicate with each other.

Proximity Cards: Cards that can be read from a short distance; mainly used for security and vehicle identification.

RAM: Random Access Memory (current memory as against stored memory on the Hard Disk).

Remote Payment: A payment carried out through the sending of payment orders or payment instruments.

Repudiation: The denial by one of the parties to a transaction of participation in all or part of that transaction or of the content of the communication.

Router: A computer system in a network that stores and forwards data packets between local area networks and wide area networks.

Scattering: The process of mixing the integrated circuit (IC) chip components so that they cannot be analyzed easily.

Search Engines: Software programs that are capable of locating specified information or web sites on the Internet.

Secure Electronic Transaction (SET): A set of standards jointly developed by Visa, MasterCard, and several technologies companies to facilitate secure credit card transactions over the Internet.

Secure Socket Layer (SSL): A protocol for providing data security during transmission using data encryption, server authentication, and message integrity.

Server: A computer that provides services to another computer (the client).

Settlement: An act that discharges obligations with respect to funds or securities transfers between two or more parties.

Settlement System: A system used to facilitate the settlement of transfers of funds.

Simple Mail Transfer Protocol (SMTP): A protocol used to transfer electronic mail between computers on the Internet.

Smart Card: A card with a computer chip embedded, on which financial, health, educational, and security information can be stored and processed.

Specification: Documents that contain basic detailed data.

Sniffing: It involves the use of a software program that is illicitly inserted somewhere on a network to capture ("sniff") user passwords as they pass through the system.

Spoofing: It refers to an attempt to gain access to a system by posing as an authorised user.

Standards: The rules under which analysts, programmers, operators, and other personnel in an information service organization work.

Stored Value Card: A card that stores prepaid value via magnetic stripe or computer chip.

Straight-through Processing: It refers to automatic transaction processing without any human intervention in the transaction process flow.

Structured Query Language (SQL): A query language used to manipulate large databases.

SVGA: Special Video Graphics Adapter (for quick display of graphics).

System Integrity: The quality that a system has when it performs its intended function in an unimpaired manner, free from deliberate or inadvertent manipulation of the system.

System Specification: A baseline specification containing all the essential computer-based business system documentation. It is completed at the end of the Development Phase.

Systemic Risk: The risk that the failure of one participant in a funds transfer system, or in financial markets generally, to meet its required obligations will cause other participants or financial institutions to be unable to meet their obligations when due.

Systems Analysis: The performance, management, and documentation of the four phases of the life cycle of a business system: study, design, development, and operation.

Tamper-evident: The capacity of devices to show evidence of physical attack.

Tamper-proof: The proven capacity of devices to resist all attacks.

Tamper-resistant: The capacity of devices to resist physical attack up to a certain point.

TCP/IP: It is a standardised communications protocol for transmitting data in packets via the Internet. TCP (Transfer Control Protocol) deals with the construction of the data packets, while IP (Internet Protocol) routes them from computer to computer.

Telecommunications: Data transmission between a computing system and remotely located devices via telephone lines, cable, or wireless technology.

Telnet: A protocol that permits users to access a remote terminal or another computer through a network; widely used on the Internet.

Threat Monitoring: The analysis, assessment, and review of audit trails and other data collected for the purpose of searching out system events that may constitute violations or attempted violations of system security.

Throughput: The total amount of useful work performed by a data processing system during a given period of time.

Topology: The arrangement of nodes usually forming a star, ring, tree, or bus pattern.

Traceability: The degree to which transactions can be traced to the originator or recipient (also referred to as auditability).

Transferability: In electronic money systems, the degree to which an electronic balance can be transferred between devices without interaction with a central authority.

Transmission Control Protocol/Internet Protocol (TCP/IP): A standard format for transmitting data in packets from one computer to another, on the Internet and within other networks. TCP deals with the construction of the data packets while IP routes them from machine to machine.

Trap Door: A concealed and unauthorized entrance into a computer operating system, designed by the programmer.

Trojan Horse: A program that appears to perform a useful function and sometimes does so quite well but also includes an unadvertised feature, which is usually malicious in nature.

Truncation: Dropping off part of a character string either to conserve space or because of limited space.

Trusted Computer System: A system that employs sufficient assurance measures to allow its use for simultaneous processing of a range of sensitive or classified information.

Trusted Third Party: A reputable entity that authenticates one or more parties to an electronic transaction. The authentication process generally involves the issuance and administration of digital certificates.

Uniform Resource Locator or Universal Resource Locator (URL): A way of specifying the location of available information on the Internet.

Upload: To transmit a file to a central computer from a smaller computer or a remote location.

Usenet: A set of many newsgroups distributed via the Internet.

VDU: Visual Display Unit.

Virtual Corporations: Corporations that have no official physical site presence and are made up of diverse geographically dispersed or mobile employees.

Virus: A program with the ability to reproduce by modifying other programs to include a copy of itself. It may contain destructive code that can move into multiple programs, data files, or devices on a system and spread through multiple systems in a network.

Vulnerability: A weakness in system security procedures, system design, implementation, internal controls, etc., that could be exploited to violate system security.

Web Page: A screen of information supporting the home page of a web site.

Web Site: The collection of an entity's home page and other proprietary pages located on the World Wide Web.

Wide Area Network (WAN): A communications network that covers a wide geographic area, such as state or country, using high speed long distance lines or satellites provided by a common carrier.

World Wide Web (web, www): A sub network of the Internet through which information is exchanged via text, graphics, audio, and video.

Worm: A program that scans a system or an entire network for available, unused space in which to run. Worms tend to tie up all computing resources in a system or on a network and effectively shut it down.

XML: XML (Extensible Mark-up Language) is a universal, flexible, text-based data format that facilitates structured data interchange on the Web.

Index

Index